Machine Intelligence and Knowledge Engineering for Robotic Applications

NATO ASI Series

Advanced Science Institutes Series

A series presenting the results of activities sponsored by the NATO Science Committee, which aims at the dissemination of advanced scientific and technological knowledge, with a view to strengthening links between scientific communities.

The Series is published by an international board of publishers in conjunction with the NATO Scientific Affairs Division

A	Life Sciences	Plenum Publishing Corporation
B	Physics	London and New York
C	Mathematical and Physical Sciences	D. Reidel Publishing Company Dordrecht, Boston, Lancaster and Tokyo
D	Behavioural and Social Sciences	Martinus Nijhoff Publishers Boston, The Hague, Dordrecht and Lancaster
E	Applied Sciences	
F	Computer and Systems Sciences	Springer-Verlag Berlin Heidelberg New York
G	Ecological Sciences	London Paris Tokyo
H	Cell Biology	

Series F: Computer and Systems Sciences Vol. 33

Machine Intelligence and Knowledge Engineering for Robotic Applications

Edited by

Andrew K.C. Wong
Systems Design Engineering, University of Waterloo
Waterloo, Ontario N2L 3G1, Canada

Alan Pugh
Electronic Engineering, University of Hull
Hull HU6 7RX, United Kingdom

Springer-Verlag
Berlin Heidelberg New York London Paris Tokyo
Published in cooperation with NATO Scientific Affairs Division

Proceedings of the NATO Advanced Research Workshop on Machine Intelligence and Knowledge Engineering for Robotic Applications held at Maratea, Italy, May 12–16, 1986

ISBN 3-540-17844-9 Springer-Verlag Berlin Heidelberg New York
ISBN 0-387-17844-9 Springer-Verlag New York Berlin Heidelberg

Library of Congress Cataloging in Publication Data. NATO Advanced Research Workshop on Machine Intelligence and Knowledge Engineering for Robotic Applications (1986 : Maratea, Italy) Machine intelligence and knowledge engineering for robotic applications. (NATO ASI series. Series F, Computer and systems sciences ; vol. 33) "Published in cooperation with NATO Scientific Affairs Division." "Proceedings of the NATO Advanced Research Workshop on Machine Intelligence and Knowledge Engineering for Robotic Applications held at Maratea, Italy, May 12–16, 1986."—T.p. verso. Includes index. 1. Robotics—Congresses. 2. Artificial intelligence—Congresses. I. Wong, Andrew K.C. II. North Atlantic Treaty Organization. Scientific Affairs Division. III. Title. IV. Series: NATO ASI series. Series F, Computer and systems sciences ; vol. 33. TJ210.3.N375 1986 629.8'92 87-12863
ISBN 0-387-17844-9 (U.S.)

This work is subject to copyright. All rights are reserved, whether the whole or part of the material is concerned, specifically the rights of translation, reprinting, re-use of illustrations, recitation, broadcasting, reproduction on microfilms or in other ways, and storage in data banks. Duplication of this publication or parts thereof is only permitted under the provisions of the German Copyright Law of September 9, 1965, in its version of June 24, 1985, and a copyright fee must always be paid. Violations fall under the prosecution act of the German Copyright Law.

© Springer-Verlag Berlin Heidelberg 1987
Printed in Germany

Printing: Druckhaus Beltz, Hemsbach; Bookbinding: J. Schäffer GmbH & Co. KG, Grünstadt
2145/3140-543210

ABOUT THE BOOK

This book consists of a collection of papers addressing the topic: "Machine Intelligence and Knowledge Engineering for Robotic Applications". It is mainly an outcome of a NATO Advanced Research Workshop held at Maratea, Italy in May 1986. The authors are internationally known and hence the book may be considered an important documentation of the state of the art in knowledge-based robotics.

The book presents and reviews the recent advances in this significant field of research. It covers: robot vision, knowledge representation and image understanding, robot control and inference systems, task planning and expert systems and integrated software and hardware systems. Each of these areas is addressed by several authors who approached the problems differently. Almost all the articles attempt not only to consider the theoretical aspects, but also to include in their presentation challenging issues such as systems implementation and industrial applications.

ACKNOWLEDGEMENT

The authors gratefully acknowledge the professional support and financial assistance provided by the late Dr. Mario di Lullo and the Scientific Affairs Division, NATO, Brussels, Belgium. Professors E. Baker, L. Pau, A. Pugh, A. Sanderson and A.K.C. Wong, members of the Organization Committee of the Machine Intelligence and Knowledge Engineering for Robotic Applications Advanced Research Workshop also deserve special gratitutde. Without their dedicated support, the Workshop could not have achieved what it did. The Editors recognize, too, the contribution of Anne Ross and Deborah Stacey, whose tireless effort made this publication possible.

PREFACE

This book is the outcome of the NATO Advanced Research Workshop on Machine Intelligence and Knowledge Engineering for Robotic Applications held at Maratea, Italy in May 1986. Attendance of the workshop was by invitation only. Most of the participants and speakers are recognized leaders in the field, representing industry, government and academic community worldwide.

The focus of the workshop was to review the recent advances of machine intelligence and knowledge engineering for robotic applications. It covers five main areas of interest. They are grouped into five sections:

1. Robot Vision
2. Knowledge Representation and Image Understanding
3. Robot Control and Inference Systems
4. Task Planning and Expert Systems
5. Software/Hardware Systems

Also included in this book are a paper from the Poster Session and a brief report of the panel discussion on the Future Direction in Knowledge-Based Robotics.

Section I of this book consists of four papers. It begins with a review of the basic concepts of computer vision, with emphasis on techniques specific for robot vision systems. The next paper presents a comprehensive 3-D vision system for robotic application. It covers various theoretical aspects and factory applications of 3-D robot vision. The last two papers deal with computer vision on moving objects. The first provides an overview on the computation of motion from a sequence of monocular or stereo images. The second describes new techniques on time-varying image analysis with a high-level system for representing and identifying time-varying characteristics of a large class of physical events.

Section II is on Knowledge Representation and Image Understanding. The first paper presents a general and flexible knowledge representation using attributed graphs and hypergraphs as the basic data structure. With these representations, model synthesis and recognition of 3-D objects based on various forms of graph morphism algorithms are described. Also presented is a knowledge directed method for recognizing and locating 3-D objects from a single 2-D perspective image. An attempt is made to show that similar representation can be used to represent the world environment of a roving robot in both path planning and navigation. As for the second paper, it describes a knowledge-based system for robotic applications. After a general discussion on knowledge-based system which is able to acquire knowledge in specified domains, store knowledge in defined structure and organize knowledge in desired format for access, retrieval, transfer, utilization and extension, it presents some recent results on a piloted vision system for roving robot navigation, label reading and 3-D object recognition. The next paper addresses image understanding for robotic application. It describes how knowledge about the robot environment could be used by an image understanding algorithm to facilitate the recovery of information from images. Two specific applications are described in the paper: one demonstrates how knowledge of ego-motion parameters of a mobile robot could be used for segmentation of a scene and the recovery of depth information, and the other shows how a hypothesize-and-test approach could be used to find road edges in real scenes for an autonomous vehicle.

Section III covers the use of machine intelligence and knowledge-based systems for robot control and target tracking. The first paper presents a hierarchical control approach for machine intelligent robots. It is based on what is defined as Hierarchical Intelligent Control and the Principle of Decreasing Precision with Increasing Intelligence. Entropy is used as a common measure for the probalistic model involved. The second paper is a critique on the application of artificial intelligence planning technique in industrial robots. It shows the limitation of some classical A.I. paradigms in industrial application and recommends directions for future development. The third paper investigates some fairly universal concepts of analogical reasoning in the context of the block world. Frames are used to represent problem situations and the three-stage underlying learning process is also described. The fourth paper is on the overall hardware and software architecture of a knowledge-based system for change detection, target tracking and threat assessment. Based on target features, numbers and maneuver patterns or changes in the scene, the system is able to assign threat level and threat scenario labels to the scene. Thus interpretation of the scene could be achieved more efficiently and reliably.

Section IV is on task planning and expert systems. The first paper covers the conceptual, the algorithmic and data structure of a task and path planning project for a mobile robot. It presents a world model which includes: a) a set of elementary task operators; b) three-level environment models, namely, geometrical, topological and semantic, and c) the functional capabilities of the machine in the form of specialized processing modules.

The next paper describes a nonlinear planning approach for task planning and control synthesis for flexible assembly systems. It proposes assembly sequences based on relational model of part contacts and attachments. The resulting plans are then consolidated into AND/OR graph representation which provides a basis for efficient scheduling of operations. A simple example is used to demonstrate the efficiency of this approach in comparison to a fixed sequence method. The last paper of this section presents a robust and practical expert robot welding system called MARS. It identifies the various relevant variables of the welding process and investigates their interrelation so as to develop a mathematical model for feedback control of a welding robot. The objective of the project is to construct a computerized hierarchical expert welder.

Section V consists of three papers describing several software/hardware robot systems. The first is on the Edinburgh Designer System which can serve as a general framework to support symbolic computing for robotics. It concludes that a) an algebra engine is required to handle temporal constructs, groups and tolerances; b) a proposed taxonomy can support activity modules, c) and an automatic plan formation would require the creation of a "specialist". The second paper describes the implementation of complex robot subsystems through distributing computational load functionally over several micro-processor systems in both tightly and loosely coupled configurations. This approach is used to explore various concepts of sensor data fusion. An autonomous mobile robot which has provided the experimental environment is also described. The third paper discusses the autonomous research robot being developed at the University of Karlsruhe. The device is able to perform simple operations in the laboratory. It contains a mobile platform, a complex sensor system, two manipulators, hierarchical controls and an expert system. The paper describes how the fundamental technology developed at the Institute is being integrated in the robot system. The last paper summarizes the research and development activities in the field of intelligent robotics at the Laboratory for Intelligence Systems for the National Research Council of Canada. It gives a brief description on the Council's objective and introduces several projects of the research laboratory on 3-D vision, sensory based control, multi-processor system architecture and applications of artificial intelligence. The last paper of this book is a closing remark based on the Panel Discussion on the Future Direction in Knowledge-Based Robotics. It summarizes the general discussions, recommendations and future directions of each of the areas covered under the five sections. The Panel Discussion concluded on an optimistic note, with researchers targeting sensor fusion, advance computer architecture and an increase in intelligence in all systems (knowledge base and sensor) as areas to be actively pursued.

A.K.C. Wong

TABLE OF CONTENTS

I. ROBOT VISION

Robot Vision
A. Rosenfeld 1

3-D Vision for Robot Applications
D. Nitzan, R. Bolles, J. Kremers,
P. Mulgaonkar 21

On the Computation of Motion from
a Sequence of Monocular or Stereo
Images - An Overview
J. Aggarwal 83

Time Varying Image Analysis
T. Huang 105

II. KNOWLEDGE REPRESENTATION & IMAGE UNDERSTANDING

Knowledge Representation for Robot Vision
and Path Planning Using Attributed Graphs
and Hypergraphs
A. Wong 113

Knowledge-Based Systems for Robotic Applications
J. Tou 145

Image Understanding for Robotic Applications
R. Jain 191

III. ROBOT CONTROL AND INFERENCE SYSTEMS

Machine-Intelligent Robots: A Hierarchical
Control Approach
G. Saridis 221

On the Application of Intelligent Planning
Techniques in Industrial Robotics
M. Lee 235

Analogical Reasoning by Intelligent Robots
N. Findler & L. Ihrig 269

Knowledge-Based Real-Time Change Detection,
Target Image Tracking & Threat Assessment
L. Pau 283

IV. TASK PLANNING AND EXPERT SYSTEMS

Task and Path Planning for Mobile Robots
R. Chatila and G. Giralt 299

Task Planning and Control Synthesis for
Flexible Assembly Systems
A. Sanderson & L.S. Homem-de-Mello 331

MARS: An Expert Robot Welding System
P. Sicard & M. Levine 335

V. SOFTWARE/HARDWARE SYSTEMS

The Edinburgh Designer System as
a Framework for Robotics
R. Popplestone 387

Implementation of Complex Robot Subsystems
on Distributed Computing Resources
S. Harmon 407

Autonomous Robot of the University
of Karlsruhe
U. Rembold & R. Dillmann............................... 437

VI. POSTER SESSION

Robotics Research at the Laboratory for
Intelligent Systems, National Research
Council of Canada
S. Elgazzar 463

VII. FUTURE DIRECTIONS IN KNOWLEDGE-BASED ROBOTICS
D. Stacey and A. Wong 477

VIII. LIST OF SPEAKERS 483

IX. LIST OF PARTICIPANTS 485

ROBOT VISION

Azriel Rosenfeld
Center for Automation Research
University of Maryland
College Park, MD 20742/USA

1. INTRODUCTION

This article reviews the basic concepts of computer vision, with emphasis on techniques that have been used, or could be used, in robot vision systems. Sections 2 and 3 discuss two- and three-dimensional vision systems, respectively, while Section 4 briefly discusses some other vision topics. References to basic papers or review papers are given in connection with each topic.

2. TWO-DIMENSIONAL VISION

The general goal of computer vision is to derive a description of a scene by analyzing one or more images of the scene. In many situations the scene itself is basically two-dimensional; for example, a robot might be dealing with flat parts lying on a flat surface, or might be looking for holes in such a surface. Vision is much easier for two-dimensional (2D) scenes, and not surprisingly, the earliest work on vision dealt with such scenes. This section outlines the basic steps in the 2D vision process, and then presents a review of some of the methods used to carry out these steps. Three-dimensional vision is more complicated; it will be treated in the next section.

2.1. The 2D Vision Process

In order for a robot to recognize parts, holes, etc. on a surface—in general: "objects"—it must first be able to distinguish the objects form the rest of the surface. In other words, it must be able to single out pieces of the image that (hopefully) correspond to the objects. This process of extracting subsets of an image that correspond to relevant parts of the scene is called *segmentation*.

Once a subset has been extracted from an image, it is usually necessary to measure various geometric properties of the subset (size, shape, etc.). These measurements can serve as a basis for recognizing the subset as representing a given object, for determining the position and orientation of the object, etc. They may also serve as a basis for further segmenting the subset; for example, if two objects touch or overlap, they may be extracted as a single subset, and it may than be necessary to break the subset into two pieces on the basis of geometric criteria (e.g., to break it into convex pieces). This stage of the computer vision process is called *geometric analysis*. Different algorithms for geometric analysis can be designed, depending on how the image subsets are represented in the computer; thus a topic closely related to geometric analysis is *geometric representation* of image subsets.

Recognition of objects by analyzing subsets of an image can vary greatly in difficulty, depending on the complexity of the objects. If the objects that might be present in the scene differ greatly from one another, simple "template matching" techniques can be used to distinguish them; in this situation it may not even be necessary to explicitly extract the objects from the rest of the image. More generally, objects can often be recognized by the fact that they have a characteristic set of geometric property values; this "feature matching" approach was used in the well-known "SRI vision module" (1). If the objects are complex, it may be necessary to break up the recognition process into stages: to first detect subobjects and recognize their properties, and then to recognize the objects as combinations of subobjects in specific relationships; this is known as the "structure matching" approach.

2.2. Segmentation

A *digital image* is a discrete array of numbers representing brightness values at regularly spaced points in the scene. The elements of a digital image are called *pixels*, and their values are called *gray levels*. (Color has not been extensively used as yet in robot vision systems; color images will be briefly discussed in the next subsection.) This section reviews basic methods of segmenting digital images into subsets. For general surveys of image segmentation see (2,3). The effectiveness of a segmentation technique depends on the properties of the class of images to which it is applied (4); approaches to defining, or "modeling", classes of images are reviewed in (5).

2.2.1. Thresholding

If an object differs significantly in brightness from its background, it gives rise to a set of pixels in the image that have significantly different gray levels from the rest of the image. (Large brightness differences between an object and its background can often be produced by controlling the illumination so as to silhouette or edge-light the object.) Such image subsets can be extracted from the image by *thresholding* the pixel gray levels, e.g., classifying a pixel as "light" or "dark" depending on whether its gray level lies above or below a specified "threshold" level.

If the illumination of the scene can be controlled and the sensor can be calibrated, it may be possible to set a threshold once and for all to correctly segment scenes of a given class; but in general it will be necessary to determine the best threshold for each individual image. If the objects occupy a significant fraction of the scene, this can be done by examining the *histogram* of the image, which is a graph showing how often each gray level occurs in the image (6). This histogram should have two peaks, one representing background gray levels and the other object gray levels; these ranges of gray levels give rise to peaks because they occur relatively frequently. Intermediate gray levels should be relatively rare, and should give rise to a valley on the histogram, between the peaks. A good gray level at which to set the threshold is evidently the level corresponding to the bottom of the valley, since nearly all object pixels and nearly all background pixels will be on opposite sides of this threshold.

If the illumination of the scene is not uniform, dark objects at one side of the scene may actually be brighter than the light background at the other side, so that the objects cannot be separated from the background by simple thresholding. One way to handle this situation (7) is to divide the image into blocks and pick a threshold for each block by analyzing its histogram. These thresholds can then be interpolated to yield a "variable threshold" that properly segments the entire image. A survey of threshold selection techniques can be found in (8).

The color at a point in a scene can be characterized by a triple of numbers representing, for example, the values of red, green, and blue "color components". Thus a digital color image is an array of triples of values. If these pixel values are plotted as dots in "color space", an object (or background) of a given color gives rise to a cluster of dots. These clusters are analogous to histogram peaks, and the image can be segmented into regions having different colors by partitioning the color space so as to separate the clusters. This approach is classically used to segment images obtained by multispectral scanners in remote sensing, but it has not yet found significant use in robot vision.

2.2.2. Edge detection

Small objects are not easy to extract from their background by thresholding, because the histogram peaks that they produce may be too small to detect reliably. Similarly, if a scene contains many objects of different brightnesses, it is not easy to extract them by thresholding, because their histogram peaks overlap. Another method of segmentation can be used in such cases, provided the objects have relatively uniform brightness and that they contrast strongly with their immediate backgrounds. This implies that the rate of change of gray level is low within the objects, but high at the borders of the objects. The objects can thus be extracted by *edge detection,* i.e., by detecting pixels at which the rate of change of gray level is high.

The classical method of detecting edges in an image is to apply an isotropic derivative operator, such as the gradient operator, to the image; such an operator will have high values at edges, no matter what their orientations (9). Many digital approximations to the gradient have been used for this purpose; an especially simple example is the "Roberts cross" operator (10), and another frequently used operator is the "Sobel operator" (11).

Several other basic methods of edge detection are the following: (a) Match the image in the vicinity of each pixel with "templates" of step functions in different orientations; if a good match is detected, an edge in that orientation is likely to be present (12). (b) Fit a polynomial surface to the image gray levels in the neighborhood of each pixel; if the gradient of the fitted surface is high, an edge is likely to be present (12). (c) Fit a step function to the image gray levels in the neighborhood of each pixel; if this step has high contrast, an edge is likely to be present (13). (d) Apply a Laplacian operator to the image; the zero-crossings of the Laplacian values correspond to edges (14). An early survey of edge detection techniques can be found in (15).

2.2.3. Texture analysis

If an object is not uniform in brightness, but rather is patterned, neither thresholding nor edge detection can be used to extract it, since its pixels do not have gray levels in a narrow range, and it has many internal edges. Nevertheless, such an object may be distinguishable from its background because of its characteristic pattern of gray levels, or "visual texture". For a general survey of visual texture analysis see (16).

Textures can be characterized by sets of local properties of their pixels, i.e., by the fact that in a textured region, certain local patterns of gray levels tend to be present in the neighborhood of each pixel. An early survey of local properties that can be used to distinguish textures can be found in (17). By computing a set of such properties at each pixel, the pixel can be characterized by a set of numbers (compare the discussion of color images in the subsection on thresholding), and the image can be segmented into differently textured regions by partitioning the "local property space" so as to separate the clusters corresponding to the regions. Since local properties tend to be more variable than colors, some degree of local averaging should be performed first in order to make the clusters more compact. Similarly, by computing average values of local properties and then taking differences of these averages, one can compute a "texture gradient" at each pixel and use it to detect "texture edges", i.e., edges between differently textured regions (18).

A powerful method of characterizing textures is by performing various types of "shrinking" and "expanding" operations on them and analyzing the results; for example, thin patterns disappear under small amounts of shrinking, while closely spaced patterns "fuse" under small amounts of expanding. This approach to image analysis has been used in a variety of applications for over 20 years; a recent comprehensive treatment is (19).

2.2.4. Tracking and region growing

The methods of segmentation discussed so far treat each pixel (or its neighborhood) independently; they are oblivious as to whether the resulting pixels constitute a connected region, or whether the resulting edge segments constitute a smooth, high-contrast boundary. Better-quality regions or edges can be obtained by requiring that the results be locally consistent, i.e., that the regions be connected or that the edges smoothly continue one another. Methods of "tracking" edges sequentially, pixel by pixel, or of "growing" regions, can be used to insure continuity. (A survey of region growing techniques can be found in (20).) A more powerful, but computationally more expensive, approach is to require (piecewise) global consistency, e.g., to search for regions that are optimal with respect to constancy or smoothness of gray level, or for edges that are optimal with respect to contrast and smoothness of direction. A useful approach to finding globally consistent regions is a split-and-merge process in which regions are split if they are inconsistent, and pairs of adjacent regions are merged if their union is consistent. For a general treatment of image segmentation by partitioning into consistent regions see (21).

2.3. Geometric Analysis

Once a region has been segmented from an image, it can be represented by a "binary image" in which pixels belonging to the region have value 1, and those belonging to the background have value 0. Various geometric properties of the region can be computed from this binary image at low computational cost. This process is sometimes referred to as *binary vision*. The following subsections discuss basic geometric properties and their measurement, as well as other, more compact ways of representing regions.

2.3.1. Connectivity and borders

If a scene contains several objects on a background, segmenting an image of the scene yields the entire set of pixels belonging to all the objects; it does not distinguish the objects from one another. In order to deal with one object at a time, it is necessary to "label" the object pixels so that pixels belonging to the same object get the same label (and conversely). This process is called *connected component labeling* (22); it assigns distinctive labels to sets of object pixels that are all mutually connected. The theory of connected regions in digital pictures is developed in (23). Connectedness is the basic principle that underlies the process of counting objects (i.e., connected regions) in an image.

The *border* of a region consists of these region pixels that are adjacent to non-region pixels. These border pixels lie on a set of curves, one representing the "outer border" of the region and the others representing the borders of its holes, if any. To label these borders individually, a "border following" process can be used (22) that, starting from any border pixel, successively visits all the pixels belonging to that border until it returns to the starting pixel.

2.3.2. Size and shape properties

The *area* of a region is (approximately) the number of it pixels (i.e., the number of pixels having a particular component label). The *perimeter* is the number of border pixels, or (for a specific border) the number of moves required to follow the border completely around. A frequently used shape measure is area/(perimeter)^2, which measures the *compactness* of the region. The *elongatedness* of a region can be defined using a process of shrinking and area measurement; a region is elongated if it has large area but disappears under a small amount of shrinking. *Distance* measures are another source of useful shape information; on measures of distance in a digital image see (24), and on the approximation of Euclidean distance see (25).

Many shape properties of a region can be derived by measuring the *curvature* (i.e., rate of change of direction) of its border. *Concavities* correspond to parts of the border where the curvature is "negative" (in the sense that direction is changing counterclockwise while the border is being followed clockwise). (On the theory of concavity in digital images see (26); on the characterization of straight line segments see (27).) *Corners* are border points where the curvature has a high (positive or negative) value. Such properties are useful in segmenting a region into parts when necessary; for example, when two objects in the scene touch or overlap, they give rise to a single connected region in the image, but they can be "cut apart", e.g., by making a cut that joins the bottoms of two deep concavities. A review of shape analysis algorithms for contours (i.e., borders) can be found in (28), and a general survey of shape analysis techniques can be found in (29).

The *moments* of a region provide useful information about its shape (30). The (i,j) moment m_{ij} is defined as $\sum x^i y^j$ summed over all pixels (x,y) of the object. (Moments can also be defined for gray-level images by weighing pixel (x,y) by its gray level.) Thus m_{00} is the area of the region, and $(m_{10}/m_{00}, m_{01}/m_{00})$ are the coordinates of the *centroid* of the region. The *principal axis* of a region is the line (through the centroid) about which the region's moment of inertia is least; its slope $\tan\theta$ satisfies the quadratic equation $\tan^2\theta + (m_{20} - m_{02})\tan\theta/m_{11} - 1 = 0$, where the m's are moments computed with the origin at the centroid. Other combinations of moments can be used to measure region symmetry, and in general, to provide a variety of properties useful for object recognition.

2.3.3. Geometric representations

A region need not be represented by the set of 1's in a binary image; other representations can be used that are more compact if the shape of the region is simple. The following are some well-known examples:

(a) *Run length code.* Each row of binary image consists of runs (= maximal sequences) of 1's alternating with runs of 0's. Thus the row is completely determined by specifying the starting value (1 or 0) and the lengths of the runs. Most geometric properties of a region can be measured directly from its run length code. Run length coding has been used for many years in data compression.

(b) *Chain code.* The sequence of moves made in following a border, together with the coordinates of the starting point, completely determine the border; this move sequence is called the *chain code* of the border. A region is determined by specifying its borders in this way, and many region properties can be computed directly from the border codes. The same techniques can be used to encode and process digitized curves (31,32).

(c) *The medial axis* (22,33) *and the quadtree* (34). Any region is the union of the maximal "blocks" of pixels (e.g., upright squares) that are contained in the region; thus the region is determined by specifying the list of centers and sizes of these squares. This representation is quite compact, but is more difficult to work with because it is an unsorted list. A less compact, but more tractable, representation, called the *quadtree,* is obtained by recursively subdividing the binary image into quadrants until blocks (sub...subquadrants) consisting of all 0's or all 1's are obtained. The recursive subdivision process defines a tree structure on the blocks, and makes it easier to perform geometric computations.

Various methods of approximating the shape of a region are also useful. The boundary of a region can be approximated by a polygon; on methods of constructing good polygonal approximations see (21). A "ribbonlike" region has a medial axis in which the centers of the maximal squares lie approximately along a curve (the "skeleton" of the ribbon); thus the region can be approximated by specifying this curve together with a "width function" that specifies how the sizes of the squares vary as one moves along the curve (35). Successive (polygonal) approximations to a curve can be organized in a tree structure; here a node represents a side of a polygon that approximates a given arc of the curve, and the children of the node represent sides of a refined polygon that approximate subarcs of this arc (36).

2.4. Recognition

This section discusses methods of recognizing objects, or parts of objects, from the regions to which they give rise in an image. The objects are assumed to be relatively flat and to have known spatial orientations (e.g., to be lying on a flat surface); this is the basic assumption in 2D vision.

2.4.1. Template matching and Hough transforms

If the shape of the desired object is precisely known, it can be recognized by a "template matching" process, without the need to extract it explicitly from the image. In this approach, the template is compared to the image, pixel by pixel, in all possible positions and (2D) orientations; comparison can be based on pixelwise differencing or on computing some type of correlation coefficient. This process is relatively expensive computationally, and is normally used only to detect small pieces of objects, such as straight edge segments or corners, using small templates consisting of only a few pixels. If a set of distinctive features of the object can be detected in this way, one can then search for combinations of features in approximately the correct relative positions, and thereby detect the object itself (37).

A somewhat less costly approach, known as the *Hough transform* (38–40), can be used for objects that are large in extent but that consist only of relatively few pixels—for example, a thin curve, or the border of an object. The basic idea is to map these pixels into a transform space defined in such a way that pixels belonging to a particular curve or border of the given shape all map into the same point in the space. If many pixels that all belong to the same curve are actually present in the image, they will give rise to a peak in the "Hough space".

2.4.2. Property matching

If the objects expected to be present in the scene are sufficiently different from one another, they can often be distinguished by measuring sets of property values (1), without the need for detailed comparison with templates. The properties used can include any of the geometric properties discussed in the previous section, as well as properties involving lightness, color, or texture that may have been used as a basis for segmenting the object from its background. A set of such property values is measured for each candidate region in the image, and compared with the "ideal" values that characterize the various types of objects. If a good match is found, an object has been recognized.

2.4.3. Structure matching

If the objects are sufficiently complex, it may not be possible to distinguish them from one another on the basis of sets of property values, since the ranges of values that characterize different objects may not be sufficiently distinct. In this situation a "structural" approach can be used. Pieces of objects are extracted from the image and characterized by sets of property values; these are the "primitive parts" out of which the objects are composed. Combinations of these pieces in given relationships are then found; these represent entire objects, or more complex object parts. This process can be repeated for several stages, if necessary, until the desired objects have been identified. This approach is generally known as "structural pattern recognition" (21). An early example of its use in robot vision is (41). Hierarchical structure matching has a general resemblance to the syntactic analysis of language, in which a sentence is "parsed" by combining words into clauses, clauses into phrases (etc.), and phrases into the entire sentence. Many "syntactic" formalisms for "languages" consisting of two- (or higher-)dimensional patterns have been developed; for a review of this "syntactic pattern recognition" methodology see (42,43).

3. THREE-DIMENSIONAL VISION

Vision is much more difficult if the scene is strongly three-dimensional—for example, if a robot is dealing with a pile of parts in a bin, or has to determine the spatial orientation of a workpiece. In such a scene the surfaces of the objects can have strongly varying orientations; as a result, a uniformly reflective object will not appear uniformly bright in the image. It is also very hard to obtain uniform illumination; shadows are unavoidable. These factors make a three-dimensional (3D) scene much harder to segment. If the effects of illumination and surface orientation could be computed, the image brightness data could be corrected so that it represented surface reflectivity, and the image could then be segmented into regions corresponding to uniformly reflective surfaces. Alternatively, if the surface orientation were known at each point, the image could be segmented into regions within which there is no abrupt change in orientation, and which therefore probably correspond to surfaces or faces of a single object. But even if correct segmentation could be achieved, object recognition would still be difficult in a 3D situation, because an object of a given (3D) shape can give rise to regions of many different (2D) shapes in the image, depending on how the object is oriented. Moreover, only one side of an object is visible from any given viewpoint, and objects can also partially hide one another, which means that recognition must be based on incomplete data.

Many techniques have been developed for recovering information about the orientations and (3D) shapes of the visible surfaces in a scene; the methods used include direct range sensing, stereomapping, and inference of surface shape from cues present in a single image. In addition, methods of representing 3D shape information have been developed, and basic work has been done on the recognition of partially visible 3D objects in unknown orientations.

3.1. Surface Shape Recovery from a Single Image

The gray level of a pixel in a digital image represents the amount of light received by the sensor from a given direction, or equivalently, the apparent brightness of a given point P on the surface of some object in the scene. This brightness value is the resultant of at least three factors: the level of illumination at P; the reflectivity of surface at P; and the spatial orientation of that surface relative to the viewing direction. It is not possible to separate the effects of these factors if only the brightness at the single point P is known. Given an entire image, however, and under plausible general assumptions about the nature of the scene and the illumination, reasonable conclusions can often be drawn.

The task of deducing illumination, reflectivity, and surface orientation from the brightness data in a single image is known as *recovery* (more fully: "recovery of intrinsic scene characteristics from an image" (44)).

If the surface orientation is known at every point of the image, the (3D) shapes of the visible surfaces in the scene are determined. For this reason, methods of recovering surface orientation information from an image are known as "shape from . . ." methods—i.e., techniques for deducing 3D surface shape from cues available in the image. Some basic "shape from . . ." concepts will be reviewed in the following paragraphs.

3.1.1. Shape from shading

In the immediate neighborhood of a point P, the illumination is likely to be approximately constant (except at the edge of a shadow). Thus if P lies on a smooth, uniformly reflective surface, the variations in brightness ("shading") in the vicinity of P are primarily due to surface orientation effects. These variations do not completely determine the surface orientation, but they do constrain it, and if it is known at some points of a region (for example, at an occluding edge of a smooth surface the orientation is edge-on to the viewer), it can be determined for the rest of the region (45–47).

3.1.2. Shape from texture

If a surface is uniformly patterned ("textured"), it gives rise to an image region in which the patterning is no longer uniform, due to surface orientation effects. The more the surface is slanted relative to the viewing direction, the more foreshortened the texture becomes, due to perspective; and the direction of foreshortening is an indication of the direction of slant. Thus the degree and direction of local surface slant can, in principle, be deduced from local measurements of texture foreshortening, or non-isotropy (48). A related idea is to illuminate the scene with a regular pattern of light, e.g., a grid; distortions in the grid as it appears in the image then provide information about the orientations of the illuminated surfaces (49).

3.1.3. Shape from shape

Clues about the (3D) shapes of surfaces in a scene can also be derived from the (2D) shapes of features (e.g., edges) in an image of the scene. An early example of this concept, using line drawings rather than real images, demonstrated that the shapes of the junctions at which edges meet provides useful information as to which surfaces meeting at a junction belong to the same object (50). Later generalizations of this work were based on the observation that an edge in an image can arise from various types of abrupt changes in the scene, including changes in illumination (shadow edges), in range (occluding edges), in surface orientation, or in reflectivity (if the edge is due to a surface marking). When edges meet at a junction, not all combinations of these cases are possible, and when all the junctions in the image are considered, the number of possibilities is greatly reduced (51). In principle, it should sometimes be possible to distinguish among the various types of edges by carefully analyzing the brightness variations in their vicinities (44). When planar surfaces meet along an edge, the spatial orientations of the surfaces are constrained, and this provides useful information about the shapes of polyhedral objects (52–54).

The shapes of edges or regions in an image can also provide quantitative information about surface shape if certain extremality assumptions are made. For example, a curve in an image could arise from any different curves in space, but humans seem to assume that the actual space curve is the one having the least possible curvature (55,56). Similarly, a region in an image could arise from many different surface patches in space, but humans seem to assume that the actual surface patch is the one having the "simplest" (e.g., most symmetric, most compact) shape (53,56). A general discussion of the detection and interpretation of image edges can be found in (57).

3.2. Stereo and Range Sensing

If two images of a scene taken from different (known) positions are available, and the two image points corresponding to a given scene point P can be identified, then the position of the point in space can be calculated by triangulation. This is the underlying principle of *stereomapping*. The chief difficulty is that corresponding pairs of image points are not easy to identify. Early work on automatic stereomapping for application to aerial photographs (58) produced systems that required considerable human interaction, and there were similar limitations on the performance of the early stereo techniques developed for 3D vision (59). Current stereo systems make use of a variety of techniques for unambiguously identifying sets of corresponding point pairs that define consistent 3D space curves and surfaces. Examples of stereo systems based on models for human stereopsis are (60–62).

3D surface shape can also be derived from multiple images taken from the same position but under different lighting conditions. The shading in each image imposes constraints on the surface orientation at each point, and the intersection of these constraints determines the orientation unambiguously (63). In this technique, which is known as *photometric stereo,* determining corresponding points is trivial, since the images are in perfect registration.

Surface shape information would be immediately available if the range to each visible surface point in the scene could be measured directly. Several different types of "range sensors" have been developed for this purpose. One approach is to illuminate part of the scene at a time, for example with a plane of light; the spatial position of an illuminated point is then completely determined from its position in the image, since it must lie along a given spatial direction and must also lie in the plane of light. An example of early work on object recognition using range data obtained in this way is (64). Another approach illuminates the scene, one point at a time, with a pulse of light, and measures the range to the illuminated point from the phase shift (i.e., the time lag) between the illuminating and reflected pulses, as in radar. An example of early work using this techniques is (65). A review of range sensing techniques for vision can be found in (66).

3.3. 3D Object Recognition

In a segmented image of a 3D scene, the regions represent surface patches. One can compute various shape properties of the patches (e.g., surface area), but this is less useful than in the 2D case, since the visible parts of the surface of a 3D object change as the object's spatial orientation changes.

Recognition of a 3D object from a 2D image is a difficult task because only one side of the object is visible (even ignoring the possibility of occlusion by other objects), and the shapes of the visible parts depend on the orientation of the object. One solution to this problem is to store descriptions of objects as seen from "every possible" viewpoint (in practice, many tens of viewpoints would usually be needed), and to recognize objects by matching their appearances in the image with a large set of stored descriptions; but this is computationally expensive.

Inherently 3D descriptions of an object can be defined in a variety of ways (67). An object is determined by specifying the shapes of its surfaces, which can be approximated by patches of various standard types, or polyhedrally. (As in the 2D case, successive approximations can be organized in a tree structure.) An object can also be regarded as the union of maximal blocks of various types. For example, in the case of a "cylinder-like" object, the centers of the blocks will tend to lie along a space curve, and the object can be approximated by specifying this curve together with a "width function" (68). Maximal blocks can also be organized into a tree structure (an "octree") defined by recursively subdividing space into octants (69). A variety of representations for 3D objects are reviewed in (70,71).

If a set of such such 3D descriptions of objects is given, the objects can be recognized in an image by a process of constraint analysis. A given edge or region in the image cannot be the projection of every possible object edge or object surface; it can only have arisen from a subset of the possible objects, and these have to be in a subset of the possible spatial orientations. If a set of edges or regions are all consistent, in this sense, with the presence of the same object in the same orientation, there is strong evidence that this object is in fact present in the scene (72). A general survey of model-based methods for image analysis can be found in (73).

4. SOME OTHER TOPICS

This concluding section briefly discusses some general vision systems topics that were not treated in earlier sections.

4.1. Time-Varying Image Analysis

This paper has considered only the analysis of static scenes, but there is a large and rapidly growing literature on the analysis of time-varying scenes using time sequences of images. An early survey of research in this area can be found in (74), and collections of papers (including other important surveys) are (75–79). Approaches based on models for human perception of motion can be found in (80,81).

4.2. Multiresolution Methods

Many types of operations on images, including edge detection, region growing, template matching, etc., can be performed efficiently using a "coarse-fine" approach in which the image is first processed at low resolution, and the results are then used to guide the processing at higher resolutions. Conversely, many types of operations can be performed using a "divide-and-conquer" strategy in which results obtained locally, at high resolution, can be recursively combined into global results. An early discussion of this multiresolution approach to image analysis can be found in (82), and an implementation using an exponentially tapering "pyramid" of images is described in (83). Two collections of papers on multiresolution and hierarchical methods in image analysis are (84,85). A general survey of the data structures used in image analysis can be found in (86).

4.3. Cooperative Computation

Many of the steps in the computer vision process can be implemented as "cooperative computations" in which local operations are performed in parallel on all parts of the data (e.g., at each pixel of an image, or at each node of a data structure). Some general classes of such computations that can be used to label parts of an image or scene (e.g., to classify pixels, to identify parts of a template, etc.) are described in (87). For reviews of this approach to image analysis and computer vision see (88,89).

4.4. Computational Constraints

Many of the image operations described in this paper involve massive amounts of computation. At present it would not be cost-effective to implement such operations in practical vision systems. The operations that can be used in real vision systems are severely constrained by the available computational resources. For this reason, many of today's vision systems are "binary", i.e., they work with thresholded images, because it would not be feasible to process grayscale (or color) data at the required speeds. However, as the cost of computer hardware continues to drop, and its power continues to increase, more types of operations will become available for practical use. Eventually, as large multiprocessor systems become available at low cost, it will become practical to implement parallel approaches to computer vision, with resulting major increases in processing speed. The coming years will see major increases in the power, speed, and versatility of vision systems.

BIBLIOGRAPHY

(1) G.J. Agin and R.O. Duda, "SRI vision research for advanced industrial automation", *Proceedings of the 2nd USA-Japan Computer Conference,* Tokyo, Japan, 1975, 113–117.

(2) E.M. Riseman and M.A. Arbib, "Computational techniques in the visual segmentation of static scenes", *Computer Graphics Image Processing 6,* 1977, 221–276.

(3) T. Kanade, "Region segmentation: signal vs. semantics", *Computer Graphics Image Processing 13,* 1980, 279–297.

(4) A. Rosenfeld and L.S. Davis, "Image segmentation and image models", *Proc. IEEE 67,* 1979, 764–772.

(5) A. Rosenfeld, ed., *Image Modeling,* Academic Press, New York, 1981.

(6) J.M.S. Prewitt and M. Mendelsohn, "The analysis of cell images", *Annals N.Y. Academy of Sciences 128,* 1966, 1035–1053.

(7) C.K. Chow and T. Kaneko, "Boundary detection of radiographic images by a threshold method", in S. Watanabe, ed., *Frontiers of Pattern Recognition,* Academic Press, New York, 1972, 61–82.

(8) J.S. Weszka, "A survey of threshold selection techniques", *Computer Graphics Image Processing 7,* 1978, 259–265.

(9) L.S.G. Kovasznay and H.M. Joseph, "Image processing", *Proc. IRE 43,* 1955, 560–570.

(10) L.G. Roberts, "Machine perception of three-dimensional solids", in J.T. Tippett et al., eds., *Optical and Electro-Optical Information Processing,* MIT Press, Cambridge, MA, 1965, 159–197.

(11) K.K. Pingle, "Visual perception by a computer", in A. Grasselli, ed., *Automatic Interpretation and Classification of Images,* Academic Press, New York, 1969, 277–284.

(12) J.M.S. Prewitt, "Object enhancement and extraction", in B.S. Lipkin and A. Rosenfeld, eds., *Picture Processing and Psychopictorics,* Academic Press, New York, 1970, 75–149.

(13) M. Hueckel, "An operator which locates edges in digital pictures", *J. ACM 18,* 1971, 113–125.

(14) D. Marr and E. Hildreth, "Theory of edge detection", *Proc. Royal Soc. B207,* 1980, 187–217.

(15) L.S. Davis, "A survey of edge detection techniques", *Computer Graphics Image Processing 4,* 1975, 248–270.

(16) R.M. Haralick, "Statistical and structural approaches to texture", *Proc. IEEE 67,* 1979, 786–804.

(17) J.K. Hawkins, "Textural properties for pattern recognition", in B.S. Lipkin and A. Rosenfeld, eds., *Picture Processing and Psychopictorics,* Academic Press, New York, 1970, 347–370.

(18) A. Rosenfeld and M. Thurston, "Edge and curve detection by visual scene analysis", *IEEE Trans. Computers 20,* 1971, 562–569.

(19) J. Serra, *Image Analysis and Mathematical Morphology,* Academic Press, London, 1982.

(20) S.W. Zucker, "Region growing: childhood and adolescence", *Computer Graphics Image Processing 5,* 1976, 382–399.

(21) T. Pavlidis, *Structural Pattern Recognition,* Springer, Berlin, 1977.

(22) A. Rosenfeld and J.L. Pfaltz, "Sequential operations in digital picture processing", *J. ACM 13,* 1966, 471–494.

(23) A. Rosenfeld, "Connectivity in digital pictures", *J. ACM 17,* 1970, 146–160.

(24) A. Rosenfeld and J.L. Pfaltz, "Distance functions on digital pictures", *Pattern Recognition 1,* 1968, 33–61.

(25) P.E. Danielsson, "Euclidean distance mapping", *Computer Graphics Image Processing 14,* 1980, 227–248.

(26) J. Sklansky, "Recognition of convex blobs", *Pattern Recognition 2,* 1970, 3–10.

(27) A. Rosenfeld, "Digital straight line segments", *IEEE Trans. Computers 23,* 1974, 1264–1269.

(28) T. Pavlidis, "Algorithms for shape analysis of contours and waveforms", *IEEE Trans. Pattern Analysis Machine Intelligence 2,* 1980, 301–312.

(29) T. Pavlidis, "A review of algorithms for shape analysis", *Computer Graphics Image Processing 7,* 1978, 243–258.

(30) M.K. Hu, "Visual pattern recognition by moment invariants", *IRE Trans. Information Theory 8,* 1962, 179–187.

(31) H. Freeman, "On the encoding of arbitrary geometric configurations", *IRE Trans. Electronic Computers 10,* 1981, 260–268.

(32) H. Freeman, "Computer processing of line-drawing images", *Computing Surveys 6,* 1974, 57–97.

(33) H. Blum, "A transformation for extracting new descriptors of shape", in W. Wathen-Dunn, ed., *Models for the Perception of Speech and Visual Form,* MIT Press, Cambridge, MA, 1967, 362–380.

(34) H. Samet, "The quadtree and related hierarchical data structures", *Computing Surveys 16,* 1984, 187–260.

(35) H. Blum and R.N. Nagel, "Shape description using weighted symmetric axis features", *Pattern Recognition 10,* 1978, 167–180.

(36) D.H. Ballard, "Strip trees: a hierarchical representation for curves", *Comm. ACM 24,* 1981, 310–321.

(37) M.A. Fischler and R.A. Elschlager, "The representation and matching of pictorial structures", *IEEE Trans. Computers 22,* 1973, 67–92.

(38) P.V.C. Hough, "Method and means for recognizing complex patterns", U.S. Patent 3069654, Dec. 18, 1962.

(39) R.O. Duda and P. E. Hart, "Use of the Hough transformation to detect lines and curves in pictures", *Comm. ACM 15,* 1972, 11–15.

(40) D.H. Ballard, "Generalizing the Hough transform to detect arbitrary shapes", *Pattern Recognition 13,* 1981, 111–122.

(41) H.G. Barrow, A.P. Ambler, and R.M. Burstall, "Some techniques for recognizing structure in pictures", in S. Watanabe, ed., *Frontiers of Pattern Recognition,* Academic Press, New York, 1972, 1–29.

(42) K.S. Fu, *Syntactic Methods in Pattern Recognition,* Academic Press, New York, 1974.

(43) K.S. Fu, *Syntactic Pattern Recognition and Applications,* Prentice-Hall, Englewood Cliffs, NJ, 1982.

(44) H.G. Barrow and J.M. Tenenbaum, "Recovering intrinsic scene characteristics from images", in A.r. Hanson and E.M. Riseman, eds., *Computer Vision Systems,* Academic Press, New York, 1978, 3–26.

(45) B. Horn, "Obtaining shape from shading information", in P.H. Winston, ed., *The Psychology of Computer Vision,* McGraw-Hill, New York, 1975, 115–155.

(46) B.K.P. Horn, "Understanding image intensities", *Artificial Intelligence 8,* 1977, 201–231.

(47) K. Ikeuchi and B.K.P. Horn, "Numerical shape from shading and occluding boundaries", *Artificial Intelligence 17,* 1981, 141–184.

(48) A.P. Witkin, "Recovering surface shape and orientation from texture", *Artificial Intelligence 17,* 1981, 17–45.

(49) P.M. Will and K.S. Pennington, "Grid coding: a preprocessing technique for robot and machine vision", *Artificial Intelligence 2,* 1971, 319–329.

(50) A. Guzman, "Decomposition of a visual scene into three-dimensional bodies", in A. Grasselli, ed., *Automatic Interpretation and Classification of Images,* Academic Press, New York, 1969, 243–276.

(51) D. Waltz, "Understanding line drawings of scenes with shadows", in P.H. Winston, ed., *The Psychology of Computer Vision,* McGraw-Hill, New York, 1975, 19–91.

(52) A.K. Mackworth, "Interpreting pictures of polyhedral scenes", *Artificial Intelligence 4,* 1973, 121–137.

(53) T. Kanade, "Recovery of the three-dimensional shape of an object from a single view", *Artificial Intelligence 17,* 1981, 409–460.

(54) S.W. Draper, "The use of gradient and dual space in line-drawing interpretation", *Artificial Intelligence 17,* 1981, 461–508.

(55) K.A. Stevens, "The visual interpretation of surface contours", *Artificial Intelligence 17,* 1981, 47–73.

(56) H.G. Barrow and J.M. Tenenbaum, "Interpreting line drawings as three-dimensional surfaces", *Artificial Intelligence 17,* 1981, 75–116.

(57) T.O. Binford, "Inferring surfaces from images", *Artificial Intelligence 17,* 1981, 205–244.

(58) G.L. Hobrough, "Automation in photogrammetric instruments", *Photogrammetric Engineering 31,* 1965, 595–603.

(59) S.T. Barnard and M.A. Fischler, "Computational stereo", *Computing Surveys 14,* 1982, 553–572.

(60) D. Marr and T. Poggio, "A computational theory of human stereo vision", *Proc. Royal Soc. B204*, 1979, 301–328.

(61) W.E.L. Grimson, *From Images to Surfaces – A Computational Theory of the Human Early Visual System*, MIT Press, Cambridge, MA, 1981.

(62) J.E.W. Mayhew and J.P. Frisby, "Psychophysical and computational studies towards a theory of human stereopsis", *Artificial Intelligence 17*, 1981, 349–385.

(63) R.J. Woodham, "Analysing images of curved surfaces", *Artificial Intelligence 17*, 1981, 117–140.

(64) Y. Shirai, "Recognition of polyhedrons with a range finder", *Pattern Recognition 4*, 1972, 243–250.

(65) D. Nitzan, A.E. Brain, and R.O. Duda, "The measurement and use of registered reflectance and range data in scene analysis", *Proc. IEEE 65*, 1977, 206–220.

(66) R.A. Jarvis, "A perspective on range finding techniques for computer vision", *IEEE Trans. Pattern Analysis Machine Intelligence 5*, 1983, 122–139.

(67) H.K. Nishihara, "Intensity, visible-surface, and volumetric representations", *Artificial Intelligence 17*, 1981, 265–284.

(68) R. Nevatia and T.O. Binford, "Description and recognition of curved objects", *Artificial Intelligence 8*, 1977, 77–98.

(69) C.L. Jackins and S.L. Tanimoto, "Oct-trees and their use in representing three-dimensional objects", *Computer Graphics Image Processing 14*, 1980, 249–270.

(70) A.A.G. Requicha, "Representations for rigid solids: theory, methods, and systems", *Computing Surveys 12*, 1980, 437–464.

(71) S.N. Srihari, "Representation of three-dimensional images", *Computing Surveys 13*, 1981, 399–424.

(72) R.A. Brooks, "Symbolic reasoning among 3-D models and 2-D images", *Artificial Intelligence 17*, 1981, 285–348.

(73) T.O. Binford, "Survey of model-based image analysis systems", *Intl. J. Robotics Research 1* (1), 1982, 18–64.

(74) W.N. Martin and J.K. Aggarwal, "Dynamic scene analysis: a survey", *Computer Graphics Image Processing 7*, 1978, 356–374.

(75) J.K. Aggarwal and N.I. Badler, guest eds., Special Issue on Motion and Time-Varying Imagery, *IEEE Trans. Pattern Analysis Machine Intelligence 2*, (6), November 1980, 493–588.

(76) T.S. Huang, ed., *Image Sequence Analysis*, Springer, Berlin, 1981.

(77) W.E. Snyder, guest ed., Special Issue on Computer Analysis of Time-Varying Images, *Computer 14* (8), August 1981, 7–69.

(78) T.S. Huang, ed., *Image Sequence Processing and Dynamic Scene Analysis*, Springer, Berlin, 1983.

(79) J.K. Aggarwal, guest ed., Special Issues on Analysis of Time-Varying Imagery, *Computer Vision, Graphics, Image Processing 21* (1,2), January and February 1983, 1–160, 167–293.

(80) S. Ullman, *The Interpretation of Visual Motion*, MIT Press, Cambridge, MA, 1979.

(81) E.C. Hildreth, *Measurement of Visual Motion,* MIT Press, Cambridge, MA, 1984.

(82) L. Uhr, "Layered 'recognition cone' networks that preprocess, classify, and describe", *IEEE Trans. Computers 21,* 1972, 758–768.

(83) S. Tanimoto and T. Pavlidis, "A hierarchical data structure for picture processing", *Computer Graphics Image Processing 4,* 1975, 104–119.

(84) S. Tanimoto and A. Klinger, eds., *Structured Computer Vision: Machine Perception through Hierarchical Computational Structures,* Academic Press, New York, 1980.

(85) A. Rosenfeld, ed., *Multiresolution Image Processing and Analysis,* Springer, Berlin, 1984.

(86) L.G. Shapiro, "Data structures for picture processing: a survey", *Computer Graphics Image Processing 11,* 1979, 162–184.

(87) A. Rosenfeld, R. Hummel, and S.W. Zucker, "Scene labeling by relaxation operations", *IEEE Trans. Systems, Man, Cybernetics 6,* 1976, 420–433.

(88) A. Rosenfeld, "Iterative methods in image analysis", *Pattern Recognition 10,* 1978, 181–187.

(89) L.S. Davis and A. Rosenfeld, "Cooperating processes for low-level vision: a survey", *Artificial Intelligence 17,* 1981, 245–263.

3-D VISION FOR ROBOT APPLICATIONS

D. Nitzan, R. Bolles, J. Kremers, P. Mulgaonkar
SRI International, Menlo Park, California, USA

I INTRODUCTION

A. **Robot vs. Human 3-D Vision**

Like human vision, robot vision should be able to perceive three-dimensional (3-D) objects, i.e., be able to detect, verify, recognize, locate, inspect, and describe different 3-D objects. Although real objects are three-dimensional, humans can perceive them on the basis of visual data that is two-dimensional (2-D) and incomplete. Utilizing a variety of range cues (e.g., geometric, perspective, texture and shading variations, and occlusion), human perception maps visual data into 3-D features and matches them with those of known models. Similar range cues are applicable to robot perception.

Human vision is based primarily on excitation of light intensity in the wavelength range of 0.38 - 0.77μm. While this type of sensory data plays a major role in robot sensing, the latter can also utilize any measurable noncontact sensory data, such as ultraviolet, infrared, sonar, range, and proximity data. In particular, sensing range data directly constitutes a major difference between human 3-D vision and robot 3-D vision. Since a human baseline (the distance between the eyes) is fixed around 5cm, the sensation of stereo disparity is effective for only nearby objects, say within 50cm. In contrast, a robot baseline can vary according to the application. Furthermore, a robot can use other schemes, besides stereo, to measure range data directly.

B. **Taxonomy of 3-D Vision Applications**

Robot applications of 3-D vision may be classified into two major categories: Man-made objects and natural objects. Scenes including man-made objects are easier to analyze, and most of the 3-D vision techniques that have been developed so far are applicable to these domains. Representation of natural objects (e.g., trees, rivers, lakes, boulders, and hills) is much more complex and three-dimensional scene analysis of natural objects is just beginning to be investigated.

Man-made objects may be classified into indoor objects (e.g., those found in manufacturing plants, storage and sorting facilities, offices, and households) and outdoor objects (e.g., roads, buildings, poles, cars, ships, airplanes, and spacecrafts). What characterizes such objects, from the viewpoint of 3-D vision analysis, is their geometry: their surfaces are well defined mathematically (e.g., by planes, cylinders, cones, etc.) or otherwise (e.g., by splines, contour lines, etc.) because their man-made construction is based on definable specifications. For this reason analysis of man-made objects of different sizes and in different domains may be based on similar 3-D vision algorithms.

Robotic application of 3-D vision in manufacturing is classified hierarchically in Table 1. At the top level are three major categories [Rosen 1979; Miller 1984]: visually-guided manipulation, visual inspection, and computer-aided design/computer-aided manufacturing (CAD/CAM). Each of these categories is divided into classes, each class may be subdivided into subclasses, and each subclass may be further subdivided into more specific tasks. Some of the applications listed in the table have been implemented by industry, especially by vision-system vendors, and have been described in the open literature. Most of these applications are marked by an asterisk (*) in Table 1.

II ROBOT SENSING

Given an environment in which a robot operates and a function it has to perform, we define *robot sensing* as the translation of relevant properties of objects in that environment into the information required to control the robot in performing that function [Nitzan et al. 1983]. Thus, for example, different types of sensing will provide the information required to control a robot that handles industrial parts, inspects printed circuit boards, arc welds workpieces, navigates outdoors or services satellites in space.

A block diagram representing robot sensing is shown in Figure 1. In the following sections we discuss these blocks separately.

Visually-Guided Manipulation

1. Part Picking and Placing
   - From Moving Conveyors
     > Flat*
     > Overhead
   - From Containers
     > Tray (semiordered; jumbled)*
     > Bin (semiordered; jumbled)*

2. Adaptive Manufacturing Processes
   - Sealing*
   - Spray Painting
   - Flash Removing
   - Routing*
   - Deburring
   - Drilling
   - Cutting
   - Grinding*

3. Assembly Operations
   - Part Presentation
   - Part Mating
   - Fastening
     > Spot Welding
     > Arc Welding*
     > Riveting
     > Bolting
     > Screwing
     > Nailing
     > Stapling
     > Gluing

------
* Have been implemented by industry and described in open literature [Miller 1984].

Visual Inspection

1. In-process
   - Material Handling
   - Assembly

2. Dimensional Measurement
   - Specified*
   - Overall Shape*

3. Surface
   - Defects*--cracks, voids, pits, burrs, warping, scratches, blemishes
   - Cosmetics--stains, color, texture

4. Part Integrity*
   - All Parts Present
   - Expected Parts
   - Expected Positions/Orientations
   - Expected Surface Image

5. Labels*
   - Position/Orientation
   - Expected Surface Image

6. Part Sorting

7. Safety
   - People
   - Robot
   - Workpieces

8. Execution Monitoring

CAD/CAM

1. Measurement of Real Parts for CAD/CAM Modeling

2. Predicting Images from CAD/CAM Models

Table 1: Robotic Applications of 3-D Vision in Manufacturing

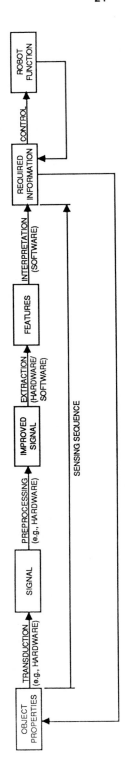

FIGURE 1 ROBOT SENSING

A. **Robot Functions**

By our definition, a robot consists of four components: effectors (arms, end-effectors, and legs), sensors (contact and noncontact), computer control (single or multilevel), and auxiliary equipment (e.g., tools and fixtures). *Robot functions* —tasks that a robot can perform—may be classified into active tasks, which require the actions of the robot effectors, and passive tasks, which do not require effector actions. The robot functions RECOGNIZE, LOCATE, FIND, VERIFY, GAUGE, and INSPECT are usually performed passively, although sometimes they may require manipulation. Active robot functions are, for example, GRASP, TRANSPORT, MOVE, PLACE, TRACK, CUT, GRIND, WELD, POLISH, SPRAY, SORT, ALIGN, INSERT, MATE, PRESS, FASTEN, ASSEMBLE, WALK, AVOID-COLLISION, AVOID-OBSTACLES, and NAVIGATE. Robot functions may be expanded hierarchically; for example, ASSEMBLE = (FIND, ACQUIRE, HOLD, MOVE, ALIGN, INSERT, VERIFY), FIND = (MOVE-CAMERA, RECOGNIZE, LOCATE), and RECOGNIZE = (TAKE-PICTURE, FIND-EDGES, EXTRACT-FEATURES, INTERPRET-FEATURES).

B. **Required Information and Object Properties**

The information required to perform a given robot function consists of entities, such as identity, location, geometry, force, and integrity, serving as input data to the robot. For example, the commands RECOGNIZE (location) and GRASP [identify (object), location (object), geometry (grasping), force (grasping)] have sufficient information for a robot to identify an object and grasp it properly. If the robot fails to perform its function because the information it received is inadequate, the robot may modify the required information (see the feedback loop in Figure 1).

In an unstructured environment robot sensing translates currently relevant object properties into the portion of the required information that is not known a priori. The *object properties* may be geometric (e.g., position, orientation, edge, corner, width, surface, volume, roughness, shape, and proximity), mechanical (e.g., weight, force/torque, and pressure), optical (e.g., reflectance, color, and texture), and material (e.g., acoustics, hardness, and temperature).

C. **Robot-Sensing Sequence**

As shown in **Figure 1**, the sequence of robot sensing consists of four steps:

(1) *Transduction*—Converting (in hardware or hardware and software) relevant object properties (e.g., surface geometry and reflectance) into a signal.

(2) *Preprocessing*—Improving the signal (usually in hardware); for example, removing irrelevant local image variation by low-pass filtering.

(3) *Feature Extraction*—Finding object features (in hardware or software) that may be compared with those of object models, including geometric features (e.g., straight and circular edges) and nongeometric features (e.g., reflectance, mass, and temperature).

(4) *Feature Interpretation*—Matching the extracted features with model features by hypothesis generation and verification, analyzing the results, and generating the required information (in software); for example, matching surface edges with those of a model of a cylindrical object, checking if such a match agrees with other objects and with the laws of physics, and determining the identity and location of the sensed object.

We do not believe that each of the four steps above can be implemented by means of a general purpose hardware or software subsystem. Instead, each step should be performed by alternative hardware/software schemes, regarded as *tools*, depending on the environmental conditions and the specified robot functions. To be able to carry out a wide variety of sensing tasks, the sensing system will consist of four sets of tools, called *tool boxes*, one for each step, and a knowledge-based "supervisor" that will select the best tools for each task.

D. **Iterative Sensing Strategy**

In some cases the information generated by a sensing sequence may be inconsistent with information based on other sources or inadequate because the transduced signal is incomplete (e.g., due to occlusion), is noisy, or has low resolution (purposely, to speed up processing). Under either one of these circumstances *iterative sensing* is performed, i.e., the sensing sequence is repeated under conditions modified by the sensing feedback. Such iterative sensing may be repeated more than once and may entail more than one sensor.

Examples of iterative sensing strategy are as follows:

- Use sparse range data to settle a disagreement between information extracted from vision and infrared sensors.

- Move a single vision sensor or use multiple ones to sense surfaces occluded by previous sensing.

- Sense the range of a dark object for a longer time than previously to cancel out the photon noise.

- Inspect image "windows," where defects may be found, with high resolution after the image of an object is found with low resolution.

III MEASUREMENT OF 3-D INFORMATION

Various techniques have been developed for measuring 3-D information. A taxonomy of these techniques is shown in a tree form in Figure 2. A distinction is first made between direct and indirect measurement of 3-D information. Direct measurement results in *range* data—values of the distances between a range sensor and a set of surface points in the scene. Indirect measurement is inferred from monocular images and results in range data or surface orientation.

A. Direct Range Measurement

Two basically different techniques can be used to measure range directly—triangulation and time-of-flight.

1. *Triangulation Techniques*

Triangulation is based on elementary geometry (see Figure 3): Given the *baseline* of a triangle, i.e., the distance between two of its vertices, and the angles at these vertices, the range from one of the vertices to the third one is computed as the corresponding triangle side. Triangulation techniques are subdivided into two schemes: stereo, using ambient light and two cameras (a passive scheme), and structured light, using a projector of controlled light and a camera (an active scheme). The plane in which the triangle lies is called the *epipolar plane* and its line of intersection with a camera image plane is called the *epipolar line*.

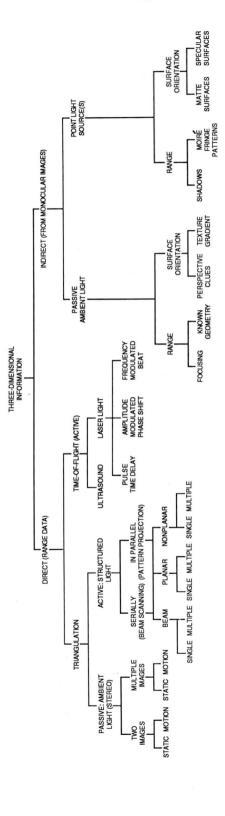

FIGURE 2 TECHNIQUES FOR MEASURING THREE-DIMENSIONAL INFORMATION

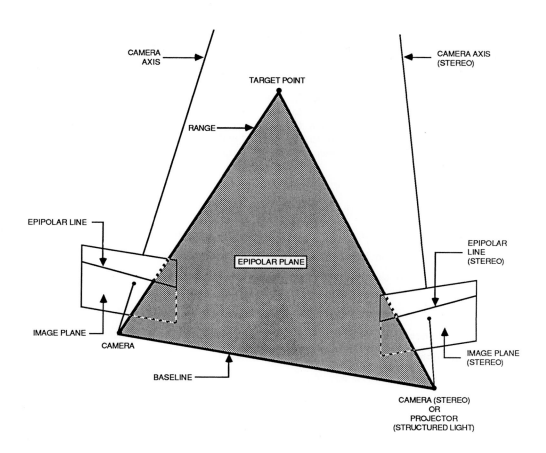

FIGURE 3   TRIANGULATION RANGE SENSING

The main drawback of any triangulation technique is *missing data* for points in the scene that are not "seen" from both vertices of the triangle. This problem can be partially solved in two ways:

- Decreasing the baseline. This remedy, however, will increase the measurement errors.

- Using multiple cameras for the stereo scheme or multiple projectors and/or cameras for the structured light scheme. This provision will also reduce the measurement errors and mitigate the problem of occlusion, including self occlusion, in machine vision, but it will increase the cost, complexity, and measurement time of the system.

a. <u>Stereo</u>

Relying on passive ambient light, triangulation *stereo* techniques use an image sensor (in particular, a TV camera) at each of two triangle vertices. A stereo pair of images can be obtained either from two static cameras (at different locations) or from one camera that is moved between two locations.

In addition to the missing-data problem, the main issue in stereo vision is the *correspondence problem*: How to match corresponding points in stereo images reliably and quickly. This problem has no solution if the two images have uniform reflectance. Conversely, the correspondence problem becomes easier as the stereo images include more intensity features, such as edges, especially if they are perpendicular to the epipolar line. These features should be extracted on the basis of micro constraints as well as macro constraints. For example, local intensity changes imply edge points, but if these points are too isolated to be linked into a continuous edge they should be disregarded. The effect of the correspondence problem is an increase in the measurement time.

b. <u>Structured Light</u>

One way to dispose of the correspondence problem is to use active light—a scheme in which one of the stereo cameras is replaced by a source of specially controlled illumination, called *structured light*. The structured light may be projected serially, by scanning a collimated light beam (usually a laser), or in parallel, either by diverging a laser beam with a cylindrical lens or by using a slit or a slide projector. The structured light may consist of single or multiple light patterns, each of which may be a straight line (a beam), planar, or nonplanar.

In addition to the missing-data problem, the structured light scheme entails two issues:

- *Specular Reflection*—Reflection from a mirror-like surface may result in no range measurement if the reflected light does not reach the camera, and false (larger or smaller) measured range values if the reflected light is subsequently reflected by other surfaces before part of it reaches the camera (see Section IV-B-1).

- *Slow Measurement*—Serial projection of multiple light planes requires too much time for data acquisition. This problem can be mitigated by projecting them in parallel but this entails determination of the correspondence between each light plane and the image of its intersection with the target. As a tradeoff between serial versus parallel projection, the following time-coded light pattern projection method was proposed [Altschuler et al. 1981]: Each plane among a set of different light planes is turned on or off during each of a sequence of time slots according to a given code and the resulting images are decoded to determine the correspondence between each plane and its image.

Another problem in structured light range measurement may arise from multiple reflections of the incident light beam if the target surfaces are simi-specular and concave (e.g., fillet joints of ground aluminum workpieces). This problem can often be solved by using a laser scanner, acting as a structured light, and a linear diode array, acting as a camera with a narrow rectangular field of view within which true target points are visible and most spurious ones are not. Such a range sensor was built at SRI [Kremers et al. 1983] for one-pass visual guidance of a robot performing arc welding of brushed or ground aluminum and steel workpieces. See Figure 4. A collimated infrared solid-state laser, A, is reflected by two steering mirrors, B and C, a scanning mirror, D, and a steering mirror, F, and hits a target point on a workpiece. A portion of the scattered light reflected from the target point enters the sensor through a window, H, is reflected by the scanning mirror, D, into a filter-lens assembly, I, and illuminates a few (e.g., 3) neighboring elements of a linear diode array, J. The total distance from the scanning mirror through the steering mirror, F, to the target point (regarded as the measured range) is determined from the average position of the illuminated elements. This measurement is repeated for a set of target points along a workpiece "slice" as the scanning mirror is rotated by a galvanometer driver, E.

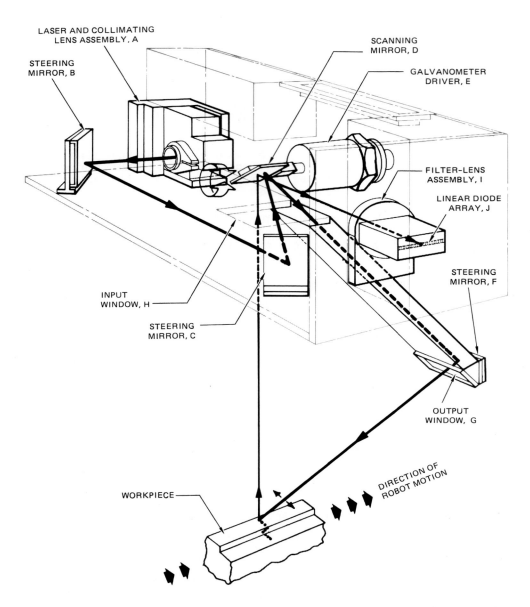

FIGURE 4   TRIANGULATION RANGE SENSOR USING LASER SCANNER AND LINEAR DIODE ARRAY

2. *Time-of-Flight Techniques*

A *time-of-flight* range sensor includes a signal *transmitter* and a signal *receiver* consisting of a collector of part of the signal reflected by the target and the electronics for measuring the round-trip travel time of the returning signal and its intensity. Two types of signal are practical: ultrasound (such as used by the Polaroid range sensor) and laser light.

A time-of-flight range sensor is an active device because it uses a signal transmitter. Any active range sensor, regardless of whether it is based on triangulation or time-of-flight or whether its signal is ultrasound or light, relies on Lambertian reflection of the incident signal and may malfunction if the target surface is too glossy. For an ideally specular surface (unless its normal happens to bisect the transmitter-target-receiver angle) the receiver detects no returning signal and there is no range data. Worse yet, if the receiver detects a returning signal that, following reflection by the target, is reflected serially by other surfaces, there is false range data. In practice, these adverse effects are significant if the signal wavelength is much larger than the average surface undulation and, hence, are worse for range sensing using ultrasound than that using light. Another drawback of ultrasonic range sensing is poor resolution (e.g., typically 4 x 4 to 10 x 10 over a 90 degree solid angle [Jarvis 1983]), which is caused by the difficulty in generating a narrow acoustic beam. For these reasons let us focus on range sensors using laser light.

Time-of-flight laser range sensors use a scanning mirror to direct the transmitted laser beam along pan-and-tilt orientations with equal angular increments in order to obtain dense range data consisting of N x M range elements, called *rangels*. Like with triangulation range sensing, by also measuring the intensity of the reflected light, we obtain an N x M array of TV-like intensity data in complete registration with the range data. However, the missing-data problem that is inherent in triangulation range sensing is eliminated by mounting the laser transmitter coaxially with the receiver's reflected light collector.

We distinguish among three schemes for measuring the length of the transmitter-target-receiver optical path in time-of-flight laser range sensing (See Figure 5):

(1) *Pulse Time Delay*—Using a pulsed laser and measuring the time of flight directly [Johnston 1973]. This scheme requires advanced electronics.

(2) *A.M. Phase Shift*—Using an amplitude modulated laser and measuring the phase shift, which is proportional to the time of flight [Nitzan et al. 1977].

(3) *F.M. Beat*—Using "chirps" of laser waves that are frequency modulated as a linear function of time and measuring the beat frequency, which is proportional to the time of flight [Goodwin 1985].

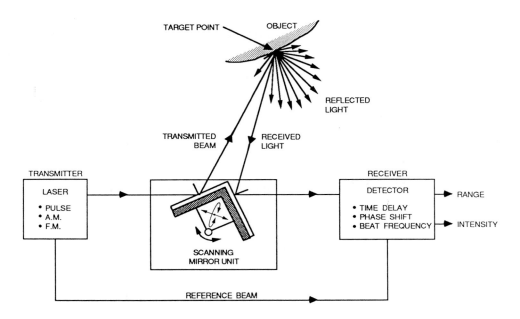

FIGURE 5   TIME-OF-FLIGHT LASER RANGE SENSING

Time-of-flight laser range sensors have the following problems:

- *Specular Reflection*—Reflection from a mirror-like surface may result in no range measurement if the reflected light does not reach the receiver, and larger measured range values if the reflected light is subsequently reflected by other surfaces before part of it reaches the receiver. See Section IV-B-1.

- *Slow Measurement*—A long integration time is required to reduce the photon noise (and other types of noise) to an acceptable level, especially if the target is dark. For given values of target reflectance, incidence angle, range, and measurement error, the integration time is inversely proportional to the product of the transmitted laser power and the area of the receiver's collector [Nitzan et al. 1977].

- *Ambiguity in A.M. Phase Shift*—If the phase shift, ϕ, between the transmitted light and the received light in an amplitude-modulated scheme may exceed 2π, then the (true) range, r, is ambiguous (see Section IV-B-2): $r = n\lambda + r(\phi)$, where n = 0,1,2, ..., λ is the wavelength of the modulation frequency, and $r(\phi)$ is the measured range assuming that $0 \leq \phi \leq 2\pi$.

B. **Indirect Measurement of 3-D Information**

A monocular image does not provide range information directly. However, under certain circumstances, geometrical constraints in the scene may be used to infer 3-D information (range or surface orientation) from a monocular image. The various techniques for obtaining range information indirectly are classified in Figure 2 into two groups, one using passive ambient light and the other using point light source(s).

1. *Passive Ambient Light Techniques*

The group of passive light techniques is further divided into two subgroups, one measuring range and the other measuring surface orientation.

a. <u>Range Measurement</u>

Two techniques have been proposed for indirect measurement of range based on a monocular image obtained under ambient light: Range from focusing and range from known geometry.

i. <u>Range from Focusing</u>

Consider the lens formula, $r = sf/(s-f)$, where r and s are the distances of an object and its image, respectively, from the lens center and f is the focal length. See Figure 6. The value of s can be determined by varying it until an image "window" is in sharp focus. This can be done only if the image window is not homogeneous. Knowing the values of s and f, the range r can then be evaluated.

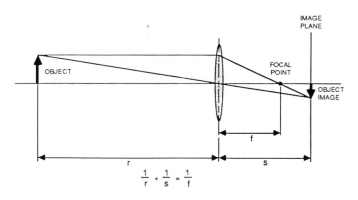

FIGURE 6   THIN LENS GEOMETRY

The depth of focus in air is

$$\Delta s = 2\lambda(f/D)^2[r/(r-f)]^2 \approx 2\lambda f^2/D^2 \qquad \text{for } r \gg f,$$

where λ is the light wavelength and D is the lens diameter. Therefore, the larger D the smaller the error in measuring s and r. Recent advances in range-from-focus techniques have been reported in [Krotkov and Martin 1986; Pentland 1985; Jarvis 1983].

ii. <u>Range from Known Geometry</u>

Although the projected 2-D image of a 3-D scene does not directly encode the distance to an object, if the identity and size of the objects in the scene and the camera parameters are known, range values can be computed on the basis of the observed scale changes. Many such *model-based techniques* have been described in the literature. In the following discussion we describe a few representative examples, assuming that we know the position of the lens center of a camera, the direction of its optical axis, and its focal length f.

Given the image of a point in space, its range (its distance from the camera center) can be computed only if the point is known to lie on a given surface (e.g., a table of a known height). Now suppose that the camera views a sphere of known radius R. If $R \ll r$, where r is the distance between the centers of the lens and the sphere, then the ellipsoidal image of the sphere may be approximated by a circle. Measuring the radius R' of this image circle, the range r can be computed from the relation

$$r \approx R\,[1 + (f/R')^2]^{1/2} \approx Rf/R'.$$

Finally, consider a known object of arbitrary shape. If the view angle of the object is narrow, then $r_j \approx r$, where r_j is the range of any jth point belonging to the object and r is the range of the center of a sphere enveloping the object (the sphere's center coincides with the centroid of the object vertices and its radius R is the largest distance between the centroid and any of these vertices). Measuring a radius R' of a circle enveloping the object image, then $r \approx Rf/R'$, where R and f are known values.

If the view angle of the object is wide, then local geometric features should be examined. If the images of six of the object's vertices (or other landmarks) are identified, then the position and orientation of the object can be computed uniquely [Fischler and Bolles 1980]. If more than six vertices or landmarks are identified, then the redundant information can be used to verify the identification of the first six vertices and to refine the computed position and orientation of the object. Having determined the position and orientation of the object, the range values of any one of its vertices or landmarks can be readily computed.

b. <u>Measurement of Surface Orientation</u>

Two techniques have been proposed for indirect measurement of surface orientation (or surface gradient) from a monocular image obtained under ambient light—one based on knowledge of perspective geometry and the other on texture gradient.

i. <u>Shape from Perspective Clues</u>

The geometry of perspective projection provides strong constraints on the possible interpretation of scenes. For example, the image of a set of n parallel lines on a planar surface, $z = f(x,y)$, in a scene consists of n straight lines that intersect at a single vanishing point (x_v', y_v'). The *vanishing point* is the image of a point at infinity in the direction of the parallel lines. The line of sight of the vanishing point is collinear with the vector $(x_v', y_v', -f)$ and is parallel to the planar surface, i.e., normal to its gradient vector $(p,q,1)$, where $p = \partial z/\partial x$ and $q = \partial z/\partial y$. Since the above two vectors are perpendicular, their dot product is zero, i.e., $x_v'p + y_v'q - f = 0$. Knowing f, we can solve for p and q by using the x_v' and y_v' values of two vanishing points corresponding to two sets of parallel lines that lie on the same plane but have different directions.

Such constraints have been used in a rule-based hypothesis generation and verification system to determine the orientations of edges and surfaces in perspective images of 3-D objects [Mulgaonkar and Shapiro 1985]. A practical problem with this technique is the difficulty in accurate detection of such edges.

ii. Texture Gradient

There are three methods for computing the orientation of a planar surface from the image of its texture. The first method is based on the shape of a texture element, called *texel*. For example, the image of a circular texel is an ellipse; the average values of its major axis orientation and minor-to-major axis ratio can be used to compute the surface rotation and tilt, respectively, with respect to the optical axis of the camera [Stevens 1979]. The second method is based on the maximum rate of change (i.e., gradient) of the texel size. The direction and magnitude of the texel-image gradient can be used to compute the surface rotation and tilt, respectively, with respect to the optical axis of the camera [Witkin 1981]. The third method is based on the vanishing points of many virtually parallel lines in the texture. For every different pair of texels, a vanishing point is established at the intersection of each of their two tangents and the line passing through their centroids. Fitting a straight line to the resulting vanishing points and representing that line by two (x_v', y_v') points, the surface orientation can then be computed, as described previously, by solving two equations of the form $x_v'p + y_v'q - f = 0$ for p and q [Ohta et al. 1981].

2. *Point-Light-Source Techniques*

As with passive light techniques, techniques for indirect measurement of 3-D information from monocular images utilizing point light source(s) are classified into two subgroups (see Figure 1)—one measuring range and the other measuring surface orientation.

a. Range Measurement

Two range-measurement techniques utilizing a light source have been proposed—one based on shadows and the other on Moire' fringe patterns.

i. Shadows

A technique using shadows to measure range of a thin object was reported by Tsuji et al. (1983). See Figure 7. Suppose that we know the position, L, of a point light source, the camera geometry (the lens-center position C, the focal length f, and the optical axis direction), and the height of a table under the light source and the camera. An object "point" (a small particle or the end of a linear object, such as a wire), P, and its shadow point on the table, S, are projected onto the camera image plane at points P' and S', respectively. By computing the intersection of Line CS' with the table we first find Position $S(x,y,z)$. By computing the intersection of Lines CP' and LS we then find Position $P(x,y,z)$. Now suppose that the system should measure the (x,y,z) values of multiple object points, such as pairs of wire ends. To do that, the correspondence between the image points P'_j and S'_j of each jth object point is established if image lines $L'P'_j$ and $L'S'_j$ are collinear, where L' is the known image point of the light source, because Points L, P, and S lie on a 3-D straight line and because such a line is mapped into a straight line in the image plane. The main limitation of this technique is that it cannot handle solid 3-D objects because self occlusion will prevent most object points and their shadows from being "seen" by both the light source and the camera.

Shadows cast by point light sources can also be used to detect and locate range discontinuities. Figure 8 shows the principle of such a system. Two diagonally opposite point-light sources alternately illuminate a stack of flat objects (e.g., letters) of different size and thickness. As a result, different shadows appear and disappear if there are range discontinuities in the scene. Two intensity images, each obtained by a single camera while one of the lights is turned on, are normalized to correct for any nonuniformity in the light sources. The shadows are extracted by subtracting the two normalized images, which are identical except for the shadows. Identifying the shadows with range discontinuities and knowing where the top of the stack is, the location of the top flat object can thus be determined.

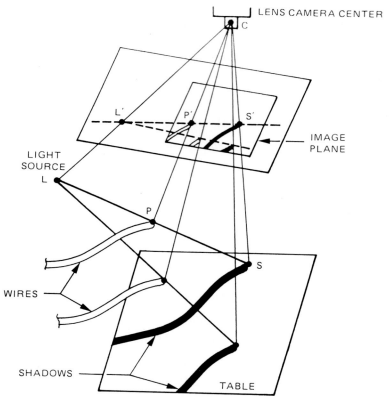

FIGURE 7   MEASURING WIRE-TIP POSITIONS FROM INTENSITY IMAGES OF THE TIPS AND THEIR SHADOWS

ii. Moire' Fringe Range Contours

A Moire' fringe interference pattern is obtained by illuminating a scene with a light source passing through a grating and viewing the scene with a laterally displaced camera through an identical second grating; e.g., see [Jarvis 1983]. The interference pattern constitutes equidistant contours, but there is no information about the corresponding range values or even the sign of the change in range of neighboring contours. The sign information can be obtained by using a phase shift in the second grating, but range recovery is a research issue. This technique can be used to measure only relative range values of continuous contours, which correspond to smooth surfaces.

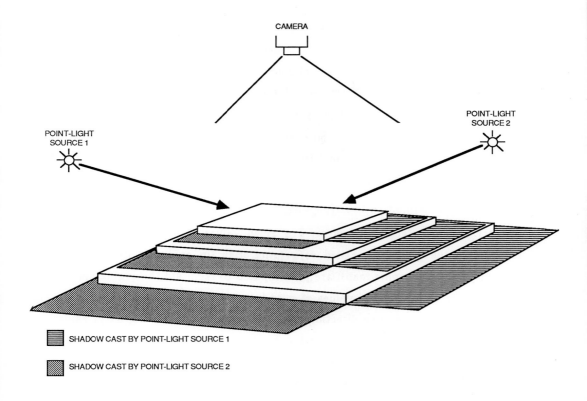

FIGURE 8   USING SHADOWS FROM POINT-LIGHT SOURCES TO DETECT AND LOCATE RANGE EDGES

b. Measurement of Surface Orientation

Techniques using distant light sources and measuring surface orientation from the resulting image brightness are classified below into techniques for matte surfaces and techniques for specular surfaces.

i. Techniques for Measuring Matte Surface Orientation

Computation of *shape from shading* was introduced by Horn (1975) and investigated further by others [e.g., Woodham 1981; Pentland 1984]. The shape of a smooth surface, $z = f(x,y)$, is represented by the gradient components $p = \partial z/\partial x$ and $q = \partial z/\partial y$, where $(p,q,1)$ are the projections (or components) of a vector normal to the surface at a point $P(x,y,z)$. See Figure 9. For a plane, $Ax + By + Cz + D = 0$ or $z = px+qy+k$, where $p = -A/C$, $q = -B/C$, and $k = -D/C$. The orientation of a point-light source is similarly represented by a vector $(p_s, q_s, 1)$. Viewing the surface from a far distance along the $-z$ direction, the image irradiance (brightness), $I(x,y)$, recorded by a camera is identified with a reflectance function $R(p,q,p_s,q_s)$. Since $R(p,q,p_s,q_s)$ is proportional to the cosine of the incidence angle, which is equal to the dot product of the vectors $(p,q,1)$ and $(p_s, q_s, 1)$ normalized, we get

$$I(x,y) = R(p,q,p_s,q_s) = r(pp_s + qq_s + 1)[(p^2 + q^2 + 1)(p_s^2 + q_s^2 + 1)]^{-1/2} ,$$

where r is the surface reflectance. For given values of p_s and q_s, constant-R contours in the p-q plane, called *reflectance map* (in analogy with constant-z contours of a topographic map), are precomputed to facilitate the evaluation of p and q. If r is known, the values of p and q can be determined by using two different (p_s, q_s) light-source directions, measuring the corresponding $I(x,y)$ values, and solving two equations for p and q by finding the intersection of the two corresponding contours in the precomputed reflectance maps (or by lookup at a two-dimensional table). However, since these equations are nonlinear, wrong (p,q) solutions may be found as well; hence, three (p_s, q_s) light-source directions and the corresponding three reflectance maps should be used to remove this ambiguity. Now if r is unknown, then an additional (p_s, q_s) direction and the corresponding reflectance map are required. Called *photometric stereo*, this technique was developed by Woodham (1978). A variation of this technique uses only a single light source but rotates the object under the camera [Kitagawa 1983].

ii. Techniques for Measuring Specular Surface Orientation

If a point-light source is used, the radiance reflected from a specular surface patch will not reach an imaging camera, unless the patch normal happens to bisect the source-patch-camera angle. Hence, point-light sources cannot be used to measure the orientation of the specular surfaces of a glossy object. To overcome this problem, Ikeuchi (1981) extended the photometric stereo technique to specular surfaces by using a nonuniform distributed light source and calculating

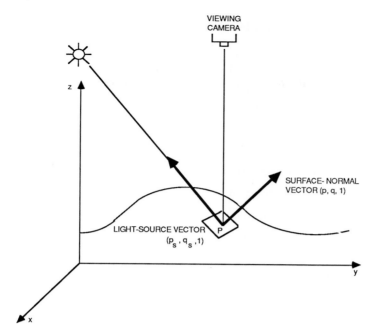

FIGURE 9   VIEWING A SURFACE ELEMENT ILLUMINATED BY A POINT-LIGHT SOURCE

the corresponding reflectance map. A large distributed light source guarantees that a specular surface patch of any orientation within a wide solid angle will reflect a portion of the distributed light, which acts as a virtual point source of orientation $(p_s, q_s, 1)$, into the camera. Note that since the object is glossy, the normal of a surface patch bisects the source-patch-camera angle and, hence, the p and q values of that patch are uniquely related to the values of p_s and q_s. Three linear lamps were placed symmetrically 120 degrees apart under the object, a planar Lambertian light reflector was placed above the object, and a camera observed the object from above through a hole in the Lambertian reflector. Virtual (p_s, q_s) point-light sources of three levels of brightness were achieved by turning on one lamp at a time. The three corresponding $R(p,q)$ reflectance maps, calculated for specular surfaces illuminated by this setup, were used to determine the surface orientation of shiny metallic objects (hooks, nuts, and bolts).

Another technique for measuring the orientation of specular surfaces used a distributed source of circularly polarized light [Koshikawa 1982]. Polarization parameters are derived from eight images taken at different angles of an analyzer retarder. The surface orientation is then determined from experimental relations between these parameters and the image intensity.

IV SIGNAL PREPROCESSING

The result of the signal transduction step, called an *image*, is a set of data values having a well-defined spatial organization. We distinguish between an *intensity image*, a rectangular array of intensity values, and a *range image*, a rectangular array of range values. For a range image produced by a time-of-flight scanner may consist of a rectangular array of distance values, with rows and columns having equal angular increments of the scanning device around the center position. Measured intensity or range data, corresponding to a spatial point in the scene, are associated with each image element. Such data contain two types of error: Error in the value of the measured data and error in the geometry of the image coordinate (i.e., error in the direction of the line of sight of the observed point). The goal of the signal preprocessing step (see Figure 1) is to transform the image produced by the sensor into an improved image in which these errors are minimized. The specific operations applied at this step depend on the characteristics of the expected errors, which in turn depend on the sensor itself. Using a sensor model, the preprocessing step is usually bottom-up or data driven in nature.

A. Preprocessing Intensity Data

There exists a large body of knowledge dealing with the improvement of the quality of intensity images. Noise minimization techniques, such as Gaussian, median, or other weighted filters in both one and two dimensions, are commonly applied in many computer-vision applications [Ballard and Brown 1982]. Depending on the nature of the application, other intensity modifications may also be used. For example, a logarithmic transformation enhances the dynamic range in the darker end of the image [Rosenfeld and Kak 1976]; histogram equalization techniques spread the most significant portion of the intensity histogram [Rosenfeld and Kak 1976]; color transforms are often used to map the machine's basis of the color space (RGB) to other tri-stimulus frames [Kender 1976; Ohta 1980; Pratt 1971], such as chromaticity or principal components; nonlinear mappings of pixel intensity values may be used to compensate for the bias and gain of each individual pixel in a CCD camera array. The key point to note is that no new information is created in these operations. In fact, most filtering operations lose information due to, say, the blurring or integration over a wide spatial region. However, the transformed values are better in other respects, such as their signal-to-noise ratio or dynamic range.

Correction of geometric errors is often very important in cases where multiple sensors are used. Such errors lead to a misregistration of the information obtained from the sensors and make sensor fusion difficult. Errors in the geometry of the image are corrected through a process of *calibration/decalibration*. By imaging a target with a known geometry the errors can be quantified, and subsequent images can then be decalibrated by applying the inverse transformation. Common geometric errors, such as barrel and pincushion distortions, skew, and rotation, can be corrected by assuming a suitable parametric form for the errors and using the calibration images to estimate the values of these parameters. Several different types of calibration targets have been used, the most common one being a field of uniformly placed lines and dots [Gennery 1982]. Decalibration too does not preserve information because geometric transformations change the spatial distribution of the image and, consequently, the spatial frequency content of the image is not preserved. Decalibration may also map a few adjacent pixels onto a single pixel and thus produce image blurring.

B. Preprocessing Range Data

Range images include errors that are affected by the reflectance and geometry of object surfaces in the scene, as well as by the scheme and mode of operation of the sensor itself. For example, "blooming" in the camera of a triangulation sensor using structured light may produce spurious range values. Both triangulation and time-of-flight range sensing are based on Lambertian reflectance and thus are adversely affected by reflectance extremes of two types—dark surfaces and specular surfaces. A dark surface reflects few photons; hence, a longer integration time is necessary to improve the signal-to-noise ratio [Nitzan et al. 1977]. In Section III-A we mentioned some range errors caused by surface specularity; in Section IV-B-1 we illustrate the sources of these errors and how to detect them. We also mentioned in Section III-A that another type of range error, which is characteristic of amplitude-modulated time-of-flight range sensors, may result from the ambiguity in the measured phase shift if it may exceed 2π; such range ambiguity is discussed in Section IV-B-2.

1. *Range Errors Caused by Surface Specularity*

Figure 10(a) shows a cross section of an object whose range image is measured by a triangulation range sensor, which includes a light-beam projector, P, and a camera, C. Suppose that Surfaces D-E and F-G are matte and Surface E-F is glossy. As the light beam hits Point j on a matte surface, where j = 1,2,6,7,8, a true range value $\overline{Cj}$ is measured. However, if a beam hits Point k on a specular surface, where k = 3,4,5, the entire beam is reflected and hits Point k', which may or may not be seen by the camera. As a result, in this example the sensor will measure no range value for Point 3 and, worse yet, wrong range values $\overline{C4^*}$ and $\overline{C5^*}$, which are smaller than the true range values $\overline{C4}$ and $\overline{C5}$, respectively. Note that the erroneous range values could also be larger than the true values. For example, if Surface E-F were matte and Surface F-G were specular, then the wrong range will be larger than the true range.

Range errors caused by surface specularity may be detected by a heuristic based on the assumption that the x values of the points observed by the camera are monotonic, where x is defined along C-P in the camera coordinate system. If all the object surfaces were matte, then these x values will change monotonically as the tilt angle of the projected light beam changes. On the other hand, since Surface E-F is specular, the x values of Point 4* (if Point 3* is seen by the camera) and Point 5* are nonmonotonic. If it is known (from either world knowledge or an intensity image) that Points 4* and 5* cannot belong to occluding objects, then the corresponding range values are regarded as erroneous and should either be measured correctly by other means or ignored.

Now suppose that the range image is measured by a time-of-flight laser scanner; hence, compared with a triangulation scheme, the camera is replaced by a photomultiplier and Point C coincides with Point P. See Figure 10(b). The laser light hitting Point k on a specular surface (e.g., k = 4,5) is reflected to Point k' on a Lambertian surface, and a portion of that light is reflected back to

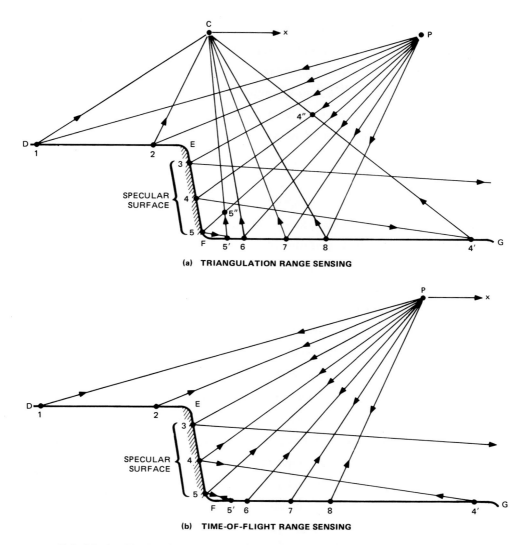

FIGURE 10   ERRONEOUS RANGE MEASUREMENT CAUSED BY SPECULAR SURFACE

Point P via Point k. Since the view angle of the sensor photomultiplier is narrow (e.g., 1 degree), the resulting measured range value is 1/2 $(\overline{Pk} + \overline{kk'} + \overline{k'k} + \overline{kP})$ = $\overline{Pk} + \overline{kk'}$, which is always larger than the true value, 1/2 $(\overline{Pk} + \overline{kP})$ = $\overline{Pk}$. Furthermore, since this scheme does not employ a camera, the x value of Point k' cannot be computed and, hence, an error in the range value cannot be detected from the raw data. Detection of such an error may require the use of world knowledge.

2. *Ambiguity in A.M. Time-of-Flight Range Data*

As described in Section III-A-2, range measurement by amplitude-modulated time-of-flight laser scanners is performed by measuring the phase shift, ϕ, between the transmitted light and the received light. As ϕ varies between 0 and 2π, such a sensor can properly measure a range value $r(\phi)$ within the wavelength, λ, of the modulation frequency. Range values beyond the modulation wavelength result in *range ambiguity* caused by a "wraparound" phase shift. In general, assuming no limitations caused by insufficient laser power, the true range between $n\lambda$ and $(n+1)\lambda$ is $r = n\lambda + r(\phi)$, where $n = 0,1,2,...$. Given ambiguous measured range data, $r(\phi)$, the function of the range preprocessing step is to detect any unreal, wraparound jumps in range values and correct them by adding $n\lambda$ to each value. A method for detecting such jumps in range is proposed as follows. Figure 11 shows, for example, an outdoor range image, where closer points are depicted darker, that was measured by using a range laser scanner made by the Environmental Research Institute of Michigan (ERIM) and whose $\lambda = 64$ feet [Hebert 1985]. Examining the data, one finds that a big jump in range (slightly larger than λ) occurs around $r = \lambda$, resulting in a jump edge across the range image. Except for the very special case in which the sensor is placed in front of a cave or a tunnel of the corresponding dimensions, such a jump edge implies that the jump is not real. This hypothesis may be verified by computing the slopes of the surfaces below and above the jump edge.

Where significant range discontinuity occurs, the scanning laser beam often splits and illuminates two surfaces. The resulting two-phase shifts, ϕ_1 and ϕ_2, where $\phi_1 < \phi_2$, are represented by ϕ, where $\phi_1 < \phi < \phi_2$. Hence, $r(\phi_1) < r(\phi)$

$< r(\phi_2)$, i.e., the measured range corresponds to a point between the two surfaces. Such a measurement is "reasonable" if both ϕ_1 and ϕ_2 are between 0 and 2π, but if $0 \leq \phi_1 \leq 2\pi$ and $2\pi \leq \phi_2 \leq 4\pi$, then the error is significant and should be corrected accordingly.

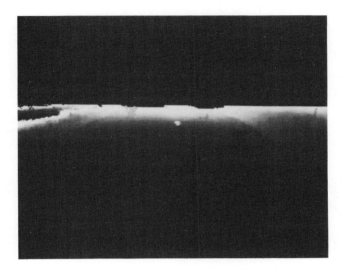

FIGURE 11 DISPLAY OF MEASURED RANGE IMAGE WITH AMBIGUITY

(Closer Points are Depicted Darker)

V FEATURE EXTRACTION

In this section we discuss and classify features of 3-D objects and describe methods for extracting them from measured images.

A. Feature Classification

A *feature* in the image of a 3-D object is an entity, consisting of one or more measurable properties, that is characteristic of the object. The aggregate of the features of an object constitutes the *model* of that object. A feature may be classified according to three symbolic categories: Feature physical properties, feature size, and feature relation. See **Figure 12**.

In 3-D vision we are concerned with two types of *feature physical properties*—geometric and photometric. For example, the normal and the area of

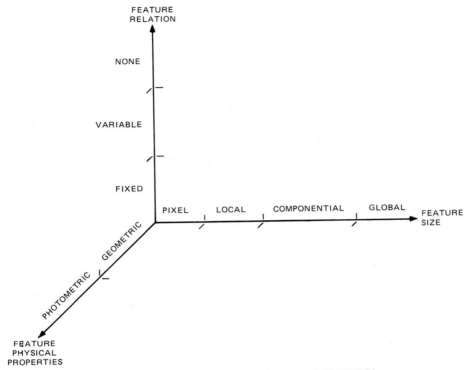

FIGURE 12   SYMBOLIC CLASSIFICATION OF FEATURES

a planar surface are geometric properties, whereas the intensity histogram and color of a label on that surface are photometric properties. *Geometric features* can be measured directly from a range image and sometimes indirectly from an intensity image. For example, the outline of an object can be found by detecting large discontinuities in range values; it can also be found by detecting edges in an intensity image if there is sufficient contrast between the object and its background. On the other hand, *photometric features* can be measured directly from an intensity image but not from a range image. Geometric features may be derived from 3-D geometric models, such as used in CAD systems. Photometric features may be derived from both 3-D geometric models and reflectance models, but the computation is more complex and the results are less realistic than those of geometric features.

The second category of feature classification is the *feature size* relative to the size of an object. Partitioning an object hierarchically, we observe that an object consists of components, each component consists of local portions, and each local portion consists of elements, the smallest of which are represented by pixels. Accordingly, we distinguish among features of four size levels: pixel, local, componential, and global features.

Pixel features are measured or computed values associated with individual pixels: Geometrical pixel features include range and location (position and orientation); photometric pixel features include intensity and color.

Local features characterize small parts of the image of an object component. If an object is partially viewed (because of a narrow view angle or partial occlusion by other objects), some of its local features are likely to be seen by a sensor because a local feature is typically much smaller than a componential feature of the same object or other objects on the scene. Examples of local features given as follows are represented (in parentheses) by lower level properties. Geometric local features include: (1) short edges, such as a small segment (location, length), a corner (location, angle), a circular hole (location, radius), and a curvature (location, radius, angle); (2) small surfaces, such as a circle (location, radius), and an arbitrary patch (location, area); (3) small volumes, such as a polygon's corner (location, number of edges, angles between edges), a sphere (position, radius), and a blob (position, volume). Photometric local features are derived from intensity properties of a small surface and include intensity histogram, average intensity, color, and texture.

Componential features characterize components of an object but usually not the entire object; they are similar to local features, except that they may be considerably larger. Geometrical componential features may include a complete edge or the face of an object part (e.g., the base or the envelope of a cylinder), as well as their combination, such as an edge contour, a straight dihedral (plane-edge-plane), or a circular dihedral (plane-edge-cylinder). Photometric componential features include those of local features as well as intensity edges and their combinations.

Global features characterize the entire image of a 3-D object. Geometrical global features are based on the topology or connectedness of the object components and the geometrical features of each component. Photometric global features are useful if the number of possible object orientations is small. The binary vision module [Gleason and Agin 1979], which was developed at SRI in the 1970s, is based on global features of silhouettes under this condition. These features include the silhouette's area, perimeter, moments, number and area of holes, radii (minimum, average, and maximum) from the centroid to the perimeter, and the like. The same features are applicable if the object outline is extracted from a gray intensity image by using one of the edge-detection methods.

So far we have discussed physical object features that are inherent. As such, they are represented in Figure 12 by two symbolic "dimensions"—physical properties and feature size. Now we wish to examine how these features are related to each other. Thus, a third category of feature classification is *feature relation* (or connectedness). Adjacent pixel features in a homogeneous image region are likely to be similar, whereas across an edge they are dissimilar. The relation between local features or between componential features of an object may be fixed, variable, or none—corresponding to strong constraints, loose constraints, and no constraints among the object features. The features of a rigid object are characterized by a *fixed relation* (e.g., the relative positions of small holes in an automobile floor or the legs of a table). A *variable relation* between features is typical of the components of flexible objects (e.g., a rubber hose), objects with both rigid and flexible components (e.g., a telephone), objects with jointed rigid components (e.g., scissors), and any combination of these classes (e.g., a robot arm with rigid and flexible links). Essentially *no relation* between features is characteristic of limp material, such as cloth, when not under tension. Finally, the relation between global features results from constraints between separate objects. Here, too, the relation may be fixed (e.g., fixtured parts), variable (e.g., parts on a bowl-feeder track or in a bin), or none (e.g., objects in zero gravity). A combination of these relational classes may also occur; for example, parts on a moving conveyor belt have a fixed relation in the z direction but no relation in the x and y directions, unless they touch each other.

B. **Feature Extraction Techniques**

Many techniques have been developed in the past for extracting photometric features (directly) and geometric features (indirectly) from intensity (gray-level or color) images. The majority of the techniques extract these features from bounding *edges* corresponding to discontinuities in the intensity image. Several *edge detectors* have been proposed, such as the Robert's cross operator [Roberts 1965], the Sobel operator [Sobel 1970], the directional zero-crossing Haralick operator [Haralick 1984], and the difference of Gaussian Marr-Hildreth operator [Marr and Hildreth 1980]. Under favorable conditions (e.g., simple 3-D vision tasks and controlled illumination) edges of interest may be found by first using a fixed intensity *threshold* to convert the gray-level image into a binary image and then applying a *run-length coding* operator to detect black-to-white and white-to-black transitions [Gleason and Agin 1979]. If the conditions are less favorable, adaptive or recursive threshold, rather than a fixed one, should be used [Ohlander 1978].

Once the bounding edge of a feature has been found, the descriptors of its interior surface (e.g., intensity histograms of red, green, and blue images) are readily measurable. Under certain conditions (e.g., an outdoor scene with textured regions), it may be more efficient to proceed in an opposite way: first use a *region growing* routine to detect a feature region and then extract its bounding edge. Recent advances have also considered optimal merging of edge information with boundaries of homogeneous regions [Levine 1983].

Edges in range images are classified into two basic types [Nitzan 1972; Bolles and Horaud 1986]: *jump edges*, where the range is discontinuous, and *convex edges* or *concave edges*, where the range gradient is discontinuous. As stated previously, geometric features can be extracted directly from range images, whereas photometric features cannot. Like intensity images, range images are single valued functions of the two image coordinates; hence, most of the techniques developed for extracting features from intensity images should be applicable to range images. The applicability of such techniques depends on the quality of the range image. For example, a triangulation range sensor may produce incomplete range images—images with no data for points not seen from both vertices of its baseline. Edge detectors, such as the Marr-Hildreth operator, may malfunction if the range image contains no-data regions. Depending on the surrounding range values, the zero crossings can stray away from the true edge into the no-data regions. To overcome this problem, the various geometric edges in range images can be detected by special routines [Duda et al. 1979; Bolles and Horaud 1986]. These routines examine the range values of each row and each column and fit a moving segment between break points, where either the range is discontinuous (indicating jump edges) or the range gradient is discontinuous (indicating convex or concave edges).

Three-dimensional objects are often modeled by polyhedra because the latter are easy to analyze, are basic components of many man-made objects, and can be used to approximate curved or irregular objects. An iterative technique for sequential extraction of planar surfaces from a range image was developed by Duda et al. (1979). Surface images (regions) are fitted by planes, each described by its unit normal vector and distance to a coordinate system origin. Surface extraction is performed in the order of its expected reliability, i.e., according to the region size (large regions are extracted before small ones) and surface orientation (first horizontal, then vertical, and finally slanted surfaces). For each significant planar surface in the scene, an appropriate histogram technique is used to find a starting plane, P_0, as a first approximation and then the plane fitting is refined in an iterative fashion. Detection of the P_0 starting planes depends on the surface orientation as follows: Horizontal planes correspond to the neighborhoods of peaks in a z histogram after its "background" has been removed by proper filtering; vertical planes correspond to the neighborhoods of peaks in the angle-distance Hough transform of linear clusters of the x-y values of the remaining image points; and slanted planes correspond to the neighborhoods of the highest peaks in intensity histograms of large regions. Recent surface fitting techniques reported in the literature use similar region growing techniques for a wide class of surfaces including planar and quadric surfaces [Yang and Kak 1986].

VI FEATURE INTERPRETATION

Feature interpretation is the process of identifying one or more features in sensory data and deducing the information required to perform a specific robot vision function, such as locating an object in a pile of objects or identifying obstacles in front of a vehicle. Interpretation is the fourth basic step in the robot sensing sequence shown in Figure 1.

The best interpretation technique for a task is a function of the goal of the task and the a priori information available for the task. In Section VI-A we describe the types of a priori information that may be available for a given task. In Section VI-B we describe some general principles of feature matching and interpretation. In Section VI-C we first define five basic classes of vision tasks in terms of their a priori information and goals, and then describe the appropriate techniques for implementing these classes of tasks.

A. A Priori Information

We distinguish among three types of a priori information: information about the scene to be sensed, information about the sensors, and information about the processing procedures.

1. *Scene Information*

Information about the scene may include models of the expected objects, estimates of the number of objects in the scene, and specifications of the expected object locations. The models may be specific CAD models, generic models, or implicit models. For example, in an obstacle detection task, the a priori scene description may include a generic model of a road at an expected location relative to the vehicle and an implicit model of the obstacles (i.e., deviations from the locally flat road). On the other hand, for an industrial part recognition task, the initial information may contain detailed descriptions of the geometry and photometry of the objects to be recognized and a statement that the objects rest on a conveyor belt in such a way that they may touch, but not overlap.

The variety of potentially useful information about a scene is enormous, which makes it difficult to codify. Not too surprisingly, however, this information and its use within the processing procedures is crucial to the performance of the task. Although progress has been made in the development of object modeling techniques, positional uncertainty representations, and the like, we feel that this area of scene description is one of the weakest links in the development of generic vision systems.

2. *Sensor Information*

The second type of a priori information is information about the sensors used to gather data from the scene. We include in this category information about the lighting as well as the sensors themselves, because many sensing techniques project special light onto the scene. Thus, for an outdoor navigation task the sensor information would include a model of the sun, a model of the sky, and a model of the light emitted by any active sensors being used, such as a laser scanner.

The geometry of range sensors is generally well understood. There are even noise models for most sensors. Some models are analytic expressions relating the noise to object parameters; others are based on measurements of the behavior of the sensor in controlled scenes. We have found both of these models to be very useful because they make it possible to tune the preprocessing steps (described in Section VI-C) to the specific sensor.

3. *Processing Information*

The third type of a priori information is a description of the processing steps to be applied to perform a specific vision task. This information may be an analytic description of the behavior of a procedure or it may be a statistical statement about the expected outcome of a procedure. For example, each edge detection procedure should include a description of precision with which it can locate edges of different types, and each feature extractor should include a statement of the probabilities of finding specific features in certain circumstances. This type of information is difficult to extract and represent, and yet it is crucial to the programming of a vision task.

We see almost a complete void in this area. As far as we know this type of information resides in the minds of a few experts, but has not been written down in a way that a person or a machine could make use of it. Another way of saying this is that computer vision is an art. We need to make it more of a science.

B.   **Feature-Interpretation Techniques**

We wish to select feature-interpretation techniques that maximize the reliability of the results and minimize the cost of obtaining them. This goal often implies a strategy that tries to locate as few features as possible and extract as much information as possible from them. Some tasks can be accomplished by identifying one feature. For example, it is possible to locate a grasp point on a part in a bin of parts by locating a slender region surrounded by open space. Other tasks require a complete configuration of features to be recognized. For example, the task of determining the 3-D location of a part in a bin generally requires the identification of two or more mutually consistent features.

The key strategy for efficiency is the ability to leverage what is known from one feature to simplify the detection and identification of the next feature. Initially the a priori information may be able to restrict the region to be searched for a feature. Once a feature has been detected and tentatively identified, it can be used to restrict the search regions further. For example, to locate obstacles on a road, the dead reckoning information can suggest the range of positions of the edges of the road. Once the edges have been found, they can constrain the search for obstacles further.

To make the best use of this leveraging strategy, feature extraction and feature interpretation need to be intimately linked together. In the road detection case, this means that it is usually more efficient to locate one side of the road and then use its position to help find the second side than it is to look for both sides of the road and match candidate pairs.

Finding one feature at a time and extracting as much information from it is only one possible strategy. There are situations in which it is more efficient to find several features and then try to find large subsets of mutually consistent features. Special-purpose hardware to detect features, for example, may sway the strategy selection this way. In general, the choice of strategy is a function of the complete task description. There are no absolute rules.

There are three different approaches to feature interpretation. The first one identifies individual features on the basis of their intrinsic properties. The second one "grows" a match from one feature to topologically connected features. The third approach applies constraints pairwise to build sets of mutually consistent features. The first approach is essentially a pattern-recognition procedure that produces a list of possible identifications by comparing the intrinsic properties of a detected feature to the properties of the model features. When all these properties are within the specified tolerances of those of the model, the model feature is reported as a possible match. An example of this approach is the procedure used in binary vision systems to recognize visible objects based on their global features. Other, more complicated, pattern-recognition techniques are alternatives for this identification process. We want to emphasize that the ability to identify a single feature on the basis of its intrinsic properties is important for vision processing in general. As mentioned earlier, the identification of one feature may be sufficient for many tasks. If not, the better this process is, the smaller the combinatorial problems are in matching sets of features.

The second approach to feature interpretation is to grow a complex set of connected features by starting with one feature and adding adjacent ones. For example, when trying to identify an object from its silhouette, a corner feature might be extended by determining the lengths of the sides forming the corner, and then examining the angles at their ends, and so on. At the end of this syntactic-type of process the system has a string of corners and arcs that is more distinctive than any individual feature. The key to this approach is some connectivity information, such as the position along a detected edge chain, to provide the connection to the next feature. This connectivity eliminates the need for searching a region for a feature. We used this type of growing strategy in the 3DPO system [Bolles and Horaud 1984] to locate moderately complex features. We first analyzed 3-D edges, such as straight lines and circular arcs, and then analyzed the adjacent surfaces, characterized them, and built surface-edge-surface descriptions of that portion of the object. These descriptions generally had enough intrinsic properties to reduce their possible interpretations to one or two.

The third general approach to feature interpretation is to use attributed relational constraints, such as the expected distance between two features, to build sets of mutually consistent features. In the road detection example, the distance between the two sides of the road (found in range data) defines pairwise constraints and is an important piece of information for eliminating spurious road edges. Many research groups have used tree searches as a flexible way of applying relational constraints to a number of features. The groups differ in the "granularity" of the features they use (i.e., the complexity of the features and the information they convey). For example, Grimson and Lozano-Perez (1984) have developed an efficient technique for point-to-plane matches, each of which contributes only a small amount of information. Bolles and Horaud (1986) have concentrated on larger features, such as the bases of a cylindrical object component, each of which determines 5 of the 6 degrees of freedom associated with the object. Mulgaonkar et al. (1984) investigated object recognition using rough models to encode relational constraints. The choice of granularity depends primarily on the sensing techniques (e.g., touch sensing versus dense range sensing) and the economics of detecting features compared to applying pairwise constraints. Another way of applying such constraints to a large number of features is to build a Hough array and accumulate "votes" for different parameter values. This approach is best suited for problems that have at most three parameters [Silberg et al. 1984].

C. **Interpretation Techniques for Classes of Vision Tasks**

As explained previously, the techniques for feature interpretation depend on the a priori information associated with a given vision task. Although each task is unique, classes of vision tasks have similar a priori information and, therefore, require similar interpretation techniques.

Five important classes of vision tasks are as follows:
- *Verification*—Given a hypothesis about the identity and location of an object, decide whether or not the hypothesis is correct.
- *Location*—Given a hypothesis of an object in a range of locations, determine the precise location of the object.
- *Recognition*—Given a set of possible objects, determine the identities of objects in the scene.
- *Inspection*—Given an object and its model (including tolerances), determine whether specific portions of the object are defective.
- *Description*—Given a set of generic models, build a structural description of a scene in terms of these models.

We view these classes as basic vision tasks. For example, finding a specific object among different objects entails recognition and location tasks. Similarly, to "find" an object in a pile of identical objects generally involves recognition as well as location because an object may look quite different at different orientations. We realize, however, that these classes are not completely disjointed. For example, both description and recognition entail identification of models. However, in recognition, the models are usually structured, i.e., their inherent and relational features are known a priori within specified tolerances. In contrast, models used to describe an object are being developed during the description process in terms of generic features, such as the mathematical definition of the object surfaces.

In the rest of this section we examine the five basic vision tasks above and discuss the characteristics that determine the techniques to be considered for implementing these tasks.

1.   *Verification*

In a pure verification task the goal is to decide whether or not the hypothesized object is at the hypothesized position and orientation. The basic approach is to predict some features of the object at the hypothesized location and then compare them to the measured data. The features could be componential (e.g. surfaces and edges), local (e.g., small holes and corners), or pixel features (e.g., raw range values). One nice thing about range data compared to intensity data is that they are relatively easy to predict, given the hypothesized position and orientation of the object. Thus, one straightforward verification technique is to predict the range data and compare it to the measured data. We used this technique in the 3DPO system [Bolles and Horaud 1984] and found it to be effective in evaluating hypotheses. Problems arose when the hypothesized positions and orientations were slightly incorrect because the range values near the edges of the hypothesized object differed significantly from the measured data. Two possible methods of solving this problem are (1) to add a location subtask for refining the location information, and (2) to ignore the data near the edges of the object. The second method was also used by Duda et al. (1979) to preclude plane-fitting errors caused by fictitious "points" along range discontinuities where two surfaces are illuminated by a split laser beam.

The key to the verification technique is the known position and orientation of the object, which makes it possible to predict the range values. If the range image produced by the sensor contains artifacts that are difficult to predict, the effectiveness of this technique would be reduced. Usually these artifacts occur at the edges of objects or at surface orientations that correspond to specular reflections.

2. *Location*

The goal of a location task is to improve the estimate of the location (position and orientation) of an object. An example of this type is visually-guided arc welding. For that task the identity and approximate location of each workpiece is known, but its location relative to the torch has to be refined by identifying visible surfaces. The key to an efficient technique for performing this task is the ability to use the a priori estimates of the surfaces' positions and orientations to reduce the number of interpretations of each detected surface.

A slightly less constrained example of this type of task is that of locating a part on a tray. Again, the identity of the object is known, but its position and orientation are to be determined. In this case, the initial search regions may be relatively large for any one feature, but the location of one feature greatly reduces the sizes of the search regions where additional features may be found.

In the Intelligent Task Automation (ITA) project for the Air Force [Honeywell 1983] we have implemented a sequential location procedure that improves the estimate of the object's two-dimensional location for each newly identified feature, and then uses this new estimate to help find the next feature. This basic idea is straightforward. In practice, however, since there are many different types of features, it is somewhat tedious to work out the contribution of each feature. In our program we worked with five types of geometrical features: line segments, corners, circular holes, circular arcs, and range regions. Each one of these features provided a different type of information. For example, a line-to-line match determined the orientation of the object (modulo 180 degrees) and a linear constraint on the origin. If the detected line segment was shorter than the model segment, it could slide back and forth by an amount that depended upon the relative lengths of the two segments.

Recently there has been an increased interest in recursive techniques, such as Kalman filtering, for updating location estimates. To be applicable to this type of vision processing, these techniques need to be extended to handle the variety of information implied by the different types of feature matches.

3. *Recognition*

The goal of a recognition task is to determine the identity of an object. This can be accomplished by one or more features. For example, it may be possible to identify one of ten different objects on a table by sensing their colors. Color is a very efficient photometric feature for object recognition.

Recognition is often required to determine the state of an object. For example, recognition of the state of an object (right-side up, sideways, or upside down) may be needed in a bin-picking task even if there is only one type of object in the bin because many objects have entirely different images at different orientations. In the 3DPO system we implemented a strategy to recognize the state of an object by recognizing clusters of distinctive features. The key to the success of such a strategy is the proper selection of the features to be analyzed. In our case we built a special CAD model for each object to be recognized, and then used it interactively to select the best features to recognize the different states. Our aim is to automate this feature-selection process as much as possible. This goal is particularly difficult to achieve if the system can extract features that contribute a wide range of constraints. For example, it is difficult to design a system that reasons about both 2-D intensity regions, which crudely constrain an object's position, and 3-D edges, which determine five of the six degrees of freedom.

The complexity of a recognition task is a function of several factors, including the number of similar and different objects, the complexity of their features, and the similarity between different objects. Most recognition strategies work well only when the objects are quite different. More research is needed to develop techniques that gracefully handle sets of similar objects.

4. *Inspection*

There are different classes of inspection tasks, including verifying the presence, identity, and location of specified components, and checking for any component defects. The detection of obstacles in front of a vehicle is one form of inspection in which the road is inspected for "defects," such as holes and protruberances in the road. The key difference between this type of inspection and the other tasks is that the models of possible damage are generic as opposed to specific. Anything sticking out of a road is an obstacle. Thus, instead of searching for specific features, inspection techniques for this class of tasks must compare measured data to model surfaces. This comparison can be done at the data level, such as comparing the expected height of the road with the measured heights, or at a higher level, such as comparing the sizes of protrusions to a nominal value. Inspection is similar to verification except that in inspection the identity is assumed to be correct and out-of-tolerance differences between the model and the data are to be classified as defects, whereas in verification the identity of the object is in question.

5. *Description*

Description of a scene in terms of generic features is generally the most difficult vision task because it starts with the least a priori information. At SRI we have performed low-level analysis of range data, which is a form of description, but have not tried to implement a system for building models of objects from sensed data. Description techniques are important for both of these tasks (feature detection and model building). The better the low-level analysis, the more descriptive are the primitives to be matched.

VII LABORATORY EXPERIMENTS

Laboratory experiments have been conducted in the past few years at SRI to demonstrate 3-D vision for robot applications. Two of these experiments are described below: visually-guided robot arc welding and locating industrial parts in a bin.

A.   **Visually-Guided Robot Arc Welding**

   1.   *Goal*

SRI has been developing a programmable 3-D vision system for real-time guidance and control of an arc-welding robot [Kremers et al. 1983,1985]. The goal of this system is to sense and analyze the 3-D geometry of weld joints of workpieces with widely varying topology, shape, dimensions, material, and surface finish (dark matte to specular) in real time, in one pass, and immediately ahead of the welding arc. The system must

- Provide guidance information for the robot without recourse to preprogrammed joint location data.
- Measure fitup and cross-sectional area of joints having variable and unpredictable geometry to allow control of welding parameters and selection of appropriate procedures to maintain adequate fusion and bead shape.

The following sections describe the implementation and operation of the existing system.

   2.   *Acquisition of 3-D Data*

A series of "slices" of light planes are projected and viewed near and ahead of the welding torch in order to recognize and verify the workpieces, locate them precisely, and measure the gap between them. Initially we tried to use a real light plane, which was generated by fanning a laser beam through a cylindrical lens. This attempt failed completely because of secondary reflections from wire-brushed or ground aluminum workpieces, as illustrated in Figure 13. For that reason we designed and built the triangulation range sensor shown in Figure 4, where a laser scanner and a linear diode array are used to locate the true slice by narrowing the field of view around it. An image of a slice of a fillet joint across a tack weld is shown, for example, in Figure 14. A set of measured slice images are stored in a computer memory for the next step: extraction of planar-surface features.

   3.   *Feature Extraction*

The workpieces are modeled by piecewise planar surfaces, each represented by the 3-D plane equation $Ax + By + Cz + D = 0$. To extract such features from the measured data, the data are segmented by gathering sufficient data

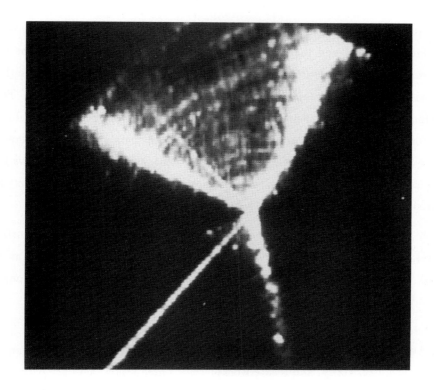

Figure 13: Reflections From Ground Aluminum Workpieces

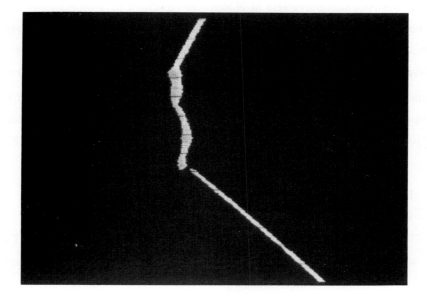

Figure 14: Fillet-Joint Slice Across Tack Weld

points (usually spanning a distance of approximately 2 inches along the weld joint) into a window and then applying a recursive procedure that partitions the range image into planar surfaces [Kremers et al. 1983]. Regions that are either too small or nonplanar (e.g., tack welds) are marked as "undefined." This procedure removes isolated noise points that may result from occasional welding sparks or any other noise. For example, Figure 15(a) depicts a perspective view of the raw 3-D data of the slices for a straight fillet joint, and Figure 15(b) shows the planar surfaces fitted to these data. As another example, Figure 16 shows the planar surfaces fitted piecewise to the surfaces of a curved fillet joint.

A variety of geometrical properties are associated with the planar-surface features. These include inherent features of each planar surface, such as its normal, width, and edge vectors, as well as relational features, such as the angle, gap, and offset between neighboring planar surfaces.

4. *Feature Interpretation and Robot Control*

Automatic control of an arc-welding robot requires that the vision system be able to recognize, verify, and locate workpieces. These vision tasks are performed by matching the model features described above with those extracted from the measured data. Such matching entails an iterative process in which certain relations between successive sets of N surfaces (specified in a user-defined model) are compared with the measured ones [Kremers et al. 1985]. The process terminates successfully when all the measured features of a given set of surfaces are found to match the corresponding model features within specified tolerances. The measured features are subsequently used to control the motion of the robot and the welding parameters. A photograph of the robot end-effector welding a fillet joint is shown in Figure 17(a), and the resulting weld is shown in Figure 17(b).

B. Locating Parts in Bin

1. *Task*

Given a bin of identical industrial parts (e.g., the castings shown in Figure 18), a 3-D part orientation ("3DPO") vision system was developed for locating individual parts in a range image so they can be picked up by a robot hand [Bolles and Horaud 1984, 1986].

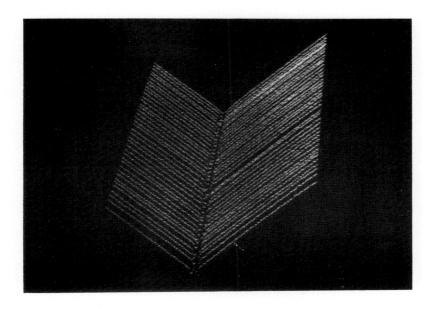

(a) Measured Data

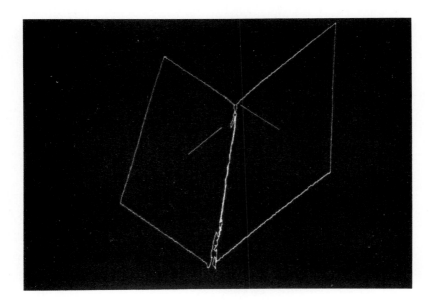

(b) Fitted Planes
Figure 15: Fitting Planes to Fillet-Joint Slice Data

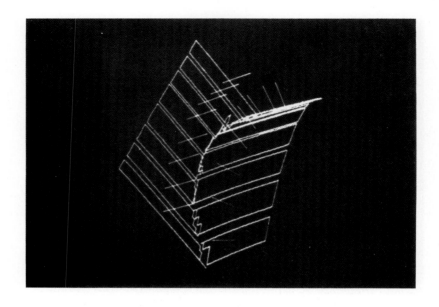

Figure 16: Piecewise Plane Fitting of Curved Fillet Joint Data

(a) End-Effector (torch and sensor) Welding a Fillet Joint

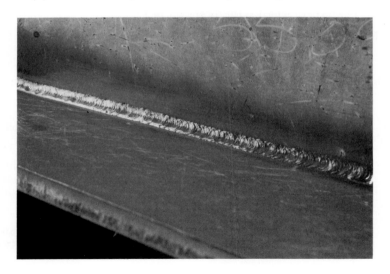

(b) Resulting Weld

Figure 17: Arc-Welding Robot Guided by Range Sensor

Figure 18: Jumbled Castings in a Bin

2. *3-D Data Acquisition*

Range images were obtained by using the Technical Arts White Scanner, a structured light range sensor that computes the (x,y,z) values and intensity of each pixel. The experimental setup is shown in Figure 19 and a height image is shown in Figure 20, where higher z values are depicted brighter.

3. *Feature Extraction*

Local geometric features, including their properties, consist of three *edge types*:

- A *straight dihedral*, including the angle between its surfaces, their widths, and the length of their intersection.

- A *circular dihedral*, including the angle between its surfaces, their widths, the radius of their circular intersection, and a binary variable indicating whether the planar surface is inside or outside the circle.

- A *straight tangential* along a cylinder, including the length and radius of the cylinder and the distance between its two tangentials.

These local features are extracted by performing the following operations:

- Detect range and range-gradient discontinuities in the range image, and classify them into jump-, convex-, concave-, and shadow-edge points.

- Discard the shadow-edge points and link the other ones into jump-, convex-, and concave-edge chains.

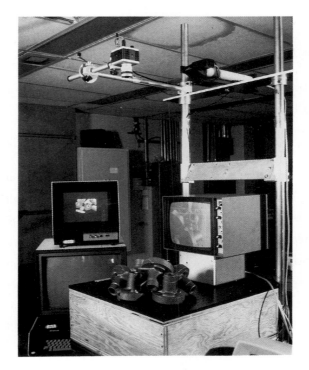

Figure 19: Structured-Light Range Measurement System

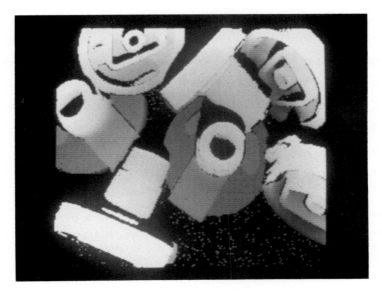

Figure 20: Measured Height Image of Castings
(Higher Elements are Brighter)

- Detect planar edge chains, segment them into (straight) lines and (circular) arcs in a 3-D space, analyze the surfaces adjacent to them, and refine the 3-D locations of the lines and arcs.

As an example, the lines and arcs extracted from the range image in Figure 20 are shown in Figure 21.

4. *Feature Interpretation*

a. Hypothesis Generation

The extracted features are compared with the model features in order to hypothesize the casting locations. Large circular arcs are sought first because they are most distinctive and, hence, constitute "focus features" [Bolles and Cain 1982]. By detecting additional edge features that are adjacent to each focus feature and consistent with the casting model, the location of the corresponding casting is thereby hypothesized. The features of the hypothesized casting are "removed" and the process is repeated, one at a time, for the remaining castings in the scene. Seven hypothesized castings resulting from this feature-matching process are shown in Figure 22.

b. Hypothesis Verification

To verify the hypothesis of each casting, the height values of all the visible image elements on the casting surfaces are first predicted by taking into account the sensor location, the hypothesized location of the casting, its model, and the image resolution. Predicted height values of elements on the surfaces of the hypothesized castings (in Figure 22) are shown in Figure 23. These values are compared with the measured data in Figure 20. Depending on the values of its predicted height, z_p, and measured height, z_m, the contribution of each image element is classified into one of three categories:

(1) *Positive evidence*, if $z_m = z_p \pm \Delta z$, where Δz is a tolerance value, which is much smaller than either z_m or z_p. The hypothesis is strengthened by this image element.

(2) *Neutral evidence*, if $z_m \geq z_p + z^*$, where z^* has a positive value comparable with a casting dimension, i.e., not negligible compared with either z_m or z_p. In this case the measured data are significantly closer to the sensor than predicted, presumably because the hypothesized casting is occluded by another casting. The hypothesis is neither strengthened nor weakened by this image element.

Figure 21: Extracted Straight and Circular Edge Features

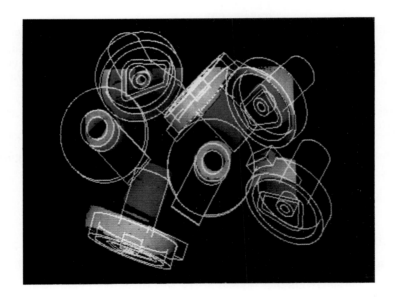

Figure 22: Seven Hypothesized Castings

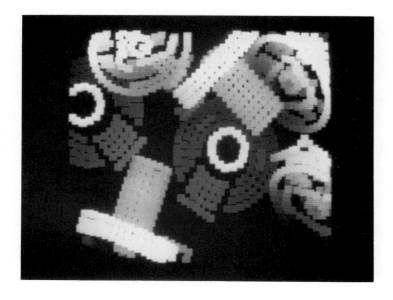

Figure 23: Predicted Height Image of Castings
(Higher Elements are Brighter)

(3) *Negative evidence*, if $z_m \leq z_p - z^*$. In this case the measured data are significantly farther from the sensor than predicted, implying that the hypothesized casting is transparent. Since in reality no casting is transparent, this case is erroneous. The hypothesis is weakened by this image element.

If an appreciable number of image elements exhibit negative evidence, the casting hypothesis is rejected; otherwise, it is accepted with a level of confidence that depends on the relative number of elements with positive evidence. In Figure 22, only the bottom-right casting hypothesis is rejected; the other six casting hypotheses are accepted.

c. <u>Configuration Understanding</u>

The 3DPO system determines which objects are on top of the pile by predicting a range image from all the verified hypotheses, tagging each of its elements with the index number (name) of the corresponding hypothesized object, and determining which of two objects whose images overlap is on top of the other. For example, the rangels that are common to Objects 1 and 2 in Figure 24 will be used to conclude that Object 2 is on top of Object 1. By comparing two castings at a time and updating such "on-top-of" relationships, the program builds a graph that shows which castings are on top and which ones are under them. The intensity image (obtained by the structured light range sensor) of the seven castings (see Figure 20) is shown in Figure 25; superimposed on this intensity

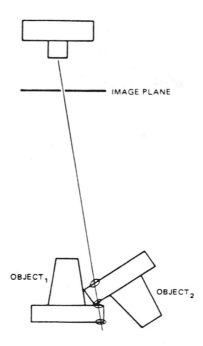

Figure 24: Determining Which Object is on Top of Another

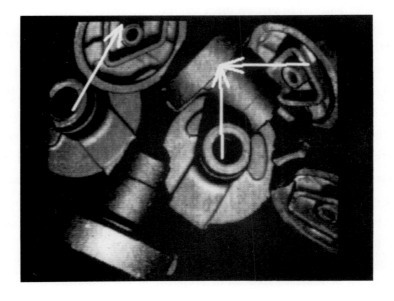

Figure 25: Arrows Pointing from Occluded to Occluding Castings

image are arrows pointing from each occluded casting to its occluding one. This information can later be used to guide a robot hand in picking up the top object(s) from the pile.

VIII CONCLUSIONS

Application of robot 3-D vision has no potential limits. Nevertheless, such application has so far been confined to manufacturing, and even there extremely few 3-D vision tasks have been automated. The major reason for this situation is that, compared with human vision, today's robot vision is still inferior and too costly. Robot 3-D vision has to go a long way before it can compete effectively with human vision.

Although we can learn from human vision how to advance robot 3-D vision, the latter can also use other types of noncontact sensing (in particular, range) and, not constrained by evolution, should be based on engineering principles. The ultimate goal of 3-D vision research is not only to enable robots to perceive as well as humans, but also to surpass them wherever possible

Four steps are distinguished in each sensing sequence: Transducing a signal, preprocessing it, extracting features from the improved signal, and interpreting the features to obtain the information required for a given vision task. Iterative sensing is represented by a major feedback loop, but there are also minor loops between the sensing steps that are yet to be explored. We do not believe that a cost-effective 3-D vision system can ever be general-purpose. Instead, such a vision system will consist of "tool boxes," each box including a variety of hardware or software tools for performing its corresponding sensing step, as well as a knowledge-based "supervisor" that selects the best tools and iterative strategies for a given vision task, including its constraints.

A variety of schemes have been proposed in the literature to obtain 3-D information in terms of range or surface orientation; Figure 2 classifies them in a tree form. None of these schemes is "best in general" as each has its own niche;

however, some schemes are clearly more useful than others (they have a bigger niche). Schemes for direct measurement of range, although very appealing, suffer from several serious problems: Specular surface errors, long measurement time, missing data (in triangulation schemes), and range ambiguity (in A.M. time-of-flight schemes). Schemes for indirect measurement of range or surface orientation, although requiring only monocular intensity images, have restricted applications and their performances have not yet been adequately assessed relative to each other and relative to the direct-measurement schemes.

We have classified object features according to the feature physical properties, size, and relation. Such classification is intended to facilitate systematic generation and application of distinguishing features in 3-D vision. Although a systematic approach to feature generation may be helpful if done manually, it is essential for automatic generation of features.

Great progress has been made in recent years in the development of techniques for extracting 3-D features from intensity images and, to a lesser extent, from range images. Despite this progress, better feature-extraction techniques are needed, especially for extracting features from combined intensity and range images.

Feature-interpretation techniques have been developed for specific classes of tasks, but very little has been done to try to integrate them into a coherent system. To do this, there must first be a task-description system, which includes a 3-D object-modeling system tailored to visual processing. Second, the system must be able to reason about its capabilities, including techniques to build and evaluate both structural and statistical descriptions.

Our final conclusion is that not enough has been done to facilitate programming of vision tasks. This situation will get worse as the number of techniques at all levels increases. If we are able to develop efficient 3-D vision techniques but are unable to simplify the programming of their applications, our efforts may fail because the cost of vision-task programming may be prohibitive.

REFERENCES

1. Altschuler, M.D., et al., "The Numerical Stereo Camera," *Proc. of Society for Photo-Optical Instrumentation Engineers Conf. on 3-D Machine Perception*, Vol. 283, pp. 15-24, SPIE, Bellingham, Washington (1981).

2. Ballard, D.H., and C.M. Brown, *Computer Vision*, (Prentice-Hall, Inc., Englewood Cliffs, New Jersey, 1982).

3. Bolles, R.C., and R.A. Cain, "Recognizing and Locating Partially Visible Objects: The Local-Feature-Focus Method," *Int. J. Robotics Research*, Vol. 1, No. 3, pp. 57-82 (1982).

4. Bolles, R.C., and P. Horaud, "3DPO's Strategy for Matching Three-Dimensional Objects in Range Data," *Proc. IEEE Int. Conf. on Robotics*, pp. 78-85, Atlanta, Georgia (March 1984).

5. Bolles, R.C., and P. Horaud, "3DPO: A Three-Dimensional Part Orientation System," *Int. J. Robotics Research*, Vol. 5, No. 2 (1986).

6. Duda, R.O. et al., "Use of Range and Reflectance Data to Find Planar Surface Regions," *Proc. IEEE Trans. on Pattern Analysis and Machine Intelligence*, Vol. PAMI-1, No. 3, pp. 259-271 (July 1979).

7. Fischler, M.A., and R.C. Bolles, "Random Sample Consensus: A Paradigm for Model Fitting with Applications to Image Analysis and Automated Cartography," A.I. Center Technical Note 213, SRI International, Menlo Park, California (March 1980).

8. Gennery, D.B., "Tracking Known Three-Dimensional Objects," *Proc. AAAI-82*, pp. 13-17 (18-20 August 1982).

9. Gleason, G.J., and G.J. Agin, "A Modular Vision System for Sensor-Controlled Manipulation and Inspection," *9th Int. Symposium and Exposition on Industrial Robots*, pp. 57-70, Washington, D.C. (March 1979).

10. Goodwin, F.E., "Coherent Laser Radar 3-D Vision Sensor," SME Technical Paper MS85-1005, Society of Manufacturing Engineers, Dearborn, Michigan (1985).

11. Grimson, W.E.L., and T. Lozano-Perez, "Model-Based Recognition and Location from Sparse Range or Tactile Data," *Int. J. Robotics Research*, Vol. 3, No. 3, pp. 3-35 (1984).

12. Haralick, R.M., "Digital Step Edges from Zero Crossings of Second Directional Derivatives," *IEEE Trans. PAMI*, Vol. PAMI-6, pp. 58-68 (January 1984).

13. Hebert, M., and T. Kanade, "First Results on Outdoor Scene Analysis Using Range Data," *Proc. of Image Understanding Workshop*, Miami, Florida (December 1985).

14. Honeywell, Inc., "Enhanced Enabling Technology Development and Feasibility Demonstration Plan, Intelligent Task Automation," Air Force Wright Aeronautical Laboratory Contract F33615-82-C-5092 (April 29, 1983).

15. Horn, B.P.K., "Shape From Shading," P.H. Winston (ed.), *The Psychology of Computer Vision*, (McGraw Hill, New York, New York, 1975).

16. Ikeuchi, K. "Determining Surface Orientation of specular Surfaces by Using Photometric Stereo Method," *IEEE Trans. on Pattern Analysis and Machine Intelligence*, Vol. PAMI-3, No. 6, pp. 661-669 (November 1981).

17. Jarvis, R.A., "A Perspective on Range Finding Techniques for Computer Vision," *IEEE Trans PAMI*, Vol. PAMI-5, No. 2, pp. 122-139 (March 1983).

18. Johnston, A.R., "Infrared Laser Rangefinder," Report NPO-13460, Jet Propulsion Laboratory, NASA New Technology, Pasadena, California (August 1973).

19. Kender, J.R., "Saturation, Hue, and Normalized Color: Calculation, Digitization Effects, and Use," Department of Computer Science, Carnegie-Mellon University, Pittsburgh, Pennsylvania (November 1976).

20. Kitagawa, H. et al., "Extraction of Surface Gradient from Shading Images," *Trans. IEEE of Japan*, Vol. J66, No. 1, pp. 65-72 (1983).

21. Koshikawa, K., "A Method of Finding Surface Orientation by Polarization Analysis of Reflected Light," *Trans. SICE*, Vol. 18, No. 10, pp. 77-79 (1982).

22. Kremers, J.H., et al., "Development of a Prototype Robotic Arc Welding Work Station," NAVSEA Report S50002-83, SRI International, Menlo Park, California (May 1983).

23. Kremers, J.H., et al., "Development of a Prototype Robotic Arc Welding Work Station," NAVSEA Report S5002-83-2, SRI International, Menlo Park, California (May 1985).

24. Krotkov, E. and J.P. Martin, "Range from Focus," *Proc. 1986 IEEE Robotics and Automation*, pp. 1093-1098, San Francisco, California (April 1986).

25. Levine, M., "Rule-Based Image Segmentation: A Dynamic Control Strategy Approach," Technical Report TR-83-9, Computer Vision and Robotics Laboratory, McGill University (1983).

26. Marr, D., and E. Hildreth, "Theory of Edge Detection," Proc. R. Soc. Lond., Vol. B-207, pp. 187-217, London, Great Britain (1980).

27. Miller, R.K., 3-D Machine Vision, SEAI Technical Publications, Madison, Georgia (1984).

28. Mulgaonkar, P.G., et al., "Matching 'Sticks, Plates and Blobs' Objects Using Geometric and Relational Constraints," Image and Vision Computing, Vol. 1 (March 1984).

29. Mulgaonkar, P.G., and L.G. Shapiro, "Hypothesis-Based Geometric Reasoning on Computer Vision: Representation and Control), Bellaire, Michigan, pp. 11-18 (13-16 October 1985).

30. Nitzan, D., "Scene Analysis Using Range Data," A.I. Center Technical Note 69, Stanford Research Institute, Menlo Park, California (August 1972).

31. Nitzan, D. et al, "The Measurement and Use of Registered Reflectance and Range Data in Scene Analysis," Proc. IEEE Trans. on Pattern Analysis and Machine Intelligence, Vol. 65, No. 2, pp. 206-220 (February 1977).

32. Nitzan, D., C. Barrouil, P. Cheeseman, and R.C. Smith, "Use of Sensors in Robot Systems," '83 International Conf. on Advanced Robotics, Tokyo, Japan (12-13 September 1983).

33. Ohlander, R., K. Price, and D.R. Reddy, "Picture Segmentation Using a Recursive Region Splitting Method," Computer Graphics and Image Processing, Vol. 8, No. 3 (December 1978).

34. Ohta, Y., T. Kanade, and T. Sakai, "Color Information for Region Segmentation," Computer Graphics and Image Processing, Vol. 13 pp. 222-241 (1980).

35. Ohta, Y. et al., "Obtaining Surface Orientation from Texels Under Perspective Projection," Proc. 7th IJCAI, pp. 746-751, VanCouver, Canada (1981).

36. Pentland, A.P., "Local Shading Analysis," IEEE Trans. PAMI, Vol. PAMI-6, No. 2, pp. 170-187 (March 1984).

37. Pentland, A.P., "A New Sense for Depth of Field," Proc. 9th IJCAI pp. 988-994, Los Angeles, California (August 1985).

38. Pratt, W.K., "Spatial Transform Coding of color Images," IEEE Transactions on Communication Technology, Vol. COM-19, No. 6, pp. 980-992 (December 1971).

39. Roberts, L.G., "Machine Perception of Three-Dimensional Solids," 315, Lincoln Laboratory, Massachusetts Institute of Technology, Cambridge, Massachusetts (1963). Also in *Optical and Electro-Optical Information Processing*, J. Tippett et al. (eds), pp. 159-197 (MIT Press, 1965).

40. Rosen, C.A., "Machine Vision and Robotics: Industrial Requirements," *Computer Vision and Sensor-Based Robots*, Dodd & Rossol, Eds. (Plenum Publishing Corporation, 1979).

41. Rosenfeld, A., and A.C. Kak, *Digital Image Processing*, (Academic Press, New York, New York, 1976).

42. Shirai, Y., *Robot Vision*, FGCS, PP. 325-352 (North Holland, 198).

43. Silberg, T.M., D. Harwood, and L.S. Davis, "Object Recognition Using Oriented Model Points," *Proc. 1st Conf. on Artificial Intelligence Applications*, IEEE (December 1984).

44. Sobel, I., "Camera Models and Machine Perception," AIM-21, Stanford AI Laboratory, Stanford University, Stanford, California (May 1970).

45. Stevens, K.A., "Representing and Analyzing Surface Orientation," in *Artificial Intelligence: An MIT Perspective*, Vol. 2, P.H. Winston and R.H. Brown (eds.) (MIT Press, Cambridge, Massachusetts, 1979).

46. Tsuji, S. et al., "Wiresight: Robot Vision for Determining Three-Dimensional Geometry of Flexible Wires," *'83 International Conf. on Advanced Robotics*, pp. 133-138, Tokyo, Japan (September 12-13, 1983).

47. Witkin, A.P., "Recovering Surface Shape and Orientation from Texture," *Artificial Intelligence*, Vol. 17 (1981).

48. Woodham, R.J., "Photometric Stereo: A Reflectance Map Technique for Determining Surface Orientation from Image Intensity," *Proc. 22nd. Int. Symp. SPIE*, pp. 136-143, San Diego, California (August 1978)

49. Woodham, R.J., "Analyzing Images of Curved Surfaces," *Artificial Intelligence*, Vol. 17, pp. 117-140 (1981).

50. Yang, H.S. and A.C. Kak, "Determination of the Identity, Position, and Orientation of the Topmost Object in a Pile: Some Further Experiments," *Proc. IEEE Int. Conf. on Robotics and Automation*, pp. 293-298, San Francisco, California (April 1986).

ON THE COMPUTATION OF MOTION FROM A SEQUENCE OF MONOCULAR OR STEREO IMAGES - An Overview *

J.K. Aggarwal

Computer and Vision Research Center
The University of Texas at Austin
Austin, Texas 78712-1084

1. INTRODUCTION

Vision and the ability to discern change due to motion are almost universal in the animal kingdom. Moving objects and changing environment surround us. Even stationary objects appear to have (relative) motion because of our own motion or the movement of our eyes. Thus, it is not at all surprising that several distinct types of investigators with widely differing backgrounds and objectives are pursuing research into motion. These backgrounds include psychophysics, neurophysiology, computer vision, computer graphics, and robotics. Briefly, psychophysicists' and neurophysiologists' interest in motion centers around understanding the part of biological visual system which senses and interprets motion. Researchers in computer graphics are concerned with the generation of images and, in particular, the generation of moving images and animation on a screen. Researchers in computer vision are interested in the analysis, processing, and understanding of images. A substantial fraction of these researchers are dedicated to the analysis, processing, and understanding of sequences or collections of images with the objective of collecting information from the set as a whole that may not be obtained from any one image by itself. Detection, computation, and understanding of motion are an integral part of these endeavors. Robotics is another discipline where researchers are intimately involved with motion. The current research scene using motion as the common element presents an interesting symbiosis between the disciplines cited above.

In computer vision, a very broad set of applications are motivating a strong interest in sensing, interpretation, and description of motion via a sequence or collection of images. The application areas include medicine, tomography, autonomous navigation, communications and television, video conferencing, dancing and choreography, meteorology, and so on. For example, the application of video conferencing and the desire to transmit sequences of images on channels with limited bandwidth serve as the strong impetus for the research on motion and compression of images. The processing of sequences of images for the recognition and the tracking of targets is of immense interest to the department of defense of every country. The computation, characterization and understanding of human motion in contexts of dancing and athletics is another field of endeavor receiving much attention. In meteorology, the satellite imagery provides the opportunity for interpretation and prediction of atmospheric processes through estimation of shape and motion

---

* This research was supported in part by the Air Force Office of Scientific Research, Contract F49620-85-K-0007.

parameters of atmospheric disturbances. The preceding examples are indicative of the broad interest in motion, time varying imagery and dynamic scene analysis.

This broad interest has been evident since the first workshop in Philadelphia in 1979 [1]. The workshop was expected to be a meeting of a relatively small number of specialists but it turned out to be a conference. Since Philadelphia, several additional meetings and special issues have contributed to the exchange of ideas and the dissemination of results. In addition, there are several sessions on motion and related issues at meetings such as the IEEE Computer Society Computer Vision and Pattern Recognition Conference and conferences of other societies interested in vision. The list of workshops and special issues devoted exclusively to motion and time-varying imagery include three special issues [2-4], two books [5-6], a NATO Advanced Study Institute [7], an ACM workshop [8], a European meeting on time-varying imagery [9] and a host of survey papers [10-15]. The extent of the breadth and depth of interest is provided by the table of contents of the book published to document the proceedings of the NATO-ASI [16]. However, this list is incomplete at best. The IEEE Computer Society workshop at Kiawah Island [17] and proposed workshop planned in Italy [18] are strong indication of the broad interest in motion at this time. The recent two volume reprint series [19] published by IEEE Computer Society includes a section on Image Sequence Analysis containing nine papers.

2. THE TWO APPROACHES TO MOTION ANALYSIS

Objects are fundamentally three-dimensional whereas images are essentially two-dimensional. However, two-dimensional sometimes models serve to capture the intrinsic nature of the problem. Also, two-dimensional problems are in general simpler and are usually studied first. This is certainly the case with motion analysis. The available techniques may be classified into two groups, namely two-dimensional motion analysis and three-dimensional motion analysis. This dichotomy is useful in the sense that two-dimensional analysis is not concerned with the complications of projection geometry. However, three-dimensional analysis depends on the projection geometry, namely, central projection or parallel projection. Further, in each case the analysis may be classified into two groups depending upon whether it uses features or optical flow. In the case where the analysis is dependent upon features, the images are segmented, features are extracted from each image and they are matched between images. Features used may be points, lines, contours, etc.. The fundamental assumption of the analysis based upon features is that correspondence between features has already been established before the computation of structure and motion. The optical flow approach depends upon the computation of the apparent velocities of brightness patterns in an image before the motion analysis stage. The computation of optical flow is, in general, difficult and highly sensitive to noise. The difficulties with both approaches are briefly reviewed in the following.

As mentioned above, the problem of correspondence (in the features-based approach) has been treated separately from structure and motion computation. The basic question is - How to establish and maintain the correspondence relationship in an image sequence? For a survey of the research on the correspondence problem in dynamic scene analysis, the readers are referred to Aggarwal, Davis, and Martin [20]. Furthermore, Ullman [6] points out that motion is one of the cues used in the human visual system to establish the correspondence relationship, or stated in another way, the problem of structure and motion computation and that of establishing correspondence are actually inter-related and to some extent inseparable. Indeed, observed 2-D motion does give clues about how to group image elements together; elements with apparently different projected motion can usually be safely hypothesized to belong to different objects in space, as shown by Yalamanchili, Martin and Aggarwal [21]. Thus motion aids in determining the correspondence relationship by eliminating implausible matches. Psychological research has collected enough evidence to support the belief that the correspondence process and structure and motion processes are closely interwoven in the human visual mechanism. But it is still not clear whether integration of these two processes will facilitate the computation of structure and motion by computers. To establish the correspondence relationship between successive frames is a challenging task and to maintain the correspondence relationship over a sequence of several images may prove to be extremely difficult. Here, we are confronted with two major obstacles: (a) observables may appear and disappear over time, and (b) occlusion may introduce extraneous observables into the field of view. The problem of extraneous observables was addressed for the 2-D case by Aggarwal and Duda [22].

The use of optical flow in the computation of motion is made difficult by the lack of methods to compute optical flow accurately. There is an elegant set of equations relating position and motion of objects in space to image positions and optical flow. However, an elegant mathematical analysis is no substitute for computational stability. The solution of these equations is not well behaved numerically and the equations are highly sensitive to noise both in image position and optical flow. The computation of optical flow depends on taking derivatives - an inherently noise augmenting process and, in general, the computed optical flow is not uniformly distributed across the image. Again, the difficulties are fundamental.

The limitations of space and time prevent an encyclopedic review. Hopefully, the two fundamental issues (i) the establishment of correspondence of features between images and (ii) the computation of reliable and uniform optical flow from images will receive attention in future. In the following, the feature based approach is discussed for both the two- and three-dimensional cases. For the three-dimensional case, the analyses of sequences of monocular images and that of stereo images are discussed.

3. TWO-DIMENSIONAL MOTION ANALYSIS

In two-dimensional analysis, a sequence of images of objects moving in a plane parallel to the image plane is given, and the basic objective of the analysis is to estimate the two-dimensional motion. Aggarwal and Duda [22] were the first to consider the motion of polygonal objects, using vertices as features to be matched between frames. The matching was based upon the angle subtended at the vertex and the length of the sides of the polygon. In this analysis, occlusion was allowed giving rise to false vertices. Martin and Aggarwal [23] considered curvilinear objects and the points of sharp changes in curvature as the feature points. Again, occlusion was allowed. The strength of the above analysis lies in its ability to consider occlusion and the lack of prior knowledge of the shapes of polygons or curvilinear figures, and the robustness of the computation of the motion parameters in the presence of noise. In the above analyses, the features used were local. In contrast, Potter [24] estimates velocity at every point in a scene by defining a skeletal template in a frame and matching it in the second frame. An analysis based upon the global properties of the objects is due to Chow and Aggarwal [25]. They consider the area, the position of the centroid, and the moment of inertia for matching objects between views, and thus estimating the velocity of objects. In general, this analysis requires prior knowledge of the shape of objects. This is the major drawback of this method. However, the strength of the method includes its ability to predict the shape of the two-dimensional object based upon the occlusion of two curvilinear objects and the ability to match the correct shape with the observed shape of the occluded object. For a more detailed discussion of two-dimensional analysis, the reader is referred to the more complete survey by Martin and Aggarwal [10].

4. THREE-DIMENSIONAL MOTION ANALYSIS - SEQUENCE OF MONOCULAR IMAGES

In principle, the observation of a certain number of points in two or more views can yield the position of these points in space and the relative displacement between the viewing systems. This line of reasoning using points as observables has been pursued by Roach and Aggarwal [26,27], Webb and Aggarwal [28], Nagel [29,30], Ullman [6], Tsai and Huang [31,34], Tsai, Huang and Zhu [32], Huang and Tsai [33], Longuet-Higgins [35,36], and Mitiche, Seida and Aggarwal [37], among many other researchers. The consensus is that the observation of five points in two perspective views is analytically sufficient to yield both structure and motion.

The use of line correspondences in the computation of structure and motion has been addressed by Yen and Huang [38-40] and Aggarwal and Mitiche [41]. Yen and Huang [40] used seven line correspondences for solving structure and motion parameters and it was shown in [41] that four lines in three views in general position can yield the orientation of the lines in space and the motion parameters. The use of line correspondences has the additional advantage over the use of point correspondences in that extraction of lines in images is less sensitive to noise than extraction of points. In the following, both point and line correspondences are considered.

4.1 Point Correspondences

Roach and Aggarwal [26,27] were among the first to consider the problem of computing 3-D structure and motion from 2-D projections using point correspondence. A scenario consisting of a static scene, a moving camera and central projection was assumed. The goal was to investigate whether it would be possible to determine the position of the points in space and the movement (translation and rotation) of the camera.

The basic equations that relate the 3-D coordinates (X, Y, Z) of a space point to its 2-D projective coordinates (x, y) are:

$$x = F \frac{a_{11}(X - X_0) + a_{12}(Y - Y_0) + a_{13}(Z - Z_0)}{a_{31}(X - X_0) + a_{32}(Y - Y_0) + a_{33}(Z - Z_0)}, \qquad (4.1)$$

$$y = F \frac{a_{21}(X - X_0) + a_{22}(Y - Y_0) + a_{23}(Z - Z_0)}{a_{31}(X - X_0) + a_{32}(Y - Y_0) + a_{33}(Z - Z_0)} \qquad (4.2)$$

Here, F is the focal length, (X_0, Y_0, Z_0) are the 3-D coordinates of the lens center, and $a_{11}, a_{12}, ..., a_{33}$ are functions of (Θ, Φ, Ψ) which specify the orientation of the camera in the global reference system.

Roach and Aggarwal show that five points in two views are needed to recover the structure and motion parameters. They relate the number of points and the number of equations available for the solution of space coordinates and motion parameters as follows. The global coordinates of each point are unknown, and the five points produce 15 variables. The camera position and orientation parameters $(X_0, Y_0, Z_0, \Theta, \Phi$ and $\Psi)$ in two views contribute another 12 variables. So this amounts to a total of 27 variables. Each point in space produces two projection equations per camera position thus forming a total of 20 nonlinear equations. To make the number of equations equal the number of unknowns, seven variables must be known or specified *a priori*. The authors chose the six camera parameters of the first view to be zero and the Z-component of one of the five points is set to an arbitrary positive constant to fix the scaling factor. The reason for fixing one variable as the scaling constant is that the information embedded in any monocular image sequence is inherently insufficient and recovery of 3-D structure and motion depends on fixing scale. For example, under central projection the observed projected motion of an object moving in space can be reproduced by another object which is twice as large, twice as far away from the camera, translating twice as fast, and rotating with the same speed around an axis of the same orientation as that of the former object. In general, the information of the absolute distance of the object from the viewer is usually lost in the image formation process.

The authors give an iterative, finite difference, Levenberg-Marquardt algorithm to solve these 18 nonlinear equations (two of the 20 nonlinear equations have no unknown variables in them). For noise free simulations, the method typically converged to the correct answer within 15 seconds on a Cyber 170/750 and hence is reasonably efficient. If noise is introduced into the point positions in the image plane, the authors report that a considerably overdetermined system of equations is needed to attain reasonable accuracy of the results. Two views of 12 or even 15 points, or three views of seven or eight points are usually needed in the noisy cases.

Webb and Aggarwal [28] have presented a method for recovering the 3-D structure of moving rigid and jointed objects from several single camera views. The fixed axis assumption is adopted to interpret images of moving objects. The fixed axis assumption asserts that every rigid object movement consists of a translation plus a rotation about an axis which is fixed in direction for a short period of time. It is shown that, under the fixed axis assumption, fixing any point on a rigid moving object causes the other points to trace out circles in planes normal to the fixed axis. Under parallel projection, these circles project into ellipses in the image plane. The structure of the rigid object can be recovered to within a reflection by finding the equation describing the ellipses.

In more detail, suppose that p_1^f, p_2^f, ..., p_n^f represent the image coordinates of n points at frame f for $1 \leq f \leq F$. A reference point p_i is chosen and $s_j^f = p_j^f - p_i^f$ is computed for all the other feature points $1 \leq j \leq n$. An ellipse is fitted to the s_j^f for $1 \leq f \leq F$. If p_i and p_j belong to the same rigid object, the error in fitting should be small according to the fixed axis assumption. Points with error under a certain threshold are then grouped as being on the same rigid part as the reference point p_i. Furthermore, it is shown that the lengths of the long and short axes of an ellipse are functions of the position of the point in space. The orientation of the axis of rotation (up to a reflection about the image plane) can then be recovered provided that the fixed axis of rotation is not parallel or perpendicular to the image plane.

A jointed object is an object made up of a number of rigid parts which cannot bend or twist. If the jointed object still moves in a way such that the fixed axis assumption holds for each rigid part, then the motion and structure of the jointed object can be recovered. It is assumed that the rigid parts are connected by joints which are treated as single points in common to two rigid parts. If the joints are visible, they can be easily identified since they satisfy two sets of motion constraints. If the joints are not visible, they can be found by solving a system of linear equations. The joints can then be used to eliminate some reflections and thus the number of possible interpretations of structure is reduced. Finally, the 3-D motion of each object is reconstructed and the next set of frames is processed.

Unlike Roach and Aggarwal [27] who solve for all the motion parameters simultaneously thus creating a large search space, Nagel [29] has proposed a technique which reduces the dimension of the search space through the elimination of unknown variables. The important observation made by Nagel was that the translation vector can be eliminated and the rotation matrix can be solved for separately. A rotation matrix is completely specified by three parameters -- namely the orientation of the rotation axis and the rotation angle around this axis. It is shown that if measurements of five points in two views are available, then three equations can be written and the three rotation parameters can be solved for separately from the translation parameters. The distance of the configuration of points from the viewer is arbitrarily fixed and the translation vector can then be determined.

Ullman in his highly acclaimed thesis has proved the following theoretical result [6]: Given three distinct orthographic projections of four non-coplanar points in a rigid configuration, the structure and motion compatible with the three views are uniquely determined up to a reflection about the image plane. The assumption of parallel projection makes it possible to find surface structure in closed form when four non-coplanar points are seen in three distinct views. However, as a consequence of the parallel projection assumption, the structure deduced could be reflected in depth to produce the same image. Further, the absolute distance from the camera to an object cannot be determined uniquely.

Tsai and Huang [31] have proposed a method to find the motion of a planar surface patch from 2-D perspective views. The algorithm consists of two steps: First, a set of eight "pure parameters" is defined. These parameters can be determined uniquely from two successive image frames by solving a set of linear equations. Then, the actual motion parameters are determined from these eight "pure parameters" by solving a sixth-order polynomial.

By exploiting the projective geometrical constraints and the notion of rigid motion, equations can be written to relate the coordinates of image points in the two frames for points on a planar surface patch $AX + BY + CZ = 1$, where A, B and C are the structure parameters. The mapping from the (x, y) space to the (x', y') space (from one image to the next image) is given by:

$$x' = \frac{a_1 x + a_2 y + a_3}{a_7 x + a_8 y + 1}, \quad y' = \frac{a_4 x + a_5 y + a_6}{a_7 x + a_8 y + 1} \qquad (4.3)$$

where, a_1 through a_8 are the eight "pure parameters" and can be expressed in terms of the focal length, the structure parameters (A, B, C), and the motion parameters N_X, N_Y, N_Z, Θ, T_X, T_Y and T_Z (**N** specifies the rotation axis, Θ is the rotational angle and **T** is the translational vector). For a particular set of pure parameters, the above equation represents a mapping from (x, y) space to (x', y') space. The collection of all these transformations has been shown to be a continuous (Lie) group of dimension eight. A linear set of equations can be derived to solve for these eight pure parameters using Lie group theory.

After the eight pure parameters are obtained, the structure and motion parameters can then be determined. Here, the Z component of the translation vector is arbitrarily chosen to fix the scale. After manipulating the equations, it is possible to get a sixth-order polynomial equation in terms of only one of the variables $T'_X = T_X / T_Z$. T'_X is solved first and then all the remaining structure and motion parameters can be easily obtained. Although potentially six real roots may result from solving a sixth-order polynomial, the authors report that aside from a scale factor for the translation parameters, the number of real solutions never exceeds two in their simulation.

The problem of a curved surface patch in motion was later investigated in [34]. Here, two main results were established concerning the existence and uniqueness of the solutions. First, given the image correspondences of eight object points in general positions, an **E** matrix can be determined uniquely by solving eight linear equations. Furthermore, the actual 3-D motion parameters can be determined uniquely given **E**, and can be computed by taking the singular value decomposition of **E** without having to solve nonlinear equations. The proofs of the claims are quite lengthy and involved and will not be discussed here.

Longuet-Higgins [35,36] worked independently to obtain results similar to those described above. He derived the **E** matrix and presented a method to recover **R** and **T** from **E** using tensor and vector analysis. The interested readers are referred to [35,36].

Mitiche, Seida and Aggarwal [37] have recently developed a different formulation to solve the structure and motion problems by exploiting the principle of conservation of distance with respect to rigid motion. The principle of conservation of distance simply states that distance between any points on an object does not change as a result of rigid motion. A scenario of a static scene with a mobile viewing system was chosen. For two points P_i and P_j in space, the projective relationship of these two points in the image frames S and S' are given as

For point P_i

$$\begin{aligned} X_i &= \lambda_i x_i & X'_i &= \gamma_i x'_i \\ Y_i &= \lambda_i y_i & Y'_i &= \gamma_i y'_i \\ Z_i &= (1 - \lambda_i) f & Z'_i &= (1 - \gamma_i) f' \end{aligned} \quad (4.4)$$

And for point P_j

$$\begin{aligned} X_j &= \lambda_j x_j & X'_j &= \gamma_j x'_j \\ Y_j &= \lambda_j y_j & Y'_j &= \gamma_j y'_j \\ Z_j &= (1 - \lambda_j) f & Z'_j &= (1 - \gamma_j) f' \end{aligned} \quad (4.5)$$

Where $\lambda_i, \lambda_j, \gamma_i$ and γ_j are the unknown parameters which determine the distance of the two space points to the two image frames S and S'. f and f' are the focal length of the two camera systems. Note that the coordinate system may be chosen such that the values of λ's and γ's are greater than 1.

The distance between points P_i and P_j expressed in frame S is therefore

$$d^S_{ij} = (X_i - X_j)^2 + (Y_i - Y_j)^2 + (Z_i - Z_j)^2 \quad (4.6)$$

Or

$$d^S_{ij} = (\lambda_i x_i - \lambda_j x_j)^2 + (\lambda_i y_i - \lambda_j y_j)^2 + (\lambda_j - \lambda_i)^2 f^2 \quad (4.7)$$

Similarly, the distance between P_i and P_j expressed in frame S' is

$$d^{S'}_{ij} = (\gamma_i x'_i - \gamma_j x'_j)^2 + (\gamma_i y'_i - \gamma_j y'_j)^2 + (\gamma_j - \gamma_i)^2 f'^2 \quad (4.8)$$

The principle of conservation of distance allows us to write the following equations

$$d^S_{ij} = d^{S'}_{ij} \quad (4.9)$$

Or

$$(\lambda_i x_i - \lambda_j x_j)^2 + (\lambda_i y_i - \lambda_j y_j)^2 + (\lambda_j - \lambda_i)^2 f^2 =$$

$$(\gamma_i x'_i - \gamma_j x'_j)^2 + (\gamma_i y'_i - \gamma_j y'_j)^2 + (\gamma_j - \gamma_i)^2 f'^2 \quad (4.10)$$

It may be observed that each point P_i contributes two unknowns λ_i and γ_i, and each pair of points (P_i, P_j) produces one second order equation. Therefore, 5 points yield 10 equations in 9 unknowns with one variable chosen to fix the scaling factor. Note that each equation involves only four of the unknowns. Note also that the formulation so far does not involve the parameters of the displacement between the two cameras. Because these motion parameters do not appear in the equations and also because only some of the unknowns of position appear in each of the equation, the authors reported that the resulting system of equations can be solved quite efficiently using existing numerical iterative algorithms.

Once the position of points in space has been computed, determining the relative position of the cameras becomes a simple matter, as follows. Take four non-coplanar points and let A and A' be the matrices of homogeneous coordinates of these points in the coordinate systems S and S', respectively. If M is the transformation matrix in the homogeneous coordinate system we have $A' = M A$.

This finding confirms what was reported earlier by Roach and Aggarwal [27], i.e., given five point correspondences in two frames, the structure and motion parameters can be uniquely determined. However, the formulation and computation in [37] are considerably simpler. It may be observed that Tsai and Huang [34] present a uniqueness of solution proof for seven points in two views. However, when enumerating the variables as reasoned by Roach and Aggarwal [27], five points in two views should be sufficient to solve the nonlinear equations. From the computational point of view, five points in two views appear adequate.

4.2 Line Correspondences

As mentioned earlier, Yen and Huang [40] use seven line correspondences for solving the structure and motion parameters. In [40], a spherical coordinate system is used and a new parameterization of lines in space is introduced. The coordinate system is chosen to be viewer centered with origin O located at the focal point. The image plane is a unit sphere centered at O. A 3-D line L is projected onto the unit sphere using central projection to create a great circle. Q denotes the normal to the plane containing the great circle. The unit normal q which is drawn parallel to Q, and from the origin to the unit circle represents the line L in the image plane. The magnitude of Q is equal to the distance from L to the origin. Note that this parameterization does not specify a unique mapping of lines in space to the image plane. There are still two possible orientations of Q with one being the negative sense of the other. The authors suggest properties such as the end of the lines or intensity gradient normal to the line to be used to determine uniquely the sense of vector Q. Once the parameterization of lines in space is obtained, geometric properties of vector q in the image plane over three successive frames can be exploited. Consider vector q^i (where i denotes frame number and ranges, say, from 1 to 3) which is the representation of a line in the unit sphere at frame i. It can be shown that if we rotate the vectors representing the same line at different frames back to a common frame, all these vectors must lie on a great circle in the unit sphere. So, for three vectors in the image plane which represent the orientation of the same line over three consecutive frames, the cross product of any two must be orthogonal to the remaining one. Equations involving only the rotational parameters (a total of six) from three consecutive frames can then be written as

$$q^2 \cdot (R_{12} \, q^1 * R_{23}^{-1} \, q^3) = 0 \qquad (4.11)$$

where R_{12} and R_{23} are the rotation matrices from frame 1 to frame 2 and from frame 2 to frame 3, respectively. An attempt to solve for the six parameters using a linear method failed. A nonlinear iterative method was used which yielded unique solutions in most cases when seven line correspondences in three consecutive frames were specified. Furthermore, if the rotational axis is fixed in space over three successive frames, a linear method can be employed to solve the rotational parameters provided that eight line correspondences are specified and not all the 3-D lines lie in a plane. However, the translation vector can not be recovered uniquely if only line correspondences are used.

Mitiche, Seida and Aggarwal [41] presented a method which utilizes the principle of invariance of angular configuration with respect to rigid motion. The principle of angular invariance states that angles between lines in a set of lines in space do not change as a result of the rigid motion of the body from which this set of lines is derived. Their approach does not require the sense of the line to be fixed as in [40] because the property of angular invariance is chosen to be the one that constrains the value of $\cos^2\Theta$ where Θ is any of the two angles Θ_1 or Θ_2 between two observed lines in space. In more detail, Let Δ_1 and Δ_2 be two lines in space. Then, for any two non-zero vectors L_1 on Δ_1 and L_2 on Δ_2 in frame S, we have

$$\cos^2\Theta = \left[\frac{L_1 \cdot L_2}{|L_1| \, |L_2|} \right]^2 \qquad (4.12)$$

If we observe the same lines Δ_1 and Δ_2 in another reference frame S', then one can write another expression for Θ:

$$\cos^2\Theta = \left[\frac{L'_1 \cdot L'_2}{|L'_1| \, |L'_2|} \right]^2 \qquad (4.13)$$

Then the principle of invariance of angular configuration for Δ_1, Δ_2 between S and S' states that

$$\left[\frac{L_1 \cdot L_2}{|L_1| \, |L_2|} \right]^2 = \left[\frac{L'_1 \cdot L'_2}{|L'_1| \, |L'_2|} \right]^2 \qquad (4.14)$$

More precisely, it can be shown that in frame S, the expressions of the unit vectors l_1 and l_2 in the direction of L_1 and L_2 are given by

$$l_1 = (l_{1X}, l_{1Y}, l_{1Z}) = \left[\frac{\mu x_j - x_i}{L_1}, \frac{\mu y_j - y_i}{L_1}, \frac{(1 - \mu) f}{L_1} \right] \qquad (4.15)$$

And

$$L_1 = \sqrt{(\mu x_j - x_i)^2 + (\mu y_j - y_i)^2 + (1-\mu)^2 f^2} \tag{4.16}$$

Where (x_i, y_i) and (x_j, y_j) are two points on the projection of L_1 in frame S. L_1 is the magnitude of the 3-D vector, and μ is the variable which determines the orientation of the line Δ_1 in space.

Similarly, we can obtain the expressions for l_2 in frame S as

$$l_2 = (l_{2X}, l_{2Y}, l_{2Z}) = \left[\frac{\Omega u_j - u_i}{L_2}, \frac{\Omega v_j - v_i}{L_2}, \frac{(1-\Omega)f}{L_2}\right] \tag{4.17}$$

Where

$$L_2 = \sqrt{(\Omega u_j - u_i)^2 + (\Omega v_j - v_i)^2 + (1-\Omega)^2 f^2} \tag{4.18}$$

Again, L_2 is the magnitude of the space vector whose projection contains image points (u_i, v_i) and (u_j, v_j). Ω is the variable which determines the orientation of Δ_2 in space.

The expressions of l'_1 and l'_2 in frame S' are quite similar to those at frame S

$$l'_1 = (l'_{1X}, l'_{1Y}, l'_{1Z}) = \left[\frac{\mu' x'_j - x'_i}{L'_1}, \frac{\mu' y'_j - y'_i}{L'_1}, \frac{(1-\mu')f}{L'_1}\right] \tag{4.19}$$

And

$$l'_2 = (l'_{2X}, l'_{2Y}, l'_{2Z}) = \left[\frac{\Omega' u'_j - u'_i}{L'_2}, \frac{\Omega' v'_j - v'_i}{L'_2}, \frac{(1-\Omega')f}{L'_2}\right] \tag{4.20}$$

The quantity $\cos^2\Theta$ measured in frame S is thus $\cos^2\Theta = (l_{1X} l_{2X} + l_{1Y} l_{2Y} + l_{1Z} l_{2Z})^2$, and $\cos^2\Theta = (l'_{1X} l'_{2X} + l'_{1Y} l'_{2Y} + l'_{1Z} l'_{2Z})^2$ in frame S'.

The principle of invariance of angular configuration for line Δ_1 and line Δ_2 between viewing systems S and S' is then

$$(l_{1X} l_{2X} + l_{1Y} l_{2Y} + l_{1Z} l_{2Z})^2 = (l'_{1X} l'_{2X} + l'_{1Y} l'_{2Y} + l'_{1Z} l'_{2Z})^2 \tag{4.21}$$

Or in expanded form

$$\frac{[(\mu x_j - x_i)(\Omega u_j - u_i) + (\mu y_j - y_i)(\Omega v_j - v_i) + (1-\mu)(1-\Omega)f^2]^2}{L_1^2 L_2^2} =$$

$$\frac{[(\mu' x'_j - x'_i)(\Omega' u'_j - u'_i) + (\mu' y'_j - y'_i)(\Omega' v'_j - v'_i) + (1-\mu')(1-\Omega')f^2]^2}{L_1'^2 L_2'^2} \tag{4.22}$$

In their formulation, each line contributes one unknown (per view) to the system of equations which is to be solved. This unknown accounts for the inability to recover the line orientation in space from a single view. With n views, $n-1$ constraints on the orientation of lines for each possible pairing of lines in the set of observed lines can be obtained. For example, if we consider four lines in space (therefore six possible different pairs of lines) and three views, then simple enumeration gives us twelve equations in twelve unknowns over the three views. The equations have the following properties: (a) The equations are of second order in each of the variables, (b) Only four of the twelve unknowns appear in every equation, and (c) The motion parameters are not involved in the equations. The authors reported that the system of equations is numerically well behaved. Once the orientation of the lines in space is solved, the rotational components of motion between the viewing systems are then readily recovered from these orientations. The translation components of the motion may also be recovered from five lines and by fixing the scale using a point on one of the lines.

Unlike the use of point correspondences which usually requires correspondence relationship to be specified between two adjacent image frames, it was shown that determining the structure and motion parameters using line correspondences from two views is in general a highly underconstrained problem [41]. Various restrictions have to be imposed on the 3-D line configuration for the existence of a unique solution [40].

5. THREE-DIMENSIONAL MOTION ANALYSIS - SEQUENCE OF STEREO IMAGES

The analyses described in the previous section determine the structure of an object given a sequence of monocular images of the object. In addition the techniques provide solutions for the motion of the projection coordinate systems from which the monocular images are derived. It was assumed that the correspondence between point or line features was already given. If one observes a sequence of stereo images, instead of a sequence of monocular images, the nature of three-dimensional motion analysis changes considerably. In the following, the problem of determining the motion of an object from a sequence of stereo images is considered. The overall analysis consists of the following steps: (i) From the sequence of stereo images, the depth map for each stereo pair is determined, (ii) the correspondence between three-dimensional features in successive depth maps is established, and (iii) the motion between the matched features is computed. This formulation of motion analysis based on sequences of stereo images has several advantages and disadvantages. In the following, the highlights of the analysis are presented, and the advantages and disadvantages are briefly discussed.

5.1 Extraction and Matching of Features - Points and Lines

Given a sequence of stereo images, the depth map for each pair may be computed using an edge-based stereo algorithm [42]. However, any other algorithm for the computation of a depth map may also be used. Having determined the depth maps, the features (consisting of lines and points) are extracted. These features are matched between successive depth maps using a two pass relaxation process. In the process of the extraction, search and matching, the search space is limited by an image differencing technique as described below.

The image difference between two intensity images taken at two different instants of time is due to either the change of illumination during the time interval or the motion of objects in the scene. If the time interval between obtaining two images is sufficiently small so that there is no detectable illumination change between the two instants of time, then we can assume that the image difference is due to the motion of objects in the scene, and represents the area of motion in the image plane. When we use the image differencing technique, we generally assume that the objects are moving against a stationary background. From the image difference between two consecutive frames in a time-sequence of images, we can find the area of motion of objects and the search space can be limited to this area of motion. The scheme of limiting the search area can be applied in the feature extraction process rather than in the matching process so that the processing time may be reduced further and also the matching process may be made more reliable. In our algorithm, the coordinate system of the left camera is used as the reference coordinate system, and hence the image differencing is performed between two consecutive left images in a sequence of stereo image pairs.

Three-dimensional lines and points are extracted by a cooperative use of intensity information, edges in the intensity images, and depth maps. The scheme of limiting the search area is applied in this feature extraction stage. In other words, only the features that lie on the surface of moving objects are extracted and used for matching.

For three-dimensional line extraction, two-dimensional lines are first extracted from the left zero crossing map of each stereo image pair using the Hough transform technique. These 2-D lines are verified to correspond to three-dimensional lines by using the corresponding depth map of the scene. The verification process is needed because a two-dimensional line on an image plane is not necessarily a projection of a three-dimensional line, e.g., a circle in three-dimensional space can appear as a line in the image plane if it is aligned in the direction of projection. The verification produces valid results because the left zero crossing and depth maps are already registered. However, one problem in this technique is that it generates clusters of lines in some areas which complicates the establishment of correspondence relationships. In order to minimize this problem, a window operator is applied on the constructed Hough transform space to select the most prominent line from a cluster of lines.

For three-dimensional point extraction, first two-dimensional points are extracted from the left intensity image of each stereo pair using Moravec's interest operator which selects the points that have a high intensity variance in its neighborhood. The intensity variance at point (i, j) is measured by a 3×3 operator and defined as follows: If $I_{i,j}$ and $V_{i,j}$ represent the intensity value and the variance measure at point (i, j), respectively, then,

$$V_{i,j} = Min\ [|I_{i,j+1} - I_{i,j-1}|, |I_{i+1,j} - I_{i-1,j}|, |I_{i+1,j+1} - I_{i-1,j-1}|, |I_{i-1,j+1} - I_{i+1,j-1}|], \qquad (5.1)$$

and if the value $V_{i,j}$ exceeds a preset threshold, then the point (i, j) is selected as a feature point for matching.

After three-dimensional lines and points are extracted from the first and second frames of depth maps, a relaxation method is employed to establish correspondences between the two frames. In our algorithm, a two pass relaxation process is applied. In the first pass of the process, the features in the first frame form a set of nodes and the features in the second frame become the labels for the nodes. In the second pass, the roles of the features in each of the images are switched, i.e., the features in the second frame form a set of nodes and those in the first frame become the possible labels of the nodes. After these two applications of the relaxation process, only those correspondences that agree in both applications are considered to be valid.

The procedure of our relaxation process is as follows. A collection of features extracted from the first frame forms a set of nodes $\{n_i\}$ for the relaxation process. Each node n_i has a set of labels $L_{n_i} = \{l_{ij}\}$ which consists of the possible correspondences (viz. a feature in the second frame which is a possible match) of node n_i. Each node also has a set of nearest neighbors, each neighbor being another node. Each element of label set L_{n_i} is associated with a probability $P(n_i, l_{ij})$ with which label l_{ij} is in correspondence with node n_i. Initially, the set of labels for each node is established by assigning to the set all the candidate features extracted in the second frame. In the second frame, a feature within the search space described earlier is considered as a possible candidate that can match a node. Each possible candidate feature has a set of nearest neighbors just as each node does.

Once a set of labels L_{n_i} is associated with a node n_i, the next step is to assign the initial probability $P(n_i, l_{ij})$ to each label l_{ij}. The simplest way of assigning the initial probabilities is to give an equal probability to every element of the label set. However, convergence of the relaxation process can be improved by using weighting functions to assign the initial probabilities. Weighting functions may be based on factors which are invariant with respect to motion. In general, many factors may be used, and each factor may be weighted according to its reliability or significance. The procedures for assigning weights, computing probabilities, updating probabilities and relaxation process can be found in [43].

At the end of each iteration in the relaxation process, if the probability of a label of a node is greater than a predefined threshold value, the label is selected as the correct one of that node. However, the label should pass the final consistency check before it is selected. When checking for line correspondence, the angle between a *reference node* (line) and the current node (line) under consideration should be approximately equal to the angle between a *reference label* (reference line in frame 2) and the current label (candidate line in frame 2). The reference node is a node which is matched already without any ambiguity and the reference label is the corresponding label of the reference node. Actually, these are the first pair of matched features and therefore the consistency check cannot be applied in determining this first correspondence. When checking for point correspondence, there are two pairs of reference nodes (the first two pairs of matched points) and two pairs of labels. Two constraints are used for the consistency check: (1) The angles formed by the lines connecting the current node (point) and the two reference nodes (points) should be approximately equal to the angle formed by the lines connecting the label (candidate point) and the two reference nodes (points), and (2) the distances between the current node and the two reference nodes should be approximately the same as the distances between current label and the two reference labels, respectively. If the current label passes this consistency check, the label is selected as the correct one for the node, otherwise, the probability of the label is reduced to half the current probability, and the other labels are entertained as being likely candidates for the node.

5.2 Determining 3-D Motion of Rigid Objects

In general, two three-dimensional line correspondences may be used to determine motion of a rigid object, assuming that the motion is small. Here, a three-dimensional line is specified by a three-dimensional direction and a point on the line. The same method can be used for three-dimensional point correspondences since two points determine a line. In general, three point correspondences, or one line and one point correspondences are sufficient to determine the three-dimensional motion parameters of a moving object. In the former case, the three points should not be collinear, and in the latter case, the point should not lie on the line.

The following notational conventions are used in the following discussion. Bold face characters are used to denote vectors and matrices while plain face characters denote scalar quantities. Unprimed characters denote the quantities at the first instant of time (t) while primed characters denote the quantities at the second instant of time (t', i.e., after motion). Scalar product is denoted by ' $\cdot$ ' and vector product by ' $\times$ '.

In general, the three-dimensional motion of an object is described by the equation.

$$\mathbf{V'} = \mathbf{R}\mathbf{V} + \mathbf{T} \tag{5.2}$$

In above equation, $\mathbf{V}$ and $\mathbf{V'}$ are the position vectors of an object point at time instants of t and t' respectively, $\mathbf{R}$ is a rotation matrix about an axis passing through the origin of a coordinate system, and $\mathbf{T}$ is the translation vector. One important fact which is often ignored in the computation of object motion is that the rotation matrix is unique regardless of the location of the rotational axis in space as long as the direction of the axis is maintained. On the other hand, the translation vector depends upon the location of the rotational axis. Based on this observation, we can find the rotation and translation components of motion separately by solving linear equations. The succeeding sections describe the determination of $\mathbf{R}$ and $\mathbf{T}$ using the correspondences established as discussed above.

5.2.1 Computation of 3-D Motion from Line Correspondences

In this section, a method is presented for estimating the three-dimensional motion of rigid objects from two three-dimensional line correspondences. Suppose two sets of nonparallel three-dimensional line correspondences are established. Here, a line correspondence means that a line L in space (not a line segment) is known to have moved to line L' over a time interval $t' - t$. Therefore, the only useful information we can get from these lines is the direction cosines of each line. This *infinite line* assumption is reasonable, because from a sequence of depth maps of a scene it is in general not possible to guarantee exact line segment correspondences between frames. This is due to object motion and possible occlusion.

From the assumptions, we have the four line equations:

At time t :

$$V_1 = V_a + \alpha A \tag{5.3}$$

$$V_2 = V_b + \beta B \tag{5.4}$$

at time t' :

$$V_1' = V_p + \gamma A' \tag{5.5}$$

$$V_2' = V_q + \delta B' \tag{5.6}$$

where, A, B and A', B' are the direction cosines of two known lines before and after motion respectively, and α, β, γ and δ are arbitrary real numbers. Also, V_a, V_b, V_p, and V_q are arbitrary points on each line and are independent of each other.

The displacement equations of these lines over the time interval $t' - t$ are as follows.

$$V_1' = RV_1 + T \tag{5.7}$$

$$V_2' = RV_2 + T \tag{5.8}$$

where, the rotation matrix R can be represented in terms of either the rotation angles Θ, Φ, Ψ about the X, Y, Z axes as

$$R = \begin{bmatrix} \cos\Phi\cos\Psi+\sin\Phi\sin\Theta\sin\Psi & -\cos\Phi\sin\Psi+\sin\Phi\sin\Theta\cos\Psi & \sin\Phi\cos\Theta \\ \cos\Theta\sin\Psi & \cos\Theta\cos\Psi & -\sin\Theta \\ -\sin\Phi\cos\Psi+\cos\Phi\sin\Theta\sin\Psi & \sin\Phi\sin\Psi+\cos\Phi\sin\Theta\cos\Psi & \cos\Phi\cos\Theta \end{bmatrix} \tag{5.9}$$

In terms of the orientation vector of the rotational axis n and angle Θ about this axis we have:

$$R = \begin{bmatrix} n_1^2+(1-n_1^2)\cos\Theta & n_1n_2(1-\cos\Theta)+n_3\sin\Theta & n_1n_3(1-\cos\Theta)-n_2\sin\Theta \\ n_1n_2(1-\cos\Theta)-n_3\sin\Theta & n_2^2+(1-n_2^2)\cos\Theta & n_2n_3(1-\cos\Theta)+n_1\sin\Theta \\ n_1n_3(1-\cos\Theta)+n_2\sin\Theta & n_2n_3(1-\cos\Theta)-n_1\sin\Theta & n_3^2+(1-n_3^2)\cos\Theta \end{bmatrix} \tag{5.10}$$

If we rewrite equations (5.7) and (5.8) in component forms, we have six nonlinear equations with eight unknowns. However, based on the discussion in the beginning of this section, we can first find a rotation matrix independent of the translation vector, and next find a translation vector given this rotation matrix. In other words, we can find **R** and **T** separately, i.e., first rotate lines V_1 and V_2 about an axis passing through the origin until they are parallel to V_1' and V_2' simultaneously, and next translate them until arbitrary points P_1 on line V_1 and P_2 on V_2 meet lines V_1' and V_2' simultaneously.

5.2.2 Finding the Rotation Matrix

The rotation matrix is denoted by **R**, and each element of **R** is treated as an unknown.

$$\mathbf{R} = \begin{bmatrix} r_1 & r_2 & r_3 \\ r_4 & r_5 & r_6 \\ r_7 & r_8 & r_9 \end{bmatrix} \tag{5.11}$$

Then, the first step described above can be represented by the following equations.

$$\mathbf{A'} = \mathbf{RA} \tag{5.12}$$

$$\mathbf{B'} = \mathbf{RB} \tag{5.13}$$

Furthermore, we can find the direction cosines of the third line in space simply by evaluating **A**×**B**. Let these third direction cosines be **C** and **C'** before and after motion, respectively. Then, we have another equation

$$\mathbf{C'} = \mathbf{RC} \tag{5.14}$$

From equations (5.12), (5.13) and (5.14), we have nine linear equations with nine unknowns. However, one should note that these are three sets of equations each set consisting of three linear simultaneous equations in three unknowns.

However, when the rotation matrix **R** is represented by equation (5.10), we need not solve any system of equations. Because the rotational axis should be perpendicular to both (**A'**−**A**) and (**B'**−**B**), we can compute the direction of the rotational axis directly.

$$\mathbf{n} = \frac{(\mathbf{A'}-\mathbf{A}) \times (\mathbf{B'}-\mathbf{B})}{|(\mathbf{A'}-\mathbf{A}) \times (\mathbf{B'}-\mathbf{B})|} \tag{5.15}$$

However, a problem arises when (**A'**−**A**) and (**B'**−**B**) are parallel to each other. This case occurs only when the rotational axis lies on the plane formed by vectors **A** and **B**. Therefore, when the computation of **n** using equation (5.15) fails, we can use the following equation.

$$\mathbf{n} = \frac{(\mathbf{A} \times \mathbf{B}) \times (\mathbf{A'}-\mathbf{A})}{|(\mathbf{A} \times \mathbf{B}) \times (\mathbf{A'}-\mathbf{A})|} \tag{5.16}$$

In equation (5.16), if $(A'-A) = 0$, we can use $(B'-B)$ instead. In fact, when $(A'-A) = 0$, $n = A$, and when $(B'-B) = 0$, $n = B$; if both $(A'-A)$ and $(B'-B)$ are zero, the motion is pure translation.

Once the orientation vector of the rotational axis is found, the rotation angle is simply the angle between the plane formed by n and A, and that formed by n and A'. Therefore,

$$\cos\Theta = \frac{(n \times A) \cdot (n \times A')}{|n \times A| \, |n \times A'|} \tag{5.17}$$

In the calculation of $\cos\Theta$ using equation (5.17), if $n = A$, we can use vectors B and B' instead of A and A', respectively.

5.2.3 Determination of the Translation Vector

Once the rotation matrix is found, we have two pairs of parallel lines in space; both pairs must be merged simultaneously by a single translation. If we have only one pair of parallel lines in space, there are infinite number of ways of translating one line to merge with the other. However, if we consider two or more pairs of such lines, there exists a unique translation which can merge each pair of lines simultaneously. From equations (5.3) to (5.8),

$$R(V_a + \alpha A) + T = V_p + \gamma A' \tag{5.18}$$

$$R(V_b + \beta B) + T = V_q + \delta B' \tag{5.19}$$

Equations (5.18) and (5.19) may seem to provide six linear simultaneous equations in seven unknowns, but note that α, β, γ and δ are arbitrary real numbers. In other words, if we select an arbitrary point on each of the lines V_1 and V_2, and if we can find the corresponding points on the lines V_1' and V_2' after a translation, that translation is the one we want to determine. Therefore, by substituting $\alpha = \beta = 0$ in equations (5.18) and (5.19), we have six linear equations in five unknowns.

$$RV_a + T = V_p + \gamma A' \tag{5.20}$$

$$RV_b + T = V_q + \delta B' \tag{5.21}$$

Solving equations (5.20) and (5.21) is straightforward.

In the case of using point correspondences, we can determine the translation vector T directly using the equation (5.2).

Using the above procedure, we compute the motion parameters for every possible combination of the matched features. A clustering or an averaging procedure is then used to eliminate the effects of false matches and bad data.

6. CONCLUDING REMARKS

Although much effort has been spent on estimating 3-D object structure and motion, a vision system which can implement this capability in real time is still far from reality. The results obtained by previous analyses are presented in the several references cited. This inability to compute structure and motion in real-time arises from several sources. One source of the problem is the nonlinear nature of the equations to be solved. Another source stems from the lack of adequate amounts of computing power and resources required to support the extensive computation required for tracking objects moving in space. Ullman [44] has introduced the idea of incremental rigidity in contrast to the rigidity considered in [37]. It appears that there is some qualitative similarity between human perception of structure from motion and the proposed rigidity schemes. It is also possible that the solution to this problem requires a scheme which exploits point, line, and other correspondences at the same time. In any case, in order to establish a structure and motion computation paradigm which begins with the input image sequence and ends with the structure and motion information of objects in space, several other issues have to be addressed and successfully resolved. The fundamental characteristic of the previous analyses is that correspondence between points or lines is established or has assumed to have been established before the computation of structure and motion is undertaken. The difficulties with this approach were briefly reviewed. However, this approach is an important initial approximation in formulating this problem and its significance should not be minimized.

REFERENCES

[1]. J.K. Aggarwal and N.I. Badler (Eds.), Abstracts for the Workshop on Computer Analysis of Time-Varying Imagery, University of Pennsylvania, Moore School of Electrical Engineering, Philadelphia, PA, April 1979.

[2]. J.K. Aggarwal and N.I. Badler (Guest Eds.), Special Issue on Motion and Time-Varying Imagery, *IEEE Trans. on PAMI,* Vol. PAMI-2, No. 6, November 1980.

[3]. W.E. Snyder (Guest Ed.), Computer Analysis of Time-Varying Images, *IEEE Computer,* Vol. 14, No. 8, August 1981.

[4]. J.K. Aggarwal (Guest Ed.), Motion and Time Varying Imagery, *Computer Vision, Graphics and Image Processing,* Vol. 21, Nos. 1 and 2, January, February 1983.

[5]. T.S. Huang, *Image Sequence Analysis,* Springer-Verlay, New York, 1981.

[6]. S. Ullman, *The Interpretation of Visual Motion,* MIT Press, Cambridge, 1979.

[7]. NATO Advanced Study Institute on Image Sequence Processing and Dynamic Scene Analysis, Advance Abstracts of Invited and Contributory Papers, June 21-July 2, 1982, Braunlage, West Germany.

[8]. Siggraph/Siggart Interdisciplinary Workshop on Motion: Representation and Perception, Toronto, Canada, April 4-6, 1983, and *Computer Graphics,* Vol. 18, No. 1, January 1984.

[9]. International Workshop on Time-Varying Image Processing and Moving Object Recognition, Florence, Italy, May 1982.

[10]. W.N. Martin and J.K. Aggarwal, "Dynamic Scene Analysis: A Survey," *Computer Graphics and Image Processing 7*, pp. 356-374, 1978.

[11]. H.-H. Nagel, "Analysis Techniques for Image Sequences," in *Proc. IJCPR-78*, Kyoto, Japan, November 1978, pp. 186-211.

[12]. J.K. Aggarwal and W.N. Martin, "Dynamic Scene Analysis," in the book *Image Sequence Processing and Dynamic Scene Analysis*, edited by T.S. Huang, Springer-Verlag, 1983, pp. 40-74.

[13]. J.K. Aggarwal, "Three-Dimensional Description of Objects and Dynamic Scene Analysis," in the book *Digital Image Analysis*, edited by S. Levialdi, published by Pitman Books Ltd., pp. 29-46, 1984.

[14]. H.-H. Nagel, "What Can We Learn from Applications?" in the book *Image Sequence Analysis*, Edited by T.S. Huang, Springer-Verlag, 1981, pp. 19-228.

[15]. H.-H. Nagel, "Overview on Image Sequence Analysis," in the book *Image Sequence Processing and Dynamic Scene Analysis*, edited by T.S. Huang, Springer-Verlag, 1983, pp. 2-39.

[16]. T.S. Huang (Editor), *Image Sequence Processing and Dynamic Scene Analysis*, Proceedings of NATO Advanced Study Institute at Braunlage, West Germany, Springer-Verlag, 1983.

[17]. IEEE Computer Society Workshop on Motion: Representation and Analysis, Kiawah Island, SC, May 1986.

[18]. The 2nd International Workshop on Time-Varying Image Processing and Moving Object Recognition, Florence, Italy, September 1986.

[19]. Digital Image Processing and Analysis: Volume 2: Digital Image Analysis, R. Chellappa and A.A. Sawchuk, IEEE Computer Society Press, 1985.

[20]. J.K. Aggarwal, L.S. Davis, and W.N. Martin, "Correspondences Processes in Dynamic Scene Analysis," *Proceedings of IEEE*, Vol. 69, No. 6, May 1981, pp. 562-572.

[21]. S. Yalamanchili, W.N. Martin, and J.K. Aggarwal, "Extraction of Moving Object Description via Differencing," *Computer Graphics and Image Processing*, No. 18, 1982, pp. 188-201.

[22]. J.K. Aggarwal and R.O. Duda, "Computer Analysis of Moving Polygonal Images," *IEEE Transactions on Computers*, Vol. C-24, October 1975, pp. 966-976.

[23]. W.N. Martin and J.K. Aggarwal, "Computer Analysis of Dynamic Scenes Containing Curvilinear Figures," *Pattern Recognition*, Vol. 2, 1979, pp. 169-178.

[24]. J.L. Potter, "Scene Segmentation Using Motion Information," *Computer Graphics and Image Processing*, Vol. 6, 1977, pp. 558-581.

[25]. W.K. Chow and J.K. Aggarwal, "Computer Analysis of Planar Curvilinear Moving Images," *IEEE Transactions on Computers*, Vol. C-26, No. 2, 1977, pp. 179-185.

[26]. J. W. Roach and J. K. Aggarwal, "Computer Tracking of Objects Moving in Space", *IEEE Transactions PAMI*, Vol. PAMI-1, No. 2, April 1979, pp. 127-135.

[27]. J. W. Roach and J. K. Aggarwal, "Determining the Movement of Objects from a Sequence of Images", *IEEE Transactions PAMI*, Vol. PAMI-2, No. 6, November 1980, pp. 554-562.

[28]. J. A. Webb and J. K. Aggarwal, "Structure and Motion of Rigid and Jointed Objects", *Artificial Intelligence*, No. 19, 1982, pp. 107-130.

[29]. H. H. Nagel, "Representation of Moving Rigid Objects Based on Visual Observations", *Computer*, August 1981, pp. 29-39.

[30]. H. H. Nagel, "On the Derivation of 3D Rigid Point Configuration from Image Sequences", *Proceedings of IEEE Conference on Pattern Recognition and Image Processing*, Dallas, TX, August 2-5, 1981.

[31]. R. Y. Tsai and T. S. Huang, "Estimating 3-D Motion Parameters of a Rigid Planar Patch, I", *IEEE Transactions ASSP,* Vol. ASSP-29, No. 6, December 1981, pp. 1147-1152.

[32]. R. Y. Tsai, T. S. Huang and W. L. Zhu, "Estimating Three-Dimensional Motion Parameters of a Rigid Planar Patch, II: Singular Value Decomposition", *IEEE Transactions ASSP,* Vol. ASSP-30, August 1982, pp. 525-534.

[33]. T. S. Huang and R. Y. Tsai, "Image Sequence Analysis: Motion Estimation", in *Image Sequence processing and Dynamic Scene Analysis,* T. S. Huang ed., Springer-Verlag, 1981.

[34]. R. Y. Tsai and T. S. Huang, "Uniqueness and Estimation of Three-Dimensional Motion Parameters of Rigid Objects with Curved Surface", *IEEE Transactions PAMI,* Vol. PAMI-6, No. 1, January 1984, pp. 13-26.

[35]. H. C. Longuet-Higgins, "A Computer Algorithm for Reconstructing a Scene from Two Projections", *Nature,* Vol. 293, September 1981, pp. 133-135.

[36]. H. C. Longuet-Higgins, "The Reconstruction of a Scene from Two Projections - Configurations That Defeat the 8-Point Algorithm", *Proceedings of The First Conference on Artificial Intelligence Applications,* Denver, CO, December 5-7, 1984, pp. 395-397.

[37]. A. Mitiche, S. Seida and J. K. Aggarwal, "Determining Position and Displacement in Space from Images", *Proceedings of IEEE Computer Society Conference on Computer Vision and Pattern Recognition,* San Francisco, CA, June 19-23, 1985, pp. 504-509.

[38]. B. L. Yen and T. S. Huang, "Determining 3-D Motion and Structure of a Rigid Body Using Straight Line Correspondences", in *Image Sequence Processing and Dynamic Scene Analysis,* T. S. Huang, ed., Springer Verlag, 1983.

[39]. B. L. Yen and T. S. Huang, "Determining 3-D Motion and Structure of a Rigid Body Using Straight Line Correspondences", *Proceedings of the International Joint Conference on Acoustics, Speech and Signal Processing,* March 1983.

[40]. B. L. Yen and T. S. Huang, "Determining 3-D Motion/Structure of a Rigid Body over 3 Frames Using Straight Line Correspondences", *Proceedings of the IEEE Computer Society Conference on Computer Vision and Pattern Recognition,* Washington, D.C., June 19-23, 1983, pp. 267-272.

[41]. J. K. Aggarwal and A. Mitiche, "Structure and Motion from Images", *Proceedings of the Image Understanding Workshop,* Miami Beach, FL, December 9-10, 1985, pp. 89-95.

[42]. Y. C. Kim and J. K. Aggarwal, "Finding Range from Stereo Images", *Proceeding IEEE Conference Computer Vision and Pattern Recognition,* San Francisco, CA, Jun 1985, pp. 289-294.

[43]. Y. C. Kim, "Structure and Motion of Objects from Stereo Images", *Ph.D. Dissertation,* Department of Electrical and Computer Engineering, The University of Texas at Austin, May, 1986.

[44]. S. Ullman, "Maximizing Rigidity: The Incremental Recovery of 3-D Structure from Rigid and Non-Rigid Motion," *Perception,* Vol. 13, 1984, pp. 255-274.

7. ACKNOWLEDGEMENTS

It is a pleasure to acknowledge the help from Mr. N. Nandhakumar and Mr. Y. F. Wang.

TIME-VARYING IMAGE ANALYSIS

Thomas S. Huang
Professor
Electrical Engineering
Coordinated Science Laboratory
University of Illinois
1101 W. Springfield Avenue
Urbana, Illinois 61801

ABSTRACT

We describe two recent projects carried out at the University of Illinois. The first is a computer lip reader, which is used to increase the recognition accuracy of an acoustic word recognizer. The second is a high-level system for representing and identifying time-varying characteristics of a large class of physical events.

INTRODUCTION

In many computer vision and image processing problems, we have to process and analyze time-varying scenes of imagery (3,4). The goals of the processing can be classified into four areas: efficient coding, enhancement, pattern detection and recognition, and event understanding.

The major difference between time-varying and stationary image processing is of course that the former involves motion. Of the four areas mentioned above probably efficient coding of TV sequences with motion-compensation is the most mature area. Motion-compensated enhancement of image sequences has been done to some extent, but mainly in reducing image noise. Work in detecting and recognizing moving objects has been limited largely to single target tracking. In event understanding, successful results have been obtained in the biomedical field where the analysis is mainly two-dimensional (5,7). Three-dimensional time-varying scene understanding is still in its infancy.

It is the purpose of the present paper to describe briefly two research projects carried out recently at the University of Illinois:

One in the area of recognition, the other in the area of event understanding.

AUTOMATIC LIPREADING

The results described in this Section are from the Ph.D. thesis research of E.D. Petajan (6) under the supervision of Professor G. Slottow.

Overview

Automatic recognition of the acoustic speech signal alone is inaccurate and computationally expensive. Additional sources of speech information, such as lipreading (or speechreading) should enhance automatic speech recognition, just as lipreading is used by humans to enhance speech recognition when the acoustic signal is degraded. We shall describe an automatic lipreading system which has been developed. A commercial device (Voterm II) performs the acoustic speech recognition independently of the lipreading system.

The recognition domain is restricted to isolated utterances and speaker dependent recognition. The speaker faces a solid state camera which sends digitized video to an MC68000 based minicomputer system with custom video processing hardware. The video data is sampled during an utterance and then reduced to a template consisting of visual speech parameter time sequences. The distances between the incoming template and all of the trained templates for each utterance in the vocabulary are computed and a visual recognition candidate is obtained. The combination of the acoustic and visual recognition candidates is shown to yield a final recognition accuracy which greatly exceeds the acoustic recognition accuracy alone.

Preprocessing

A GE TW2500 CID camera is used to image and digitize the face of the speaker. The data rate is 30 frames per second, with 244x248 picture elements per frame, and 8 bits per picture element. Figure 1 shows a typical image frame, but with only 4 graylevels.

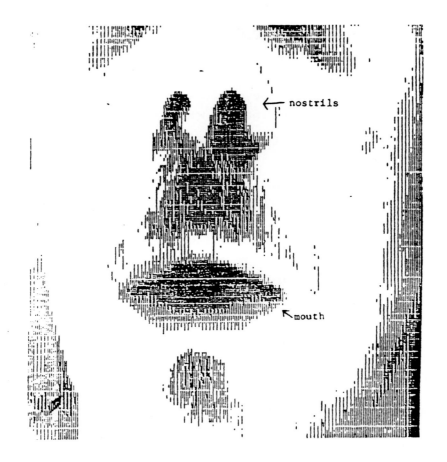

Figure 1. Facial Image (4 graylevels)

The image frames are made binary by thresholding, and then smoothed. Regions in the images corresponding to the nostrils, teeth, lips, and tongue are extracted by contour tracing-using the Predictive Differential quantization algorithm (2) to speed up computation.

Recognition

For each region, several parameters (features) are measured: area, perimeter length, hortizontal extremum positions, and vertical extremum positions. Matching between an image sequence corresponding to an unknown utterance and a prototype image sequence is done by comparing the region parameters of the two sequences. A simple distance measure between the two parameter sets is defined for this purpose.

Combining Visual and Acoustic Recognition

Based on the results of the acoustic recognizer (Voterm II) and the lipreading system, a final decision on the word is made in the following way. Among the candidates proposed by the acoustic recognizer, we choose the one with the best lipreading score (smallest distance).

Experimental Results

Tests were carried out using three vocabularies: the digits 0 to 9; the letters a to z; a 100-word vocabulary including the digits, the letters, and 64 words listed in the Voterm manuals. The digits were uttered 10 times; the letters 4 times, and the 100 words 2 times. The recognition accuracies are summarized in Table 1. The processing time of the lipreading system is about 1 second for each word.

Table 1

Percentages of Correct Recognition

Method	Vocabulary		
	Digits	Letters	100-Word Vocabulary
Acoustic alone	95	64	65
Lipreading alone	99	65	60
Combined	100	66	78

EVENT REPRESENTATION AND IDENTIFICATION

The results described in this section are from the M.S. thesis research of G.C. Borchardt (1) under the supervision of Professor D.L. Waltz.

Overview

This work presents a formal mechanism for specifying the steady and time-varying characteristics of a large class of physical events. The rules which embody these specifications are then used to identify occurrences of particular events in a changing scene, starting from low-level data such as the positions and orientations of objects as they vary over time. In this manner, a high-level description of changes occurring in a scene is formed.

This mechanism has been implemented as a package which runs in the Interlisp environment on a VAX 11/780 computer. A knowledge base has been constructed for the application of this system in a simplified assembly-line context, including robot arms, bolts, nuts and other simple objects, and events such as transportation of objects in different manners, stacking objects, fastening objects together with bolts, and so forth.

The process of recognizing physical events occurs at such a low level in humans that much of its inherent complexity is hidden from view. When this process is duplicated computationally, this complexity becomes apparent. Some of the finer points which must be dealt with include the following:

a) Time-related parts of an event. Many events contain parts or sub-events which may or may not be required to fit together in a particular manner in time. All instances of these events must nevertheless be recognized.

b) Parts of wholes. If a hand is seen to be grasping an object, for example, it is necessary to have some idea of whether or not that object is part of a larger object. If so, then the event should be taken in a more general sense of the hand grasping the larger object, and so forth.

c) Avoiding redundancy. If a simple event is found to occur (for example, a hand carrying an object) then if later a higher level event is found which contains the first event as a component (for example, stacking that object on another object) the first event should be excluded from a final description of changes in the scene, as a statement that the latter event has occurred is sufficient.

An Example

Figure 2 illustrates the sequence of actions for a particular example involving a robot arm and two blocks. The robot hand first makes a false grasping motion at BLOCK1 followed by an actual grasping of the object, a lifting of the object, and a setting down of the object on top of BLOCK2.

Input to the implemented system, however, is provided at a level which is considerably less refined than the verbal description

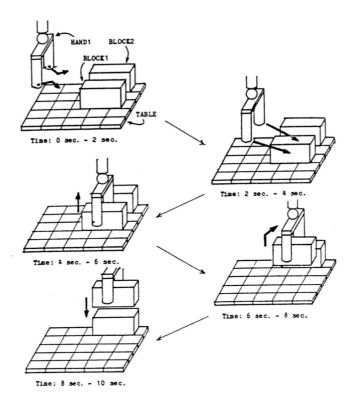

Figure 2. Sequence of Actions for the Example

above. The input file contains only data which falls into one of the following three categories.

1. Static attributes and relationships of objects. This includes dimensions of objects and relationships of parts of an object to the whole (e.g. the parts of the robot hand). All composite objects are represented using constructive geometry based on simple objects such as spheres, cylinders and blocks.

2. Time-varying positions and orientations of objects. These are input as graphs of values over time. For positions, each object is traced along the "world" coordinate axes "X", "Y" and "Z". For orientations, each object is associated with zero to three "object axes". Spheres have no object axes, cylinders have one ("P" = primary, "S" = secondary, "T" = ternary, all mutually orthogonal). Components re-

lating vectors along these axes to the world axes "X", "Y" and "Z", comprise the input for orientations of objects.

3. Simple spatial relationships of the objects. For this example, these include notions of "betweenness" (one object directly between two others), touching (or more precisely, very close proximity) and support (when one object touches a part of another which faces downwards).

Given input of the form described above, the resulting descriptions of changes occurring in the scene as produced by the implemented package is as follows:

    (HAND1 IS PICKING BLOCK1 UP FROM TIME 0 TO TIME 6)
    (HAND1 IS GRASPING BLOCK1 FROM TIME 2 TO TIME 4)
    (HAND1 IS HOLDING BLOCK1 FROM TIME 4 TO TIME 6)
    (HAND1 IS STACKING BLOCK1 ON BLOCK2 FROM TIME 6 TO TIME 9)
    (HAND1 IS HOLDING BLOCK1 FROM TIME 9 TO TIME 10)

(This high-level description should be contrasted with the low-level description in the input file, which is some 550 lines long.)

A number of observations can be made about this resultant description. First of all, the implemented package has correctly identified the major events occurring in the example. Of equal importance however, it has also attached reasonable boundaries in time to these events. Secondly, although the system must identify many lower-level events in the course of finding the high-level events, only the top-most events are included in the final description. For instance, HAND1 is holding BLOCK1 from time 4 to time 10, but for those times at which a higher-level event occurs which explains the holding (e.g. stacking) the event of holding is not mentioned.

Without going into details, we mention very briefly that the implemented system can be seen as the integration of three parts: a knowledge base of rules for determining the occurrences of particular events, an interpreter for evaluating these rules, and a supervisory program which forms a high-level description of changes in a scene from the input low-level description by a process of successive refinement.

CONCLUDING REMARKS

In the majority of time-varying image analysis problems, real-time (typically at video rate) processing is essential. Examples include: efficient coding for TV sequences, target tracking, and mobile robot navigation. Because the processing of time-varying images involves a tremendous amount of data, real-time processing has been possible only for relatively simple algorithms.

Video-rate processing systems have been constructed for TV sequence coding and simple target tracking. The lipreading system described in this paper does not quite operate in real time, but can probably be upgraded to do so. The most difficult area to achieve real-time processing is probably event understanding. Here, we have to do not only low-level image processing, but also high-level symbolic reasoning. And in between, there is the perhaps most difficult task of linking the two levels. To build real-time event understanding systems is a major challenge for the next decade.

ACKNOWLEDGEMENT

The preparation of this paper was supported by National Science Foundation Grant NSF-ECS-83-19509.

REFERENCES

1. Borchardt, G.C., "A Computer Model for the Representation and Identification of Physical Events", Report T-142, May 1984, Coordinated Science Laboratory, University of Illinois, Urbana-Champaign.
2. Huang, T.S., "Coding of Two-Tone Images", IEEE Trans. Communications, 25:11, November 1977.
3. Huang, T.S. (ed.), Image Sequences Analysis, Springer-Verlag, 1981.
4. Huang, T.S. (ed.), Image Sequence Processing and Dynamic Scene Analysis, Springer-Verlag, 1983.
5. Levin, M.D., P.B. Noble and Y.M. Youssef, "A rule-based system for characterizing blood cell motion", in Image Sequence Processing and Dynamic Scene Analysis, Springer-Verlag, 1983.
6. Petajan, E.D., "Automatic Lipreading to Enhance Speech Recognition", Ph.D. thesis, 1984, Dept. of Electrical and Computer Engineering, University of Illinois, Urbana-Champaign.
7. Tsotsos, J.K., J. Mylopoulos, H.D. Currey and S.W. Zucker, "A framework for visual motion understanding", IEEE Trans. on PAMI, 2; 6, Nov. 1980, pp. 563-657.

KNOWLEDGE REPRESENTATION FOR ROBOT VISION AND PATH PLANNING USING ATTRIBUTED GRAPHS AND HYPERGRAPHS

Andrew K.C. Wong

Department of Systems Design Engineering
University of Waterloo

ABSTRACT

This paper presents a general and flexible knowledge representation system using attributed graph representation (AGR) and attributed hypergraph representation (AHR) as the basic data structure. Based on these representations, object recognition and interpretation can be achieved by a hypergraph monomorphism algorithm and a knowledge directed search procedure. A graph synthesis procedure is used to combine the AGR's or AHR's obtained from images of different views of an object into a unique AHR. For recognition and location of 3-D objects in 2-D perspective images, another form of AHR, known as Point Feature Hypergraph Representation (PHR) is introduced. With PHR, a constellation matching algorithm can be used to compare images and models as well as to derive 3-D information from stereoscopic images. From the PHR of 3-D objects, procedural knowledge can be formulated and used to search for features in a 2-D perspective image for the recognition and location of 3-D objects in 2-D images. Further, the AGR can also be used to represent the geometric and topological information of the world environment of a mobile robot. A special search algorithm converts the AGR into a AHR from which a compact road map is derived for path and trajectory planning as well as navigation. The proposed method renders greater tolerance to local scene changes.

I. Introduction

In robot vision and task planning, an adequate representation of a real world object should enable a system to model the object or scene automatically by synthesizing data derived from the image of the real world. The challenge to machine intelligence and knowledge engineering in the development of visual based robotics is to provide general and flexible knowledge representation systems for effective communication and contol of robots or robot workstations.

To represent 3-D objects in data form suitable for manipulation and recognition, various approaches have been made. The most common ones are boundary representation [14], constructive solid geometry representation [18], sweep representation [4] and decomposition representation [10]. More recent development in computer vision can be found in [3,15]. Most of these methods are feasible to acquire the geometric information from the object image, yet they lack the flexibility for effective recognition if the orientation of the object varies, or certain parts of the object are occluded, or when the class of prototypes is very large. Furthermore, with most of these systems, the knowledge of the prototype objects has to be input by the users.

In this paper, we propose a general yet flexible knowledge representation system using Attributed Graph Representation (AGR) and Attributed Hypergraph Representation (AHR) as the basic data structure [17,19,20]. For the recognition of objects in images with range data, we use a special form of AHR, known as Edge Feature Hypergraph Representation (EHR), the vertices of which are made up of edges or curve or line segments (observed or derived). The complete AGR or AHR of an object can be obtained directly through synthesizing the AGR's or AHR's derived from the scanned images of the various views of the object. Based on this knowledge representation, we have developed various graph morphism algorithms for object recognition and interpretation.

For the derivation of 3-D information from objects and the recognition and location of 3-D objects in 2-D images, we use another form of AHR, known as Point Feature Hypergraph Representation (PHR), the vertices of which consist of point features. The procedural knowledge used to search for features in a 2-D image is directed by a special AGR known as rule network using information from the partial subgraphs of the PHR of the objects. With regard to the world environment of mobile robots, the AGR is used to represent the acquired geometrical and topological information. A special search algorithm then converts the AGR into a AHR from which a compact road map is derived for path and trajectory planning as well as navigation [17].

The advantage of this is that it furnishes a more direct and simpler approach to the representation, recognition and location of objects. It also renders greater tolerance to local scene change due to partial occlusion of objects. Another advantage is that the proposed representaion complies with some newly developed structural pattern recognition methodologies [9,21,22], knowledge directed search algorithms [16], and random graphs method [23] which allows structure variation of scenarios.

II. Attributed Hypergraph Representation

Basic Definitions and Notations

First the basic concepts and notations of attributed graphs and hypergraphs are introduced.

Definition1 An *attribute pair* is an ordered pair (A_n, A_d) where A_n is the attribute or property name of the object and A_d is the attribute or property value.

Definition2 An *attribute set* is an m-tuple $[p_1, p_2, \ldots, p_i, \ldots, p_m]$ where each element in the tuple is an attribute pair.

For our 3-D model representation, the attribute set can be used to record the properties of a surface. For example an attribute set for describing a red triangular surface is

$$S : [(\text{type, planar}), (\text{shape}, G_1), (\text{colour, red})]$$

where G_1 is a shape descriptor or an attributed graph whose vertices and edges represent the sides and angles respectively of the triangle.

Definition3 A *graph* is a pair of sets $G = (V, A)$ where $V = [v_1, \ldots, v_p, \ldots, v_q, \ldots, v_m]$ is a set of vertices, and $A = [\ldots, a_{pq}, \ldots]$ is a set of arcs. The arc a_{pq} connects vertices v_p and v_q.

Definition4 An *attributed vertex* is a vertex associated with an attribute set called vertex attribute set. An *attributed arc* is an arc associated with an attribute set called arc attribute set.

Definition5 An *attributed graph* of an object is a graph $G_a = (V_a, A_a)$ where $V_a = [v_1, \ldots, v_p, \ldots, v_q, \ldots, v_m]$ is a set of attributed vertices and $A_a = [\ldots, a_{pq}, \ldots]$ is a set of attributed arcs. The arc a_{pq} connects vertices v_p and v_q.

In order to introduce an effective scheme for grouping parts of objects or scenes into a meaningful component and to provide a flexible hierarchical structure in object representation, attributed hypergraphs is introduced.

Definition 6 The _hypergraph_ [2] is defined as an ordered pair $H = (X, E)$ where $X = \{x_1, x_2, \ldots, x_n\}$ are the vertices and $E = \{e_1, e_2, \ldots, e_m\}$ are the hyperedges of the hypergraph and

(1) $e_i \neq \emptyset$,($i = 1, \ldots, m$)

(2) $\bigcup_i^n e_i = X$.

Definition 7 An _attributed hypergraph_ $H_a = (X_a, E_a)$ consists of a set of attributed vertices X_a and a set of attributed hyperedges E_a. The attribute of the hyperedges is a mapping that maps the set of vertices into a range which provides a description of the set. The range can be: 1) an attributed graph defined on the vertex set; 2) a set of nominal or ordinal values; 3) a geometrical configuration of that set or 4) other desirable representation.

Definition 7 renders a very general representation of objects or scenes with high complexity. It also provides a means to group components or parts of an object in different manners according to the relation that induces the hyperedges. Thus, the same object can be described in different ways depending on how the hyperedges are formed. For instance, 'the colour red' can serve as a predicate relation that groups all components with attribute value red into a hyperedge. In the representation of the floor plan model for a roving robot, hyperedges can be used to represent vertex groups that form free spaces made up of convex polygons. Such hyperedges can be induced by rules based on a search procedure guided by the bounding walls [17].

III. 3-D Object Recognition and Attributed Hypergraph Synthesis

Attributed Hypergraph Representation of 3-D Objects

The purpose of introducing the AHR is to reduce the cost of finding monomorphisms during the recognition phase and to guide the graph synthesis process. The complexity of finding the monomorphism between graphs largely depends on the number of vertices and arcs in these graphs. The use of AHR results in the reduction of the number of vertices and arcs, and hence the computational cost of finding monomorphisms. It also enables the recognition of the spatial configuration of objects in 2-D images using knowledge directed search.

To simplify the definition of AHR for 3-D objects, we use the block model. It shall later be shown that such representation can be generalized to represent various 3-D objects in robotic applications as well as in other real life scenarios. What is described here below is an object attributed hypergraph representation of block models.

Definition8 An *elementary area attributed graph* $G_e = (V_e, A_e)$ is an attributed graph for representing a surface, where 1) V_e is a set of attributed vertices representing the set of boundary segments of the face and 2) A_e is the set of attributed arcs representing a set of relations between the segments.

As an example, the elementary area graph of a face can be represented by:

$$G_e = (V_e, A_e) \quad where \quad V_e = [v_1, v_2, v_3] \quad A_e = [a_{12}, a_{23}, a_{13}]$$

e.g. $v_1 = [(type, line), (length, 5)]$ and $a_{12} = [(type, angular-relation), (angle, 60°)]$.

Definition9 A *primitive block* of an object is a block bounded by surfaces such that there is no concave angular relation between any pair of the surfaces in the block.

Hence a pyramid, a column, a cylinder and a ball can be a primitive block of an object or a model.

Definition10 A *primitive block attributed graph* is an attributed graph $G_b = (V_b, A_b)$ representing the primitive block of an object. The set of attributed vertices V_b represents the set of distinct and well-defined surfaces and the set of attributed arcs A_b represents the set of angles or angular descriptors between them.

As an example, the primitive block graph of a block in Figure 1 can be represented by:

$$G_b = (V_b, A_b)$$

where $V_b = [v_1, v_2, v_3, v_4, v_5]$ and $A_b = [a_{12}, a_{23}, a_{34}, a_{14}, a_{15}, a_{25}, a_{35}, a_{45}]$

Here, we notice that one could set

$$v_1 = [(type, plane), (area, 5), (\# of edges, 3)] \quad and \quad a_{12} = [(type, line), (angle, 47°)]$$

To provide a high level object representation while retaining a general structural relation among the object components, the object attributed hypergraph is introduced.

Definition11 An *object attributed hypergraph* $H_o = (X_o, E_o)$ consists of a set of attributed vertices X_o and a set of hyperedges E_o. Each vertex corresponds to an elementary area attributed graph representing a distinctly bounded surface, and each hyperedge corresponds to a primitive block attributed graph representing a primitive block of the object.

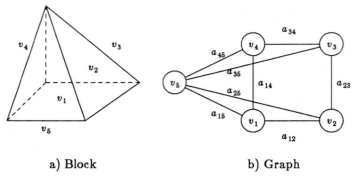

a) Block b) Graph

Figure 1 Primitive Block Attributed Graph

The following example (Figure 2) illustrates the object AHR. This object consists of one main block and two side wedge blocks.

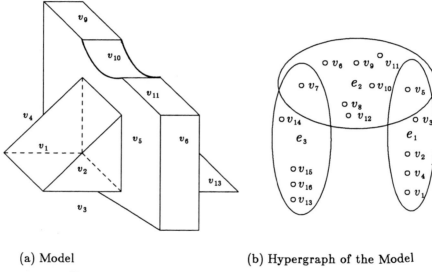

(a) Model (b) Hypergraph of the Model

Figure 2 A Model and Its Attributed Hypergraph

In summary, we can describe the AHR of 3-D objects as follows:

1. Each face of an object can be represented by a vertex in the AHR. The attribute value of each AHR vertex is an elementary graph, with the boundary segments of the faces as vertices and the relation between these segments as arcs.

2. Each primitive block of the object is represented as a hyperedge in the AHR. Each hyperedge, in turn, can assume an attribute value which is a primitive block attributed graph defined on the subset of vertices representing faces with arcs as the relations between the faces or surfaces.

3. A hidden face of a primitive block is a vertex representing the touching face of that primitive block and its adjoining block(s). In an AHR, it is the intersection of the hyperedges each of which is a primitive block adjoining to each other with faces representing by the intersecting vertices. The attribute value assigned to both surfaces is the largest of the elementary AGR of the two adjacent surfaces with the position of the smaller one relative to the larger one as part of the attribute value.

Since the object AHR is defined on a vertex set which is essentially made up of edges of the object, it is also referred to as EHR of the object.

Construction of Attributed Hypergraph Representation

When constructing the AHR for 3-D objects or their images, we can proceed in three stages. First we construct the elementary surface attributed graph for each surface in the image. We then construct the primitive block attribute graph for each component block. By considering each surface as a vertex and each set of vertices associated to a primitive block as a hyperedge, we obtain the AHR for the object(s) or the image. To relate regions in an image to the surfaces of a 3-D object, the following correspondence is observed.

(1) The regions in the image of the object generally correspond to the surfaces of an object.

(2) Some of the edges in the image correspond to the boundary segments of object surfaces.

(3) Certain groups of regions in the image may correspond to the surfaces of a certain block of the object.

Figure 3 illustrates the application of the above rule in constructing an AHR for the object image.

Once a view of an object in an image is represented by an AHR, an attributed hypergraph monomorphism algorithm [13,19] can be applied to compare the AHR with those AHR's of different prototypes. For our purpose in this paper, we will not present the AHR morphism algorithm. We introduce only some basic definitions and show the result of a 3-D object recognition experiment.

Definition 12 For an object hypergraph $H_o(V_1, E_1)$ and a model graph $H_m(V_2, E_2)$, there exists a _hypergraph monomorphism_ of H_o onto H_m if the following necessary conditions are satisfied:

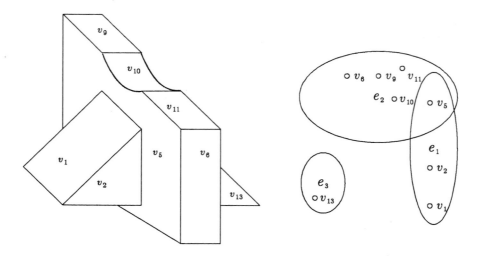

(a) Image of object after preprocessing   (b) Hypergraph of the object
Figure 3   Attributed Hypergraph for a View of an Object in an Image.

All vertices in H_o are matched or partially-matched by the vertices in H_m.

1) Vertex matching: $v_i \rightarrow v_j$ (all vertices, say v_i in H_o are partially-matched by vertices, say v_j in H_m).

2) Hyperedge matching: e_i is monomorphic to e_i' where $e_i \in E_o$ and $e_i' \in E_m$.

To find the monomorphism of H_1 onto H_2 is actually to find a partial subhypergraph of H_2 which is isomorphic to H_1.

Object Recognition Based on AGR and AHR Monomorphism

An object can be recognized by finding the monomorphism of its AHR onto a model AHR in the database.

The proposed object recognition procedure is as follows:

1   Construct all primitive attributed graphs for the blocks of the candidate object.

2   Construct all elementary area attributed graphs for the areas of the candidate object.

3   Identify all attributed primitive block graphs of the object and to form the object hypergraph.

4   Search the model database and identify H_m with the vertex set V_m containing the vertex set V_o obtained from the object hypergraph H_o. In other words, each vertex in the object hypergraph corresponds to an elementary area

attributed graph. Determine if there exists a monomorphism from the vertex (an elementary area graph) in the object hypergraph onto a vertex in the model hypergraph.

5   Find the hypergraph monomorphism from the object hypergraph to the model hypergraph.

6   Find the monomorphism from the hyperedges (primitive block graph) in the object hypergraph onto one of the matched hyperedge in the model hypergraph.

7   If the result is unique then the object is recognized, otherwise obtain AHR of another view and synthesize it with the previous view AHR.

8   Go to the step 4.

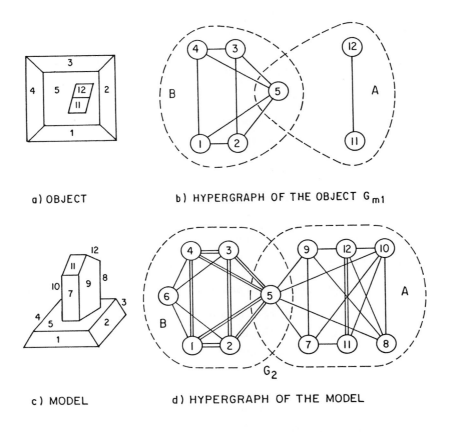

Figure 4   A 3-D Object Recognition Experiment

Figure 4 give an example of hypergraph monomorphism for object recognition from range image. The top view, Figure 4(a), of an object with its AHR shown in Figure 4(b) is compared to the AHR's of different models in the database. The AHR of the object in the image, Figure 4(b), is found to be monomorphic to the complete AHR (Figure 4(d)) of the object model (Figure 4(c)). Once the correspondence between object and model surfaces has been established, we can interpret the image with respect to the model. Figure 5 shows a collection of objects recognized by our system. Among these, only the object, C6, in one image cannot be identified due to the presence of a sizeable sensory noise that breaks up the flat rectangular surface in the range image.

Figure 6 shows another experiment. The lower left hand corner is a laser range image with one object being partially occluded by another object. A newly developed algorithm is able to obtain the AHR for the image and relate the hyperedges to the AHR's (upper middle of the diagram) of the object models (upper right corner of the diagram) stored in the database. Our algorithm is able to recognize both objects and interpret their spatial relation. The details of the algorithm will be reported in another paper.

Attributed Graph Synthesis

From the image of each view of an object, we obtain an attributed graph which represents the edges and faces of the object visible only to the vantage point of the camera. We call that graph an image graph of the object. To gather more information, several images obtained from different views of a 3-D object should be used. We have developed a method by which attributed graphs obtained from different views can be combined (synthesized) to form a single attributed graph that yields a graph representation of the entire object.

We introduce the graph synthesis for two purposes:

1) To combine two view graphs of a candidate object in the recognition process.

> When an image graph of a candidate object is monomorphic to two or more model graphs in the database, in order to resolve the ambiguity, another image graph with different view of the object should be obtained for further recognition. We then synthesize the two image graphs into one which contains the information of both graphs. The comparison of the combined graph with model graphs in the database may yield a unique monomorphism.

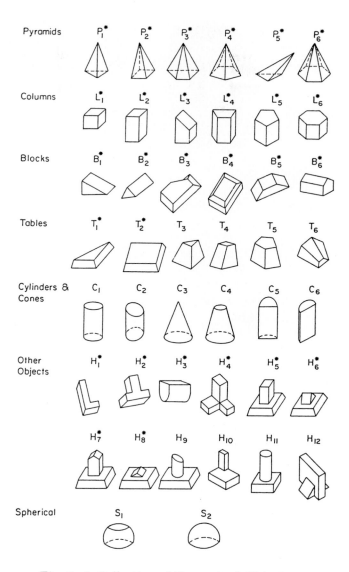

Fig. 5 A Collection of Recognized Objects

2) To build a model graph for a new object in the learning phase.

The model graphs in the database could be constructed from direct physical measurements or through learning using information derived from images of different views. Several image graphs of a model can be synthesized into a model graph.

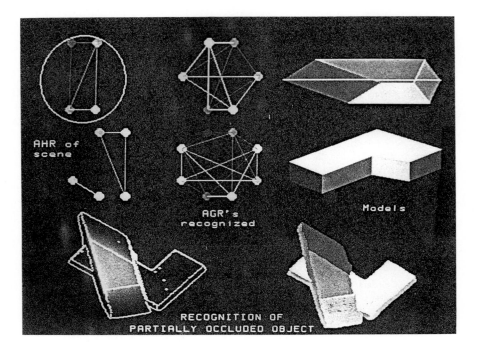

Figure 6 Recognition of Partially Occluded Objects

Using the graph synthesis process, a unique model graph can be obtained for an object. This approach is superior to the alternative approaches which require, for a single object, several representations corresponding to its various views. As a result it is difficult for them to determine whether or not the representation is complete.

To facilitate the graph synthesis process, two basic operations on graphs, namely the union and the intersection, are introduced. They are used to guide the synthesis process.

The procedure of synthesis is as follows:

1   Obtain the AHR for the image. Mark the hidden face vertex from the AHR if they exist.

2   Mark each of the marginal edges (those between the region of a face and the background) in the image graph.

3   Find the normal of the left-most (or the right-most) face. Determine the angle between the normal and the camera axis. Rotate the object such that the new view contains more of the previously hidden faces and reveals more of their relations with the known faces.

4   Obtain the AHR for the new image and mark the marginal edges.

5   Form the union of the two attributed graphs

6   Check the marginal edges in the union graph. If the marginal edge is represented in two view graphs of the object, replace them with the inner edge.

7   If there is no marginal edge in the union graph, then the graph represents the entire object and no more view is needed for further synthesis.

Figure 7 demonstrates a graph synthesis process. The object is shown at the upper left of the figure. The laser images for the object are obtained from several views and displayed at the upper right. The corresponding image view graphs are shown at the lower right and the synthesis result at the lower left.

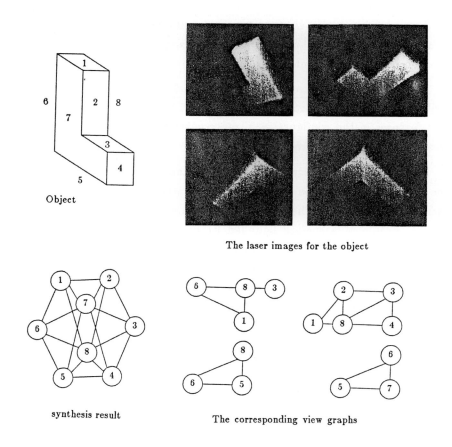

Figure 7   Graph Synthesis Experiment

IV. Point Feature Representation of Objects

In the object AHR, curve or line segments form the attributed vertices at the atomic level. For complicated objects or scene, the size of the vertex set can become very large. Thus, a brute force algorithm for finding graph morphisms is computationally infeasible. However, both in the recognition and the interpretation of 3-D objects, if certain specific feature points together with their spatial relations (even a subset) can first be detected and used to direct the search, the search would be drastically reduced. In view of this, point feature representations, also in the form of attributed graphs and hypergraphs, are introduced. They are abbreviated as PGR and PHR respectively. In these representations, point features are represented by vertices and spatial relations (such as Euclidean distance) or other attributed relations (such as edge and line features) between points by attributed arcs. It can be defined as follows.

Definition 13 The *point feature graph representation* of an object is an attributed graph, denote by $G_p = (V_p, E_p)$, such that the set $V_p = v_i$ represents point features in the configurational representation and $E_p = e_j$ represent the set of relations between pairs of point features.

Definition 14 The *point feature hypergraph representation* of an object is an attributed hypergraph, denoted by $H_p = (V_p, E_p)$, such that the set V_p represents point features in the configurational representation of the object and E_p represents the set of attributed hyperedges.

For object model description, the coordinates of the point feature with reference to the object can be used as a special attribute for that point. For object description in an image, the image coordinates can also be used as point attribute. The 3-D profile of a surface bounded by or associated with a subset of feature points can be assigned as an attribute to the hyperedge defined by the subset of vertices corresponding to the subset of feature points. Thus, once the coordinates of the feature points are determined, the 3-D profile of any convenient form of an object surface can be related to a local reference system induced by those feature points.

More specifically, let three points in an hyperedge be selected as control points. We introduce a transformation that relates these points to a convenient reference system. Thus a set of points on the surface (or a surface profile) depicted by an attributed hyperedge can be expressed in this coordinate system. When a comparison of a 3-D image with a prototype is required, the control points in the image are first identified. Then, the transformation is then obtained to relate the 3-D information derived from the image to those relative to the local reference of the prototype. If one of the control points is not found in the image, its position can be inferred from another set of control points and the same process can be used to relate the image surface profile to the prototype profile. Such representations are of great significance to industry application when the geometry and dimension of the prototype are known.

Figure 8 shows the dual representations EHR and PHR of an automotive panel. Figure 8(a) shows a 3-D projected view of the panel. Various point features labelled by numerals are marked on the diagram. The edges of the panel are labelled by upper case alphabet letters. The line separating two distinct curved surfaces are shown in dotted line and labelled by lower case alphabet letters. Figure 8(b) shows an EHR of the panel. The Greek symbols are used here as labels to the hyperedges. Here each hyperedge either represents a distinct 3-D surface of the panel or a groups of holes (or fixtures). Figure 8(c) is the PHR counterpart of the panel. Here all vertices are given as point features, and the surface normal can also be used as a specific attribute for the hyperedge.

V. Constellation Matching

When point features and their spatial relationship are used for object recognition and comparison, a special procedure for finding an optimal match between two sets of points defined on two isometric spaces can be employed. Such a procedure is generally referred to as a constellation matching. The constellation matching algorithm [20] utilizes a complete edge-attributed graph representation of a constellation constructed by treating the vertices as points and the values assigned to the edges as distances between points. It is a special case of PGR.

Let C be the complete edge-attributed graph representation of a constellation C'. The function $d(c_i, c_j)$, where c_i and $c_j \in V(C)$ and $i \neq j$, denotes the value on the edge (c_i, c_j) (i.e. the distance between points c'_i and c'_j in C'). The graph of any sub-constellation in C' occurs in C as a complete subgraph.

Let A and B be the PGR's of constellation A' and B' respectively. Then $A' = V(A)$ and $B' = V(B)$. A point to point matching of size n between A' and B' corresponds to an incidence preserving one-to-one mapping

$$f: V(A) \rightarrow V(B).$$

Thus, the edge (a_i, a_j) with $a_i, a_j \in V(A)$ corresponds to the edge $(f(a_i), f(a_j))$ with $f(a_i), f(a_j) \in V(B)$. Graphs A and B are (n,k)-similar (with the n point matched and with all the differences of distances between corresponding points in the matched pairs within the tolerance threshold k) if there exists $f: V'(A) \rightarrow V'(B)$, where $V'(A)$

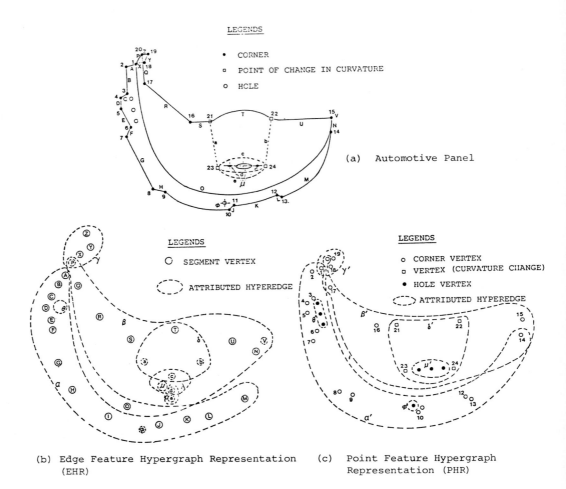

Figure 8 The EHR and PHR of an Automotive Panel

and $V'(B)$ are subsets of $V(A)$ and $V(B)$ respectively with size = n, such that for all $a_i, a_j \in V'(A)$, $i \neq j$, $|d(a_i, a_j) - d(f(a_i), f(a_j))| \leq k$.

The cost of a mapping f is defined as:

$$C(f) = \sum_{\substack{a_i, a_j \in V'(A) \\ i \neq j}} |d(a_i, a_j) - d(f(a_i), f(a_j))|.$$

As was mentioned previously, if f satisfies the condition for (n,k)-similarity then $C(f) \leq kn(n-1)/2$. The constellation matching problem is defined formally as the search for f such that:

1) n is maximum;

2) (n,k)-similarity is satisfied for a given k;

3) for all f satisfying (1) and (2), choose the one with min $C(f)$.

In brief, the algorithm conducts a search through a tree which represents every possible matching as a path from the root to a leaf. Initially, the maximum value of n is chosen as $n = min(V(A),V(B))$, and a search for matchings that satisfy condition (2) is systematically conducted. If this fails, n is decremented and the search is repeated until for a certain value of n, appropriate matchings are found. Of these, the matching with the least cost (condition(3)) is chosen as the final solution.

Constellation Matching for Stereoscopic Vision

Figure 9 shows a pair of stereoscopic views of an automotive part acquired by the left and right cameras of a steroscopic vision system. To derive the 3-D representation of an object, the points in each pair must be first correlated and constellation matching provides a general methodology for this purpose. The constellation extracted for special feature representation are shown by dots and small circles. The dark dots in each view denoted the matched sub-constellations. Some of the feature points are deliberately removed to demonstrate that our general constellation matching algorithm is able to tolerate incomplete or extraneous information. From the correspondence of the matched points the three dimensional coordinates of the special features are computed based on stereo disparity.

The CPU is of the order of 100 to 200 msec in running in Pascal on a timesharing VAX 11/750 system. To further speed up the search through the tree of possible matchings for larger constellations, additional information can be used in conjunction with the constellation matching strategy. For instance, since the two stereosopic views of an object have no vertical translation with respect to each other, the nodes in the search tree can be ranked according to their likelihood of representing a correct match. In other words, the nodes representing a match between points with a large vertical displacement are ranked lower than those with a smaller vertical displacement. In general, if a problem has certain natural or artificial constraints, specific heuristics such as the one shown above, can be incorporated to speed up the constellation matching.

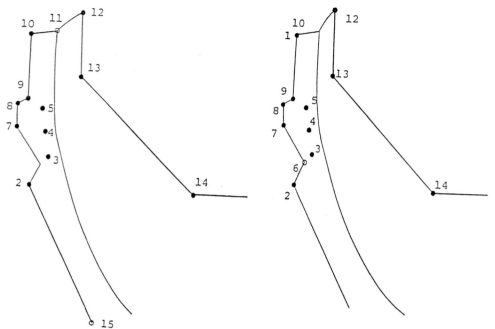

Fig. 9 Stereoscopic Views and Point Feature of an Automotive Panel

VI. Recognition and Location of 3-D Objects from a Perspective Image

Using PGR as the declarative knowledge, a new method is developed [16] which is capable of identifying and locating the position and orientations of objects with known 3-D models in a single perspective image. In this method, the interpretation of the observed information in the image is guided by three basic forms of knowledge: procedural knowledge, declarative information and contextual information. The declarative and procedural knowledge represent knowledge about objects and the effective methods of searching for those objects. Contextual information is information that can be inferred on the basis of the procedural and declarative knowledge and the information contained in the observed image.

Procedural knowledge is maintained as a set of search rules that determine: 1) how to expand or direct the search, 2) when to terminate the search and 3) how to control the analysis of declarative and contextual information to form and test hypotheses about the image, and the objects or patterns to be identified. To control the application of the search rules, the procedural knowledge is maintained as a rule network which determines the order of application of the rules and identifies the useful declarative information at each stage of the search.

The knowledge-directed search is organized as a set of search activations each of which uniquely associates a detected feature on the observed image, a node of the rule network and the context of search activation (Figure 10). The complete set of search possibilities could be represented as a list of search activations. Multiple search activations may record and test various hypotheses of object identity and location.

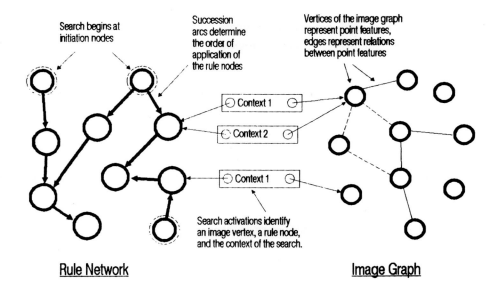

Figure 10 Image AGR and Rule Network

Observed Image

The observed image is represented as a PGR, $G_p = (V_p, E_p)$. The set of vertices $V_p = \{v_i\}$ represents the point features. The set of edges $E_p = \{e_j\}$ represents relations between pairs of image point features. Each v_i is associated with its position (x_i, y_i) in image coordinates as well as any additional attributes extracted from the image. Each edge is associated with an attribute that defines the nature of the relationship between the connected vertices; for example, it could denote a straight or curved line segment joining the image points.

Rule Network

The rule network is a directed graph $G_r = (R, S)$ with a set of rule nodes $R = \{r_i\}$ where each r_i is a collection of search rules and declarative information. The nodes are connected by a set of succession arcs $S = \{s_i\}$ which define the sequence in which the rule nodes may be applied.

The context of a search activation is that information which is dependent on the past and present history of the search. The context information of a search activation usually includes:

1) Observed data recorded by ancestor search activations; for example, the location of image vertices already encountered by the search, and

2) Information inferred by the rules and observed data of ancestor search activations, such as the hypothesized image transformation, the suspected target object, and the predicted characteristics of the application environment.

The search rules associated with each node of the rule network perform one of the following functions:

Termination Rules

> Termination rules evaluate the context information of the search activation, the observed data, and the declarative information of the rule node to determine whether or not a search activation should be terminated or propagated.

Incidental Rules

> Incidental rules specify a change in the context of the search activation and its decendents.

Growth Rules

> Growth rules select a set of candidate image features to be examined next by descendent search activations of the rule node.

A subset of the nodes of the rule network are designated as initiation nodes. The search begins by specifying an initial context; for example, specifying the camera height and elevation angle. The search activation with the initial context is created for each initiation node of the rule network. The incidental rules are then applied to establish all possible context information. If for any inferred information there exists more than one possible hypothesis, then multiple context records are produced. The termination rules are then applied to determine whether the search activation should be terminated or propogated. If the search activation is to be propogated, then the growth rules are applied to select the next set of image features to be considered. Descendent search activations are created for every combination of successor rule node, context record, and candidate image feature. If the search activation is to be terminated, then termination is positive if the desired object has been successfully detected, or negative if the search has determined that a possible match is impossible. Combined termination results are possible. Termination may be defined as positive-propagate if an object has been located which may provide information to guide the detection of an associated object. For example, detection of some part of a robot arm can be used to guide the search for less easily identified components. The search then continues with any new or remaining search activations.

The efficiency and reliability of the analysis depend highly on the quality of the information and knowledge coded in the rule network. Since the search is directed by image data acquired in a piecemeal fashion, it is essential to apply as much knowledge as possible to the selection of initial image features for examination.

In general, the exact location of a 3-D object may be determined with a minimum of four corresponding 2-D image features and their 3-D model coordinates. Actually, it only requires three point features to identify a finite number of 'possible' image transforms (up to a maximum of four unique image transforms depending on the geometry and the object position relative to the camera). Multiple image transform hypotheses can be tested by creating different context records for descendant search activations.

To verify a model hypothesis, the object features which should be visible are compared to image features positioned near the projected location of the object features given the hypothesized image transform. If the image feature is within the measurment tolerance of the corresponding object features, the value of the match is incremented by a specific amount. If the total value of all matched features exceeds a predefined normalized threshold value, the object is successfully identified and located.

Application Examples

The knowledge-directed search is used in this sample application to identify the keyboards and monitors (VT220 and VT240) in an image of our terminal room (Figure 11) and to return the three-dimensional position and orientation of each object in the camera or room coordinate systems. The image was obtained by a camera with known focal length and image plane (16 mm focal length, CCD array 5.8x8.8 mm).

An object model consists of the three-dimensional coordinates of prominent features measured relative to the object centroid and with convenient orientation. It is a special case of PHR. All measurements are in centimetres with a stated tolerance of one centimetre.

Figure 11 A Perspective Image of Computer Monitors and Keyboards in a Laboratory

The rule network is constructed to first identify the VT220 and VT240 terminals and then locate keyboards as an associated component. The rule network uses subsets of point features created by the corners of the dark screen surface as the chosen three object point feature used to obtain the image transform. Terminals and keyboards will only be recognized if their front surface is visible.

Analysis of Point Feature Representation

From the PGR of the image and a rule network designed to identify VT220 and VT240 monitors, the search has correctly identified the nearest ones. Two additional monitors are ignored because too many of the features required for recognition are distorted or hidden by other objects.

The search implemented in Pascal on a VAX/750 running a VMS operating system required 310 seconds of processing time. This is considered excessive for use with real-time systems such as a roving robot or factory inspection system.

In order to reduce the search two possible solutions are:

i)   to increase the number of features used to identify the object, or

ii) to add additional attributes to the point features of the image.

The second solution is preferred since it also offers a means of reducing the overall search complexity.

Analysis with Point Feature Attributes

The addition of any attribute to the point features of the image is certain to reduce the cost of the search, and can potentially increase the reliability of the search (provided the feature attribute can be reliably extracted from the image). In this example, the number of edges incident on the image point feature is used as the only attribute in addition to the image coordinate. The attribute values of the point features created by various line junctions are illustrated in Figure 12

Figure 12   Point Feature Attributes

The application of the rule network of the previous example with the addition of the junction information was then applied to the attributed image point feature representation. The results of the analysis are illustrated in Figure 13. The addition of a single attribute has eliminated the false keyboard detections, it has also identified and located the correct objects in only 5.9 seconds. This is a dramatic improvement over the 310 seconds previously required.

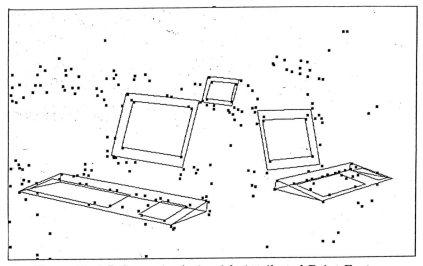

Figure 13 3-D Scene Analysis with Attributed Point Features

Analysis of PGR with Line as Arc Attributes

The knowledge-directed search is easily applied to a PGR with line as arc attributes by changing the rule network of the preceding example so that the growth rules of first two levels search nodes can only select image features that are adjacent to the current image feature (i.e. add a new growth rule that requires an edge relation between the observed image feature and the candidate features). The analysis of the line drawing with such a rule network requires only 780 ms. of processing time. In addition, the required objects are reliably identified, as shown in Figure 14.

Figure 14 3D-Scene Analysis with Line Drawing

The next example shows how the rule network is expanded to represent various objects commonly found in an office environment. Models for a standard size door, a light-switch, two different chair styles and a waste basket are added to the rule network. The image was obtained using a camera with focal length of 16mm and image plane 5.8x5.8 mm. The camera was positioned at a height of 93 cm with approximately one degree elevation angle.

The identification and location of objects in the scene are relatively successful with the notable exception of failure to identify two of the chairs visible in the image (Figure 15). In general, the rule network demonstrate considerable immunity to missing or occluded features. For example, the door frame was identified despite being hidden behind other objects. The analysis required 5.4 seconds to apply the search to the 443 point features in the image.

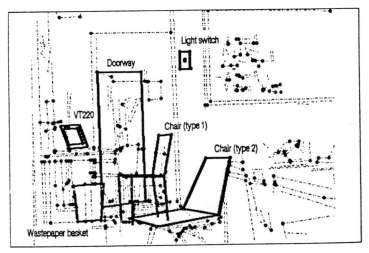

Figure 15 Analysis of Office Scene

VII. AHR Representation of Free Space for Mobile Robots

In a mobile robot path planning task [17], free space of the robot's environment is structured into a set of overlapping convex regions ideally suited to path planning and navigation tasks. The boundaries of each unobstructed convex region are constraints posed by obstacle walls. The structure of the free space environment is maintained as an AHR with each convex region represented by an attributed hyperedge identifying the boundary walls of the region.

Obstacle boundaries or walls are represented by the vertices of the graph (EGR). The attribute values for each vertex describe the end points of the wall and the direction of the wall faces. Edges between the wall vertices indicate the connection of boundaries which define obstacles or rooms (Figure 16). The edge attribute value is the angle attained by connected walls.

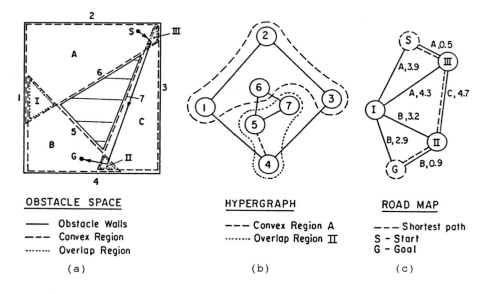

Figure 16   The AHR and Road Map of an Obstacle Space

With the basic information of the free space environment represented in an AGR, we have developed algorithms which are able to reveal the structure of free space and construct the AHR through a directed search for a set of fundamental circuits in an abstract graphical representation of the environment geometry. In the AHR, the unobstructed convex regions are represented by hyperedges which are the sets of wall vertices that form region boundaries. The overlaps of the convex regions are used to determine their topological properties. Each overlap is represented by a hyperedge with attribute values identifying the intersecting regions and the centroid of the overlap (Figure 16(b)). The algorithm is able to obtain the unique set of overlap regions each of which intersects the maximum number of convex regions. This can be achieved through two fundamental control processes in the depth first search in the generation of the fundamental cycles in the following manner:

1. The graph is constructed such that every possible directed cycle must correspond to a convex region.

2. The spanning tree is obtained by a depth first search with a strictly specified order for searching edges and with restrictions on which combinations of vertices may be included in the search tree.

Once the obstacle space is represented in this way, the information in the graph is abstracted to form a 'road map' graph used to plan feasible paths on a symbolic level. The road map is formed with the overlaps as the nodes or vertices, and the common convex regions joining each pair of overlaps as connecting roads or edges. The edges have attribute values specifying the length of the path between overlaps, an estimate of the path width and a pointer to the original convex regions of the low level representation (the set of critical walls for obstacle avoidance and navigation).

A feasible path is obtained by finding the shortest path in the road map from start to goal locations along paths of sufficient width. Figure 16 illustrates the simple obstacle space, and the hypergraph representation of a convex region and an overlap region of free space. The resulting road map with three overlap vertices, the start and goal positions, and the connecting convex regions is shown, along with a dashed line indicating the shortest path between start and goal vertices.

Figure 17 illustrates the floor plan of a 238 wall environment, its free space representation, and the paths found for differently sized objects. Note that the floor plan represents only permanent walls and obstacles (room walls, fixed partitions, study areas, and permanent fixtures such as computing equipment or bookshelves). Temporary or moveable obstructions such as doors or chains are not included.

Approximately 35 seconds of processing time was required to generate the region graph. Locating the fundamental circuits of the maximal overlap regions and primary convex regions required an additional 16 seconds. Construction of the road map required 147 seconds with the majority of processing time devoted to calculating the minimum path width associated with each edge in the road map. Once the road map is complete, path planning is comparatively rapid. For example, the paths for the smaller object were found in approximately one second. Obtaining the path of the larger object required only 600 ms.

Vehicles with complex movement constraints may be accommodated by varying the minimum path width requirement to provide room to maneuver. Dynamically altering the apparent object width during the search for a path creates no additional computational overhead.

Conclusions

This paper presents a knowledge representation system for robot vision and path planning. The generality, flexibility, and effectiveness of such a system have been demonstrated through its successful application to various tasks and task environments. Geometric information from range data image of any arbitrary view of an object now can be transformed into an AHR. The graph synthesis algorithm we introduce is capable of synthesizing the AHR's derived from these images to form an AHR that represents the entire object. The complete AHR's for a collection of objects to be classified are then stored in a database. Through the use of our new hypergraph monomorphism algorithm, it can now be determined which model in the database would correspond to the object.

From point feature representations, we have developed an effective constellation algorithm for matching spatial configurations of partial views of an object to those of the candidate models. We have also demonstrated that 3-D point feature configuration of an object can be derived and synthesized from stereoscopic images by constellation matching methods. With the knowledge-direct-search approach, multiple objects can now be identified and located from a single perspective image. The PHR and the flexible way of organizing the procedural knowledge rule network, have shown that the method is capable of working with various levels of image feature quality, including a simple point feature representation of an image. It has been demonstrated that such a method is capable of tolerating high noise levels and local variations. We also illustrate that an effective use of knowledge such as the combination of point and line features can drastically reduce the search and enhance the reliability of recognition. The AHR does provide a flexible data structure for using knowledge flexibly and effectively.

Using the same representation, a complex world model for a mobile robot can be acquired as an AHR from which a compact road map in the form of AGR can be derived. This representation of free space permits a path planning methodology capable of representing obstacle environment of considerable complexity. Experiments with simulated environments have shown the efficiency of our approach. With the AHR, local changes of the environment can now be easily coped with.

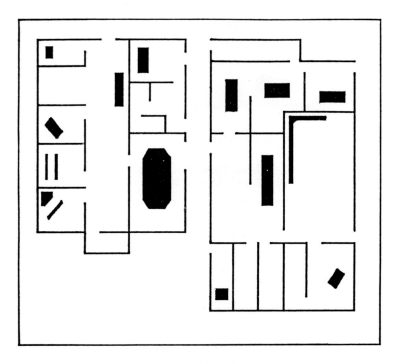
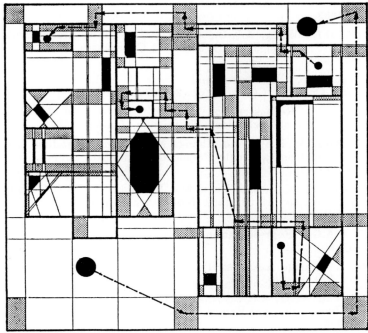

Figure 17 Free Space Representation of a Simulation Robot Environment

Thus, it is obvious that with this new representation method, a great variety of tasks and task environment can now be modelled. Its generality, flexibility and effectiveness, as observed earlier, is highly encouraging.

REFERENCES

1. Barnard, S.T., "Interpreting Perspective Images", Artificial Intelligence 21, 1983.
2. Berge, C., "Graphs and Hypergraphs", North-Holland Publishing Company, Amsterdam.
3. Besl, P.J. and Jain, R.C., "Three-Dimensional Object Recognition", Computing Survey, Vol. 17, No. 1, pp. 75-145, March 1985.
4. Binford, T.O., "Visual Perceptions by Computer", IEEE Conf. on Systems and Control, Miami, Dec. 1971.
5. Brooks, R.A., "Symbolic Reasoning Among 3-D Models and 2-D Images", Artificial Intelligence, 17, 1981.
6. Brooks, R.A., "Solving the Find-Path Problem by Good Representation of Free Space", IEEE Trans. on Systems, Man and Cybernetics, Vol. SMC-13, No. 3, March/April 1983.
7. Bugihara, K., "An Algebraic Approach to Shape-from-Image Problems", Artificial Intelligence, 23, 1984.
8. Giralt, G., Chatila, R., and Vaisset, M., "An Integrated Navigation and Motion Control System for Autonomous Multisensory Mobile Robots", First International Symposium on robotics Research, The MIT Press, Cambridge, Massachusetts, 1983.
9. Grahraman, D.E., Wong, A.K.C., and Au, T., "Graph Monomorphism Algorithms", IEEE Trans. on Systems, Man and Cybernetics, Vol. SMC-10, No. 4, pp. 181-189, 1980.
10. Jackin, Tanimoto, "Oct-trees and Their Use in Representing Three-Dimmensional Objects", CGIP, 14, pp. 249-270, 1980.
11. Langeland, N.J., "Reconstruction of 3-D Objects from 2-D Image", The Norwegian Institute of Technology, Trondheim, January 1984.
12. Lee, S.J., Haralick, R.M., and Zhang, M.C., "Understanding Objects with Curved Surfaces from a Single Perspective View of Boundaries", Artificial Intelligence 26, 1985.
13. Lu, S.W., Wong, A.K.C., and Riuox, M., "Recognition of 3-D Objects in Range Images by Attributed Hypergraph Monomorphism and Synthesis", Proc. of International Symposium on New Directions in Computing, pp. 389-394, 1985.
14. Requicha, A.A.G., "Representation of rigid Solid Objects", Computer Survey, 12.4, pp.437-464, 1980.
15. Rosenfeld, A., "Image Analysis: Problems, Progress and Prospects", Pattern Recognition, Vol. 17, No. 1, pp. 3-12, January 1984.
16. Rueb K.D., and Wong, A.K.C., Analysis of Point Feature Representation of a Perspective Image, Internal Report, Systems Design Engineering, University of Waterloo, 1986.
17. Rueb K.D. and Wong, A.K.C., "Structuring Free Space as a Hypergraph for Roving Robot Path Planning and Navigation", to appear in IEEE Trans. on PAMI, 1987.
18. Voelcker, H.B., and Requicha, A.A.G., "Geometric Modelling of Mechanical Parts and Processes", Computer, 10, pp. 48-57, 1977.

19. Wong, A.K.C., Lu, S.W., and M. Riuox, "Recognition and Knowledge Synthesis of 3-D Object Images based on Attributed Hypergraphs", Internal Report, Systems Design Engineering, University of Waterloo, 1986.
20. Wong, A.K.C., and Salay, R., "An Algorithm for Constellation Matching", Proc. of the 8th International Conf. on Pattern Recognition, Paris, pp. 546-554, 1986.
21. Akinniyi, F.A., Wong, A.K.C. and Stacey, D.A., "A New Algorithm for Graph Monomorphism based on the Projections of the Product Graph", IEEE Trans. on SMC, Vol. SMC-16, No. 5, pp. 740-751, September, 1986.
22. Wong, A.K.C., and Akinniyi, F.A., "An Algorithm for the Largest Common Subgraph Isomorphism Using the Implicit Net", IEEE Proc. of the International Conference on Systems, Man and Cybernetics, Bombay, 1983, pp. 197-201.
23. Wong, A.K.C., and You, M., "Entropy and Distance of Random Graphs With Application to Structural Pattern Recognition", IEEE. Trans. on Pattern Analysis and Machine Intelligence, Vol. PAMI-7, NO. 5, pp. 599-607, September 1985.

KNOWLEDGE-BASED SYSTEMS FOR ROBOTIC APPLICATIONS

Julius T. Tou
Center for Information Research
University of Florida
Gainesville, FL 32611
U.S.A.

INTRODUCTION

Today, quality design and high productivity in engineering and manufacturing are often synonymous with the use of computers, robots, expert systems, and other computer-based technologies. The greater the degree of computer-based automation exploited and implemented, the greater a nation's ability to survive in tomorrow's extremely competitive world market. Among the computer-based technologies, knowledge-based systems for robotic applications is of paramount importance in manufacturing industry.

An intelligent robot is made up of four fundamental components: machineware, hardware, software, and brainware. By machineware we mean mechanical structures and electro-mechanical or hydraulic devices in a robot. Hardware includes electronic, optical and sonic gears for sensing, computation, interfacing, communication, and command generation. Software implies the required computer subroutines and programs for driving the robotic electronic gears. Brainware provides the central nerve system for the robot in order to make it intelligent. An approach to the design of robotic brainware is to make use of knowledge-based systems. This paper is concerned with the brainware aspect of intelligent robots.

In a factory of the future[1], we expect to see around-the-clock operations with robots roving on the factory floor, handling materials, performing parts assembly and packaging. To accomplish these robotic applications, we introduce knowledge-based system approaches. In this paper we will discuss the design of knowledge-based systems to enhance the mobility of roving robots, to provide recognition capability for robots to handle parts and materials, and to enable robots to understand schematic diagrams for trouble shooting and repairing.

A knowledge-based system is a computer-based system which acquires knowledge in specified domains, stores knowledge in defined structures, and organizes knowledge in desired formats for ready access, retrieval, transfer, utilization, and extension[2]. The concept of knowledge-based systems opens up new approaches to the design of robotic applications. The incorporation of knowledge bases provides an effective design of intelligent robots. The major advantage of using the knowledge base is to enable the robot to get access to desired knowledge, to do reasoning tasks, and to accumulate past experience which is needed for improving the performance.

In this paper, we will present some of our research results in this area of machine intelligence. Specifically, we will discuss our computer vision system for roving robots, the VIREC system for reading labels and recognizing 3-D objects, and the AUTORED system for understanding schematic diagrams[3-6].

KNOWLEDGE-BASED SYSTEM DESIGN CONCEPTS

Among the problems of fundamental importance in knowledge engineering[2] are:

1) Preservation of knowledge
2) Proliferation of knowledge
3) Dissemination of knowledge
4) Application of knowledge

The first problem is the outcome of the fact that what masters really know or is normally not included in the textbooks written by masters, nor can it often be found in their papers. How to make this type of expert knowledge available is a very important but difficult problem. The second problem has raised the question of how a person should read and acquire information every day in order to meet his needs. The third problem is concerned with what are the most efficient, expedient and economical ways of disseminating and transferring knowledge. The fourth problem is concerned with the question of how to make full use of experts' knowledge when the need arises. An approach to this problem is to make use of knowledge-based expert systems.

A knowledge-based expert system generally consists of four fundamental components: knowledge base, knowledge acquisition, mechanism, recognition/inference mechanism, and user interface scheme. The knowledge base is composed of a knowledge sketch and knowledge details. The knowledge sketch represents the relational structure and the problem-oriented hierarchies of the knowledge stored. Knowledge of various levels of specificity in a domain is distributed throughout the hierarchy. The structure of the knowledge base provides useful information to facilitate knowledge classification, new knowledge acquisition by inserting it at the right location in the hierarchy, identification of possible problems, and efficient seeking of the appropriate knowledge when needed. The knowledge hierarchy may be linked to knowledge details via semantic pointers[7].

The knowledge acquisition mechanism is a knowledge-base generation system which automatically generates the knowledge base after a body of knowledge on a subject matter is acquired and entered into the computer. The recognition/inference mechanism interprets the request for knowledge to solve a specified problem which has arisen, performs pattern matching, issues specific answers, and generates recommendations and inferences. The decision-making process may be accomplished by discriminate analysis, Bayesian statistics, clustering techniques, feature extraction, syntactic rules, and production rules. The primary distinction between knowledge-based systems for robotic applications and those for human users is in the user interface scheme. For robotic applications, the knowledge-based system is an autonomous intelligent system which is designed with the capability of generating requests for knowledge on the basis of measurements, observations, and inferences.

For instance, MYCIN[8] and MEDIKS[9-11] are knowledge-based expert systems which have been designed for human users. The user enters queries and manifestations into the system and works interactively with the system to seek satisfactory solutions to a problem. The user interprets the outcomes of the expert system and successively generates new queries for the system to respond until an answer is found. Consequently, the human is a part of the integrated knowledge-based system and interactions between human intelligence and machine intelligence play a very significant role in this environment. On the other hand, knowledge-based systems for robots are autonomous intelligent systems, the inputs to which are measurements, observations, and self-generated inferences. The robot has no other intelligence than the knowledge-based systems. In other words, the knowledge-based system is an integrated part of the robot who has to interpret the outcome of the knowledge-based system, to monitor the sensory devices, and to perform inference functions. No additional human intelligence is available other than the one which has already been built in the knowledge-based system of the robot. Thus, robotic knowledge-based systems should be designed with self-checking, self-governing, self-evaluation, and self-verification capabilities.

The primary contents in a knowledge base are definitions, characteristics, methodologies, casual relations, rules of thumb, transfer functions, performance goals, and reasoning procedures. The major tasks of an inference/

recognition mechanism are diagnostic analysis, data validation, consistency analysis, violation detection, decision simulation, and pattern interpretation. These tasks are performed by a combination of pattern matching, content searching, rule interpretation, mathematical operations, inference heuristics and programmed algorithms.

A knowledge-based system which is designed to make full use of experts' knowledge whenever the need arises is often referred to as an expert system. In simple language, an expert system is a computer-based system that searches for a solution within a set of statements or a body of knowledge formulated by the experts in a specific area. Thus, an expert system is designed to do various tasks that an expert in that area would normally perform. The major approaches to the design of knowledge-based expert systems are the rule-based approach and the pattern-directed approach. Frequently a combination of both approaches is cleverly employed in the design.

For robotic applications, the knowledge-based systems are employed to enhance the intelligence level by filling the system database with experts' knowledge in specified domains. Human knowledge and inference capability are transferred to the machine to make it intelligent and to improve its performance. The architecture of the proposed knowledge-based system for robotic applications is illustrated in Figure 1. The knowledge-based system consists of four major components: knowledge base, knowledge-base generation unit, know-

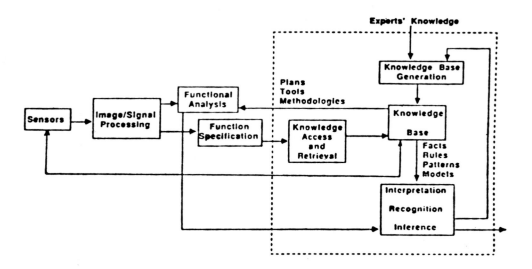

Figure 1: Architecture of a Knowledge-Based System for Robotic Applications

ledge access and retrieval unit, and interpretation, recognition and inference unit. The robotic system is equipped with sensors, image/signal processors, functional analysis, and function specification. This design is based upon the UNIPS[12] concept which was developed by Tou some eight years ago.

KNOWLEDGE-BASED SYSTEM FOR ROVING ROBOTS

The basic requirements for the guidance and navigation of roving robots are directional guidance, obstacle avoidance, orientation determination, range finding, and object identification[3]. To meet these requirements, we have developed a computer vision system for the roving robot. The vision system architecture consists of the camera as the input device, the image processing unit, the scene analysis and interpretation unit, the robot memory unit, and the robot motion control unit as shown in Figure 2. A knowledge-based system for scene analysis and interpretation has been designed. The knowledge base stores the algorithms for directional analysis, obstacle analysis, orientation analysis, range finding, and object identification as well as the rules for interpretation, recognition, and inference, as shown in Figure 3. In addition, the knowledge base stores a plan for scene analysis which provides a directory for the monitoring and management of analysis and interpretation routines. The scene analysis plan describes a structure for knowledge sketch of the knowledge base. A partial plan for scene analysis is shown in Figure 4. The algorithms and program modules constitute the knowledge details in the knowledge base. Our proposed vision system is designed to perform three phases of image processing:

1) the planning phase,
2) the "walking" phase,
3) the warning phase.

The integration of these phases of operation provides the necessary guidance for the roving robots.

The planning phase is initiated before the robot starts to "walk" or rove. During this phase the vision system determines a safety path for the robot to follow by analyzing scenes in its vicinity. The safety path is generated by the directional analysis algorithms in the knowledge-base. If the roving robot is doing repetitive work, the various safety paths may be retrieved from the database. A fundamental task in the planning phase is the directional guidance analysis. It is assumed that the roving robot is initially

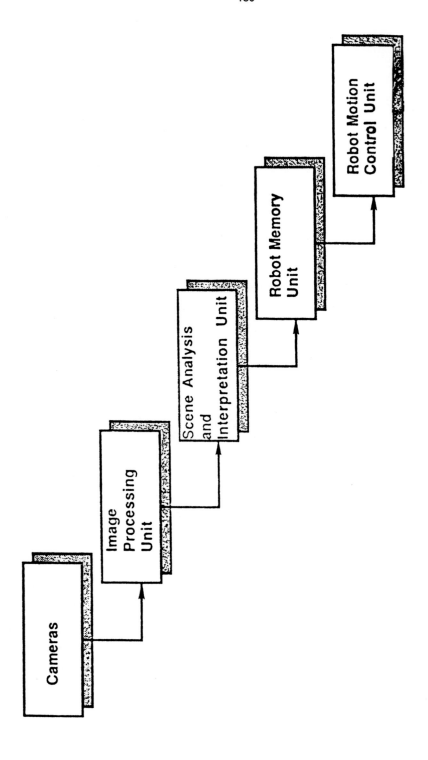

Figure 2
Vision System Architecture for Roving Robots

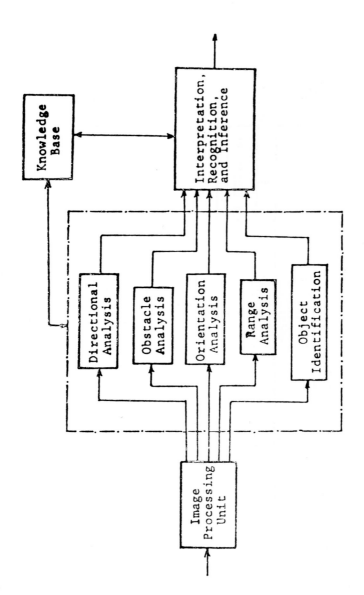

FIGURE 3

Organization and Scene Analysis and Interpretation Unit

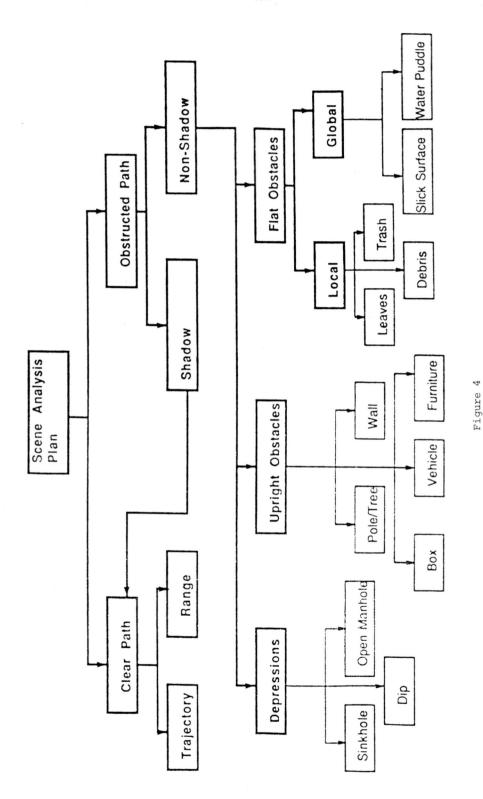

Figure 4
Hierarchy of Design for Scene Analysis

located in a safe area. To facilitate this analysis, the image in front of the robot is partitioned by a virtual grid. We have developed an iterative procedure to determine an optimal regional grey level by eliminating grid cells with abnormal average grey levels. The optimal regional grey level is used as a reference grey level for the determination of obstacle-free regions.

After the planning phase, the vision system enters the "walking" phase which determines a direction of travel and analyzes the image taken in this direction to generate a partial footprint of the scene layout. Path tracing is performed by processing one row of the image cells at a time. Each step along the direction of travel is declared either as a clear step or as a potential obstacle. As the robot roves, the vision system takes more pictures and updates the footprint of the scene layout and the region of safety. When a potential obstacle is detected, the warning phase is initiated.

When an obstacle is detected in the observed scene, the system informs the robot to move more slowly or to take a pause so that the vision system is allowed sufficient time to complete image analysis and interpretation. The main objective is either to issue directional cues for avoiding the obstacle or to invoke the recognition algorithms to identify the obstacle which can be a shadow, a flat object, an upright object, or a depression.

Obstacle avoidance analysis involves shadow discrimination, detection of depression and water puddles, recognition of slick surfaces, and identification of flat and upright objects. We have developed the concept of texture invariance under shadow for the identification of shadows from detected obstacles. This approach is conducted on the basis of the fact that the texture of the shadowed area is not changed by the presence of the shadow. Texture invariance may be measured and characterized by histogram analysis or by correlation analysis. In each analysis, performance parameters are determined and then compared with a set of thresholds for shadow identification. Analysis algorithms and programs are stored in the knowledge base.

The discrimination of upright objects from flat objects may be accomplished by two approaches: 1) using sonic waves, and 2) shading analysis. A sonic wave is swept on the detected obstacle in the verticle directions and the variances in range and the viewing angle are recorded. Roughly speaking, small range variations and large viewing angle signify an upright object, large range variation and small viewing angle suggest a flat object. The viewing angle is measured from the baseline of scene layout footprint. Interpretation rules for range variance, viewing angle, and shading analysis are stored in the knowledge base. We define shade by average texture. The edges of an object are the boundaries between areas of dark and light shade. Via image processing, the vision system determines the front view of the object. Tall front view suggests the presence of an upright object. This information is used in conjunction with the range and angle information fron sonic waves by the interpretation and recognition unit.

Depressions or drop-offs create a dangerous obstacle for the roving robot. However, the detection of depressions is a very complex image understanding problem. In the human vision system, many visual cues such as stereopsis, occlusion cues, context in the viewed scenes, textural variations, shading, etc. are acquired and made available for interpretation and integration to accomplish the task of perception. Under controlled or pre-determined environment, we may introduce markets or make use of landmarks in the detection of depressions.

Orientation information is very important for the guidance and navigation of roving robots. In our design the orientation information is represented by scene layout footprint and 3-D map of landmarks. Most objects in the real world can be characterized by polyhedral configurations. The bottom of a polyhedral object carries most useful information for the guidance of a roving robot. Such information is integrated to form scene layout footprint and 3-D map of landmarks.

Referring to Figure 1, the Function Specification Unit issues signals to retrieve various methodologies to perform functional analysis including directional analysis, obstacle analysis, orientation analysis, range analysis and object identification. The results of functional analysis are fed into the Interpretation, Recognition, Inference Unit to generate instructions and commands for the roving robot.

KNOWLEDGE-BASED SYSTEMS FOR PARTS HANDLING ROBOTS

A fundamental task in robotic assembly, inspection, material handling, packaging and distribution is automated visual recognition of industrial parts. An industrial part may be identified from its labels or it may be recognized from 2-D pictures taken from the 3-D part. Knowledge-based systems for automated visual recognition provide the brainware for parts handling robots. During the past several years, we have designed a visual recognition machine for industrial parts, known as VIREC[4]. The VIREC is a knowledge-based system which conducts two modes of operation: <u>Mode I</u> is designed for label reading and <u>Mode II</u> performs 3-D object recognition. Under Mode I operation, the VIREC enables roving robots to fetch parts from storage bins which are labelled with pictoral descriptions or to handle boxes of finished products for distribution. Under Mode II operation, the VIREC enables robots to visually recognize industrial parts for handling or for automated assembly. Mode I operation makes use of knowledge on topological transformations and Mode II operation is based upon the concepts of orthographic projection transformation. Shown in Figure 5 are typical labels and sample pictures of 3-D objects. The VIREC system is designed to read the labels and to recognize 3-D objects from their pictures.

The system architecture for VIREC is illustrated in Figure 6. It is composed of two channels. The cameras are the input devices for both channels. The camera of Channel I acquires the images of pictorial labels which are digitized and preprocessed before being encoded into a recognition language representation. The VIREC system performs topological transformations on the pictorial labels to convert them into canonical forms. The VIREC then performs interpretation and feature extraction. Knowledge about various labels which are under examination is stored in the knowledge base to facilitate topological transformation, interpretation, and feature extraction. The interpretation results and the extracted features form the contents of the database.

The cameras of Channel II acquire the images of 3-D objects. The images are digitized, processed and converted into pictorial drawings. The generation of pictorial drawings is assisted with spatial understanding which makes use of the necessary knowledge in the knowledge base. The VIREC system then performs orthographic projection transformations on the pictorial drawings to generate three or more orthographic projections of the 3-D object. The orthographic projections for known objects are stored in the database as references. The feature matching unit identifies objects by comparing orthographic projections of the unknown object with the database. The orthographic projections are used as invariant 2-D features for positive identification of a 3-D object.

<u>Label Reading</u>

The label reading function is to identify geometrical and topological properties of the pictorial labels, such as symmetry, inclusion relationship,

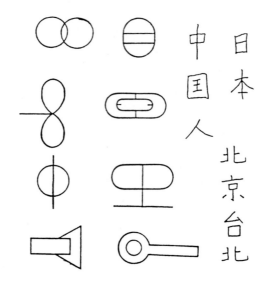

Figure 5a
Typical Pictorial Labels

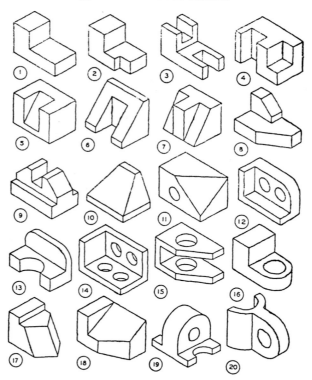

Figure 5b
Sample Pictures of 3-D Objects

Figure 5b (continued)

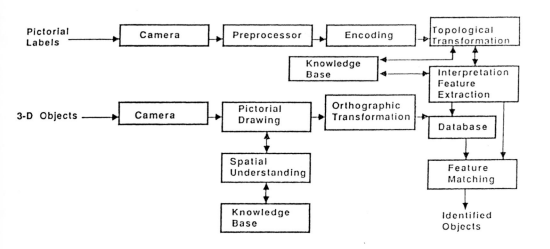

Figure 6

VIREC System Architecture

ridges, bridges, fins, pigtails, appendices, number of loops, touching contours, folding, shape description, stroke sequence, convexity and concavity. The interpretation rules are developed to facilitate the determination of geometrical and topological properties of a label from its octal chain-vector representation. Topological transformations are utilized to make hidden relationships between chain sequence and figure topology more visible to the computer. The interpretation rules for the number of loops, ridges, overlapping areas, inclusion relations, bridges and fins are summarized as follows:

1) Number of Loops

   The number of loops in a multiple-loop configuration is determined from the octal chain-vector representation by applying the coordinate-matching algorithm[13, 14]. The number of coordinate-matches determines the number of independent loops. Repeated coordinates are excluded, because they represent overlapping contours.

2) Ridges and Overlapping Areas

   The ridge between two loops is the segment joining two junction points and separating these loops. Let the two adjacent loops of a geometrical configuration be represented by the canonical form

   $$| c_{11}c_{12}\cdots c_{1(m-1)}c_{1m} \overset{C_1}{} \quad | \quad c_{21}c_{22}\cdots c_{2(n-1)}c_{2n} \overset{C_2}{} |$$

   which consists of two CL segments, C_1 and C_2. We define code sets

   $\{C_{11}\}$ = a set of head codes of segment C_1

   $\{C_{1m}\}$ = a set of tail codes of segment C_1

   $\{C_{21}\}$ = a set of head codes of segment C_2

   $\{C_{2n}\}$ = a set of tail codes of segment C_2

   According to topological properties, we classify the ridges into five types:[13]

   (a) Type 1 Ridge

   If $\{C_{11}\} = \{C_{21}\}$, and loop C_1 and C_2 are in the same direction, then this is an external ridge.

   If $\{C_{11}\} = \{C_{21}\}$, and loop C_1 and C_2 are in opposite directions, then this is an internal ridge.

(b) **Type 2 Ridge**

If $\{C_{1m}\} = \{C_{2n}\}$, and C_1 and C_2 are in the same direction, then this is an external ridge.

If $\{C_{1m}\} = \{C_{2n}\}$, and C_1 and C_2 are in opposite directions, then this is an internal ridge.

(c) **Type 3 Ridge**

If $\{C_{11}\} = \{\bar{C}_{2n}\}$, and C_1 and C_2 are in the same direction, then this is an internal ridge.

If $\{C_{11}\} = \{\bar{C}_{2n}\}$, and C_1 and C_2 are in opposite directions, then this is an external ridge.

(d) **Type 4 Ridge**

If $\{C_{1m}\} = \{\bar{C}_{21}\}$, and C_1 and C_2 are in the same direction, then this is an internal ridge.

If $\{C_{1m}\} = \{\bar{C}_{21}\}$, and C_1 and C_2 are in opposite directions, then this is an external ridge.

(e) **Type 0 Ridge**

If the above conditions do not hold, then the ridge degenerates into a touching point. Furthermore, if there is an OCL segment between two CL segments, then this OCL segment describes the overlapping area between these two loops represented by the CL segments.

3) **Inclusion Relations**

To facilitate computer interpretation of inclusion relations, we introduce a measure of curvature[14]

$$C(c_i, c_{i+1}) \triangleq c_i \theta c_{i+1}$$

for the region tracing in clockwise direction, where

$$c_i \theta c_{i+1} = c_i - c_{i+1}, \text{ if } |c_i - c_{i+1}| \leq 4$$

$$c_i \theta c_{i+1} = c_i - c_{i+1} + 8, \text{ if } c_i - c_{i+1} < -4$$

$$c_i \theta c_{i+1} = c_i - c_{i+1} - 8, \text{ if } c_i - c_{i+1} > 4$$

For the region tracing in counterclockwise direction, we use the curvature given by

$$C(c_i, c_{i+1}) \triangleq c_{i+1} \theta c_i$$

where

$$c_{i+1} \theta c_i = c_{i+1} - c_i, \text{ if } |c_{i+1} - c_i| \leq 4$$

$$c_{i+1} \ominus c_i = c_{i+1} - c_i + 8, \text{ if } c_{i+1} - c_i < -4$$

$$c_{i+1} \ominus c_i = c_{i+1} - c_i - 8, \text{ if } c_{i+1} - c_i > 4$$

If $C(c_i, c_{i+1})$ is positive, the curve $c_i c_{i+1}$ is convex. If $C(c_i, c_{i+1})$ is negative, the curve $c_i c_{i+1}$ is concave.

To design computer interpretation of inclusion relations, we divide the analysis into two cases:

Case I Touching Contours. The algorithm is (a) determine the directions of regions under consideration, (b) find the touching points, and (c) determine the strategic curvatures $C(c_{1n}c_{11})$, $C(c_{1n}c_{21})$, $C(c_{2m}c_{21})$, and $C(c_{2m}c_{11})$. The geometrical configuration is represented in a canonical form.

The interpretation rules are

(1) If $C(c_{2m}c_{21}) > C(c_{1n}c_{11})$, $C(c_{1n}c_{21})$ and $C(c_{2m}c_{11})$ are convex, then $C_1 \supset C_2$.

(2) If $C(c_{2m}c_{21}) > C(c_{1n}c_{11})$, $C(c_{1n}c_{21})$ and $C(c_{2m}c_{11})$ are concave, then $C_2 \supset C_1$. Region C_2 is external to region C_1.

Case II Folding. The algorithm is (a) determine the directions of regions under consideration, (b) identify external ridges, and (c) perform interpretation. An external ridge is not a bridge, because its deletion will reduce the number of loops. The removal of a double ridge will change the topology of the configuration. The removal of a bridge will not reduce the number of loops.

(3) Bridges. To determine the inclusion relationship for geometrical configurations with bridges, we have developed the following algorithms. First the directions of the regions under consideration are determined. Then applying the ridge-finding rules to identify the bridge. Let the canonical forms of the bridged contours be

(1) $\{ c_{b1} \} c_{11}c_{12} \cdots c_{1n} \{ c_{b2} \} c_{21}c_{22} \cdots c_{2m}$

or (2) $c_{11}c_{12} \cdots c_{1n} \{ c_{b1} \} c_{21}c_{22} \cdots c_{2m} \{ c_{b2} \}$

Then the strategic curvatures for the configuration are determined:

(1) $C(c_{b1}c_{11})$, $C(c_{b2}c_{21})$, $C(c_{1n}c_{b2})$, $C(c_{2m}c_{b1})$

(2) $C(c_{1n}c_{b1})$, $C(c_{b1}c_{21})$, $C(c_{b2}c_{11})$, $C(c_{2m}c_{b2})$

The interpretation rules are:
For Case I,

(a) If $C(c_{b1}c_{11})$ or $C(c_{1n}c_{b2})$ is concave and $C(c_{b2}c_{21})$ or $C(c_{2m}c_{b1})$ is concave, then c_{b1}/c_{b2} is an external bridge and region C_2 is external to region C_1.

(b) If $C(c_{1n}c_{b1})$ or $C(c_{b2}c_{11})$ is concave and $C(c_{b1}c_{21})$ or $C(c_{2m}c_{b2})$ is concave, then c_{b1}/c_{b2} is an external bridge and region C_2 is external to region C_1.

For Case II,

(a) If $C(c_{b1}c_{11})$ and $C(c_{1n}c_{b2})$ are convex, then bridge c_{b1}/c_{b2} is inside region C_1 and $C_2 \subset C_1$.

(b) If $C(c_{1n}c_{b1})$ and $C(c_{b2}c_{11})$ are convex, then bridge c_{b1}/c_{b2} is inside region C_1 and $C_2 \subset C_1$.

(5) <u>Fins</u>. Fins are open-loop segments attached to a closed-loop configuration, which can readily be identified from the chain-vector representation. Let a fin be identified as c_i or $\{c_i\}$ for a one-way fin, or as $c_i \bar{c}_i$ or $\{c_i \bar{c}_i\}$ for a two-way fin. We have developed the following rules for determining whether the fin is a pigtail or an appendix by checking the strategic curvatures:

(a) If the strategic curvature of a one-way fin is positive, it is an appendix.

(b) If the strategic curvature of a one-way fin is negative, it is a pigtail.

(c) If the strategic curvatures of a two-way fin are negative, it is a pigtail.

(d) If the strategic curvatures of a two-way fin are positive, it is an appendix.

<u>Illustrative Examples</u>

(1) The following pictorial label is shown to VIREC which converts the image into an octal vector-chain representation as

```
        |       C L      |
    0 6 6 6 4 3 4 2 0 1 0 6 6 6 4 3 0 2 4
                    |       O C L       |
```

The ECC representation is

$$0\ 6\ 6\ 6\ 4\ \dot{3}\ \dot{4}\ 2\ \dot{0}\ 1\ 0\ 6\ 6\ 6\ 4\ \dot{3}\ \dot{0}\ 2\ 4\ \dot{1}$$
$$|\qquad\ \ \text{C L}\qquad\ \ |\qquad\quad\text{C L}\qquad\ \ |$$

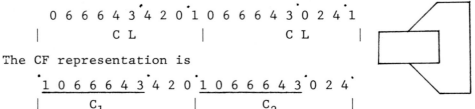

The CF representation is

$$\dot{1}\ 0\ 6\ 6\ 6\ 4\ \dot{3}\ \dot{4}\ 2\ \dot{0}\ 1\ 0\ 6\ 6\ 6\ 4\ \dot{3}\ 0\ 2\ 4\dot{\ }$$
$$|\qquad c_1 \qquad\quad |\qquad\quad c_2 \qquad\ \ |$$

Invoking rules for ridge finding, VIREC has found that segment 1 0 6 6 6 4 3 is an external ridge. Then VIREC determines the strategic curvatures:

$C(34) = -1$ and $C(30) = 3$

since $C(30) > C(34)$, VIREC concludes that segment 0 2 4 is an internal ridge and segment 4 2 0 is an external ridge. Applying the symmetry rules, VIREC has found that this configuration has x-symmetry.

(2) A Master Card logo is shown to VIREC which converts the image into an octal chain-vector representation as

| O C L | S C L |
7 1 7 6 6 5 3 3 1 7 5 5 3 2 2 1
       | C L |

The ECC representation is

        | C L |
$\dot{1}$ 7 6 6 5 3 3 $\dot{1}$ 7 5 5 3 2 2 1 7
| O C L | O C L

The CF representation is

| C L | C L | C L |
3 1 1 7 6 6 5 3 3 1 7 5 5 3 2 2 1 7 7 5

Invoking interpretation rules, VIREC finds the following descriptions of the Master Card logo:
* multiloop closed curve
* x-symmetry and y-symmetry
* overlapping area
* two internal ridges
* two junction points

(3) A Chinese character 中 is shown to VIREC which converts the image in octal chain-vector representation as

```
           |           C L              |
  6 6 6 0 0 6 6 4 4 6 6 6 2 2 2 4 4 2 2 0 0 6 6
           |   O C L   |       O C L    |
```

VIREC identifies one-way fin 6 6 6 and two-way fin 6 6 6 2 2 2. The strategic curvatures are C(60) = -2, C(46) = -2, and C(24) = -2. Thus, the fins are pigtails, which are vertical with equal length and located at (0,0), (0,-3) and (0,-5), (0,-8). By dropping the pigtails, the octal vector-chain representation reduces to

```
               |     O C L     |
  0 0 6 6 4 4 4 4 4 2 2 0 0 6 6
  |           C L               |
```

The ECC and CF representations are

```
  0 0 6 6 4 4 4 4 2 2 0 0 6 6 4 4 2 2 0 0
  |       C L         |       C L       |
```

Invoking interpretation rules, VIREC has found that this configuration is symmetrical and has two vertical pigtails and two loops separated by an internal ridge 6 6 located at (0,-3), (0,-5).

Parts Recognition

The design principles underlying 3-D parts recognition are based upon the concept of spatial understanding which determines the orthographic projections from the knowledge on camera system geometry and on pictorial drawings derived from 2-D images of an object. The orthographic projections are the 2-D features of the object, which provide the footprints for the representation of 3-D industrial parts. The footprints of a 3-D object consist of the top view, the bottom view, the front view, the rear view, and the side views. These views are the projections of the object on three orthographic planes. Via spatial understanding the VIREC system generates the footprints which are used for the recognition of 3-D objects.

The inputs to the VIREC system are grey-level pictures of the objects taken from arbitrary viewing angles. From the 2-D images, the VIREC system determines spatial information characterizing the objects. For a given camera setting, the image of a

3-D object lies in the focal plane which we call the picture plane. The viewing direction is normal to the picture plane. The coordinate system (x,y,z) for the picture plane and the viewing axis is chosen as the picture-plane coordinates. The VIREC system obtains measurable parameters from the 2-D images on the picture plane (x,y) and interprets these parameters to generate 3-D descriptions in (x,y,z)-coordinates. From this characterization, the VIREC system determines 3-D descriptions of the object in terms of the base-plane coordinates (x*,y*,z*). The origin remains unchanged. The new x*-y* plane is parallel to the base plane of the object.

Let a surface of the 3-D object be characterized in picture-plane coordinates by

$$z = f(x,y) \tag{1}$$

Taking differentials yields

$$\frac{\partial f}{\partial x}dx + \frac{\partial f}{\partial y}dy - dz = 0 \tag{2}$$

which may be written as

$$(p,q,-1) \begin{bmatrix} dx \\ dy \\ dz \end{bmatrix} = 0 \tag{3}$$

where $p = \frac{\partial f}{\partial x}$ and $q = \frac{\partial f}{\partial y}$. The vector

$$\underline{n} = (p,q,-1) \tag{4}$$

is normal to an elementary area defined by dxdydz, and the surface gradient at point (x,y) is

$$\underline{G} = (p,q) \tag{5}$$

Let $\underline{A}_1$ and $\underline{A}_2$ be two orthogonal vectors in a plane of the 3-D object, which will be called object vectors. These two vectors will generate two image vectors $\underline{a}_1$ and $\underline{a}_2$ in the picture plane. The VIREC system will establish the relationship between the object vectors and the corresponding image vectors based upon the following reasoning: An image vector $\underline{a}$ is characterized by

$$\underline{a} = (\cos \alpha, \sin \alpha) \tag{6}$$

where α is the angle between the image vector and the x-axis.

Then the object vector $\underline{A}$ is given by

$$\underline{A} = (\cos \alpha, \sin \alpha, \underline{G} \cdot \underline{a}) \qquad (9)$$

For two orthogonal object vectors $\underline{A}_1$ and $\underline{A}_2$ in an object plane, the orthogonality condition leads to

$$\cos(\alpha_1 - \alpha_2) + (\underline{G} \cdot \underline{a}_1)(\underline{G} \cdot \underline{a}_2) = 0 \qquad (8)$$

Based upon the measured values of $\underline{a}_1$, $\underline{a}_2$, α_1 and α_2, the VIREC system will determine the gradient vector $\underline{G}$.

Consider three adjacent object surfaces A, B, and C meeting at one vertex, with surface gradients given by $\underline{G}_a$, $\underline{G}_b$, and $\underline{G}_c$, respectively. For each surface, we have the relationship

$$\cos(\alpha_1 - \alpha_2) + (\underline{G}_a \cdot \underline{a}_1)(\underline{G}_a \cdot \underline{a}_2) = 0 \qquad (9)$$

$$\cos(\beta_1 - \beta_2) + (\underline{G}_b \cdot \underline{b}_1)(\underline{G}_b \cdot \underline{b}_2) = 0 \qquad (10)$$

$$\cos(\gamma_1 - \gamma_2) + (\underline{G}_c \cdot \underline{c}_1)(\underline{G}_c \cdot \underline{c}_2) = 0 \qquad (11)$$

where $\underline{a}_1$ and $\underline{a}_2$ are image vectors from object surface A, $\underline{b}_1$ and $\underline{b}_2$ are image vectors from object surface B, $\underline{c}_1$ and $\underline{c}_2$ are image vectors from object surface C.

To determine these surface gradient vectors, the VIREC system takes more measurements which lead to three additional relationships:

$$\underline{G}_a \cdot \underline{e}_1 = \underline{G}_b \cdot \underline{e}_1 \qquad (12)$$

$$\underline{G}_b \cdot \underline{e}_2 = \underline{G}_c \cdot \underline{e}_2 \qquad (13)$$

$$\underline{G}_c \cdot \underline{e}_3 = \underline{G}_a \cdot \underline{e}_3 \qquad (14)$$

The image vectors $\underline{e}_1$, $\underline{e}_2$, and $\underline{e}_3$ are derived from three intersecting edges of the image in the picture plane. These image vectors are expressed in terms of their directional angles with respect to the x-axis of the picture plane.

From the above six equations, the VIREC system determines the surface gradient vectors $\underline{G}_a$, $\underline{G}_b$, and $\underline{G}_c$, which yield three normal vectors for intersecting surfaces A, B, and C:

$$\underline{n}_a = (\underline{G}_a, -1) \qquad (15)$$

$$\underline{n}_b = (\underline{G}_b, -1) \qquad (16)$$

$$\underline{n}_c = (\underline{G}_c, -1) \qquad (17)$$

To understand the spatial relationships of a 3-D object from its 2-D images, the VIREC system conducts the above computational analysis of all the visible surfaces of the 3-D object under consideration.

After the characterization of the visible surfaces on a 3-D object in terms of picture-plane coordinates (x,y,z) has been determined, the VIREC system will perform orthographic projection transformation. The spatial information describing the 3-D object is first transformed from the picture-plane coordinates to the base-plane coordinates (x^*, y^*, z^*), and then from the base-plane coordinates to the absolute coordinates (x^{**}, y^{**}, z^{**}). The object base plane is the footprint on the ground level. The x^*-y^* plane of the base-plane coordinates is parallel to the base plane. With knowledge about the camera system available, the VIREC system performs transformation from the picture-plane coordinates to the base-plane coordinates via a rotation, and transformation from the base-plane coordinates to the absolute coordinates via a translation. The origin of the absolute coordinate system is the strategic corner which is at the lowest position of the picture plane. The VIREC determines the orthographic projections from the absolute coordinate characterization of the 3-D object.

To identify the footprint of a 3-D object from its 2-D image, the VIREC system follows the following procedure: (1) use the grey-level histogram thresholding to segment the object and the background, (2) obtain the silhouette of the object, (3) identify the $\vee$-type vertex from the picture plane as a starting point for base tracing which moves in two directions and terminates when an $\llcorner$-type vertex is met, and (4) connect all line segments that have been traced to create a partial footprint.

The orthographic projections which serve as 2-D features for the characterization of 3-D objects are represented by their primitives and relations. The primitives consist of line segments, angles, circular arcs, holes, center of circles, aspect ratios, and their structural relationships. For a given class of industrial parts, we are developing a database in terms of the primitives and their relationships. During the learning phase, the VIREC system acquires the knowledge and the orthographic projections of sample industrial parts to be recognized and stored in the database. During the recognition phase, the VIREC system extracts the orthographic projections of unknown industrial parts for comparison with the descriptions in the database.

Experimental Results

Mark I of the VIREC system has been implemented on a VAX 11/750 minicomputer. As an illustration, VIREC looks at a telephone intercom box from different viewing directions. The orthographic projections for each case are generated by the VIREC system, as illustrated in Figure 7. Since the three sets of footprints are the same, VIREC concludes that these pictures describe the same object, the telephone intercom box. As another illustration, a 3-D object is shown to VIREC from two different viewing angles depicted in Figure 8. VIREC concludes that these two different views describe the same object, since the two sets of footprints generated by VIREC are the same.

Concluding Remarks

Under the framework of the system architecture shown in Figure 1, the Function Specification Unit issues signals to retrieve various methodologies to perform functional analysis which includes encoding into octal chain-vector representations, conversion to equivalent closed chain, conversion to canonical form, and determination of strategic curvatures for Mode I operation, and generation of pictorial drawings, determination of surface normals and gradients, coordinates transformations, and generation of orthographic projections for Mode II operation. The interpretation and Recognition Unit determines the number of loops, the symmetry property, the inclusion relationship, ridges, bridges, fins, pigtails, appendices, touching contours, folding, shape description, convexity and concavity properties for Mode I operation, and generates feature database from orthographic projections for Mode II operation. Reading of labels and recognition of 3-D objects are based upon feature matching.

KNOWLEDGE-BASED SYSTEMS FOR UNDERSTANDING DRAWINGS

When we wish to design robots to do trouble-shooting of electrical and electronic systems, a fundamental task is to teach the robot to read electronic schematic diagrams and drawings. The

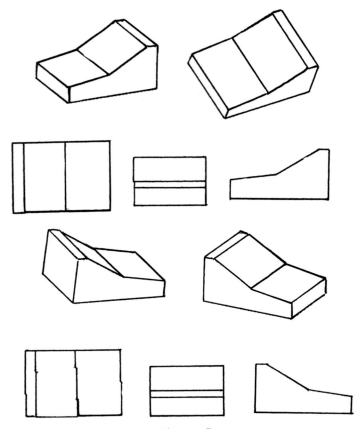

Figure 7
Orthographic Projections of a Telephone Intercom Box Generated from Various Viewing Angles

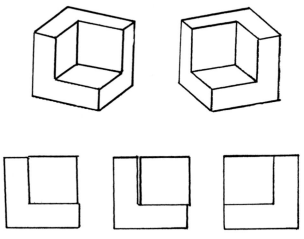

Figure 8
A 3-D Object with Different Views but Same Orthographic Projections

AUTORED system which we conceived and developed some four years ago may be used for this type of robotic applications. AUTORED is a knowledge-based system which automatically reads electronic circuit diagrams and schematics.

Electronic circuit diagrams may be considered as a graphic language to express and communicate design ideas among electrical, electronic, computer, and production engineers. Such diagrams are line drawings which consist of junction dots, line segments, symbols for circuit elements, and denotations for labeling. Thus, an electronic circuit diagram is characterized by functional elements, connecting elements, and denotations. The functional elements are represented by symbols consisting of two or more terminals for connection to other elements. Resistors, capacitors, and diodes are two-terminal elements. Transistors and amplifiers are three-terminal elements. Flip-flops are four-terminal elements. VLSI is a multi-terminal element. The connecting elements are represented by junction dots and horizontal and vertical line segments which are used to connect functional elements. The denotations are used to describe physical properties or names of the functional elements in a diagram. For instance, the denotation 10K attached to a functional element implies that the element is a resistor of 10,000 Ω resistance. The denotation C1 attached to a functional element indicates that the element is capacitor #1 in the diagram. The use of denotations in a circuit diagram will simplify automatic recognition and interpretation tasks. The major problems in automatic reading of circuit diagrams are similar to those in automatic recognition of hand-written text. The difficulties are due to the interconnections of functional elements. Thus, the classical pattern recognition techniques do not apply.

To expedite the recognition and understanding tasks, we classify the functional elements into six categories according to their geometrical shapes. They are (1) junction-pair shape, (2) triangular shape, (3) circular shape, (4) rectangular shape, (5) curved shape, and (6) non-closed boundary shape. The junction-pair shape category consists of capacitors, diodes, crystals, and ground symbols. The triangular-shape category consists of operational amplifiers, NOT gates, and driver. The circular shape category is further divided into Bipolar Transistors, Junction

FET, and MOS FET. The rectangular shape category is further divided into two major subcategories: the vertical bar shape, and the horizontal bar shape. The former includes flip-flops, adders, multiplexers, and various VLSI circuits, and the latter includes symbol representations commonly used in Europe. The curve shape category consists of the logical gates subcategory and the inductor/transformer subcategory. The non-closed boundary shape category consists of two-terminal elements which are resistors and inductors, and three-terminal (or more) elements which are Bipolar Transistors, Junction FET and MOS FET. From the above classification scheme, we create a symbol tree structure which is built into the knowledge base to facilitate recognition and interpretation of functional elements. Typical symbols for functional elements are shown in Figure 9.

Automated Interpretation

The fundamental procedures in circuit diagram interpretation and understanding are the recognition of functional elements, the reading of denotations and labels, the tracing of connection line elements, and the determination of interconnecting relationships. Before a functional element can be accurately and completely recognized, a prerequisite is to segment functional elements and denotations from connection lines. This is accomplished in AUTORED by the technique of multiple-pass pattern extraction which was proposed by Tou[15]. Each pass of pattern extraction generates a page of information of one kind through multiple-pass pattern extraction. The electronic diagram is segmented according to the nature of the elements. The reading of isolated electronic symbols is then treated as a "character" recognition problem. The connecting line elements provide the relational information for interpretation and understanding.

The AUTORED system decomposes an electronic diagram into five sets of drawings, one on each page, via a multiple-pass pattern extraction process. These drawings contain junction dots, connecting line elements, functional symbols, denotations, and unrecognizable elements, respectively. The first page stores a drawing for the junction dots, the second page for connecting line elements, the third page for functional symbols, the fourth page for denotations, and the fifth page for unrecognizable elements.

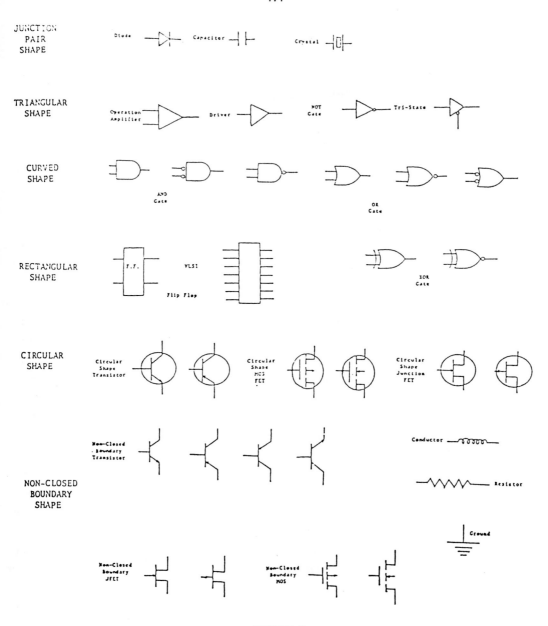

FIGURE 9

TYPICAL SYMBOLS FOR FUNCTIONAL ELEMENTS IN ELECTRONIC
AND LOGIC CIRCUIT DIAGRAMS

To accomplish automatic interpretation and understanding of electronic circuit diagrams, the AUTORED system performs the following five tasks:

Task 1 -- image enhancement, binary image generation, and binary image smoothing.

Task 2 -- extraction of symbols, denotations, and line elements from the circuit diagram.

Task 3 -- interpretation of graphic functional symbols.

Task 4 -- reading of characters and labels of the denotations.

Task 5 -- generation of database for CAD and CATS (computer-aided trouble-shooting)

Task 1 makes use of image processing techniques which are well documented in the literature[16].

Task 2 separates each functional symbol, denotation, and connection line element. This task performs five functions: (1) junction and corner extraction, (2) line segment tracing and linking, (3) line segment classification, (4) connection line segment removal and blocking, and (5) symbol locating and denotation character grouping[17]. The main purpose of junction and corner extraction is to locate all the possible junctions and corners that exist in the circuit diagram. They represent the special features of the connectivity between the vertical and horizontal connection lines and between the connection line and the functional symbol. The second step is to find all the line segments and to link the overlapping line segments into one line segment. However, these line segments can be either real connection line segments or the component line segments that are part of the symbols. The classification procedure is carried out in the third step. After all the possible connecting line segments having been identified, they are removed from the image. The blocking routine is called to isolate all the components and the denotations. The final step is implemented to classify the blocks and to group the neighboring denotation blocks into meaningful blocks.

Task 3 interprets each isolated functional symbol. The interpretation process follows the hierarchical structure of the symbol classification to recognize the geometrical shape, the orientation and the type of each functional symbol[6]. Task 4 is designed to read the characters and labels adjacent to the functional symbols or inside the symbols. Task 5 determines the interconnecting relationships between functional elements and generates circuit element database for computer-aided design and for computer-aided trouble-shooting. The database contains the names and functions of the circuit elements, the interconnecting relationships between elements, the names and functions of a group of circuit elements as a subsystem, and the interconnecting relationships between subsystems.

Shown in Figure 10 are some experimental results of AUTORED system. The inputs to AUTORED are electronic circuit and logic diagrams. The outputs are the functional symbols in the diagrams.

Recognition of Functional Elements

The recognition of segmented functional elements is performed in several steps. Through shape and structure recognition, the system automatically determines various categories of transistors (bipolar, channel, and MOS) from the segmented elements. The next step is to distinguish among these three categories of functional elements and to recognize their orientations. The last step is to determine the specific type in each category, such as PNP, NPN, P-channel, N-channel, P-MOS, and N-MOS. The knowledge hierarchy is shown in Figure 11.

(A) Category Identification

To identify transistor categories from circuit diagrams, we make use of the knowledge on the structural properties of functional symbols. A bipolar transistor is characterized by one vertical line, one horizontal line, and two slant lines. Regardless of its orientation, a bipolar transistor may be represented by the feature vector $V_b = [1,1,2]$. A junction FET transistor is characterized by three vertical lines, three horizontal lines, and no slant lines. Regardless of its orientation, a junction FET transistor may be represented by the feature vector $V_j = [3,3,0]$. A MOSFET transistor may be represented by two kinds of combinations of primitives. They are (a) six vertical lines and four horizontal lines for east-west and west-east orientations, and (b) four vertical lines and six horizontal lines for north-south and south-north orientations. Thus, a MOSFET transistor may be characterized by feature vectors $V_m = [6,4,0]$ or $V_m = [4,6,0]$. These feature

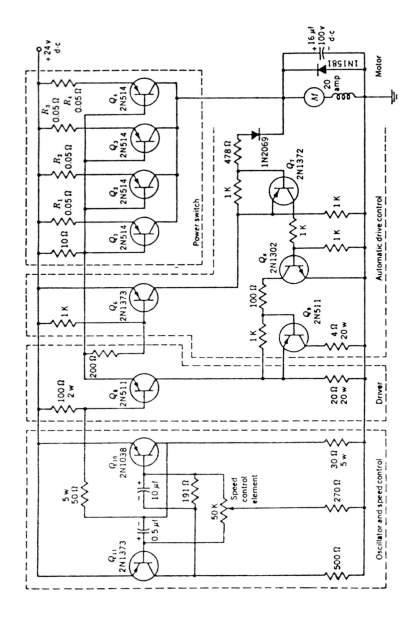

FIGURE 10(a)   Experimental Results of AUTORED - A Transistor Circuit Diagram

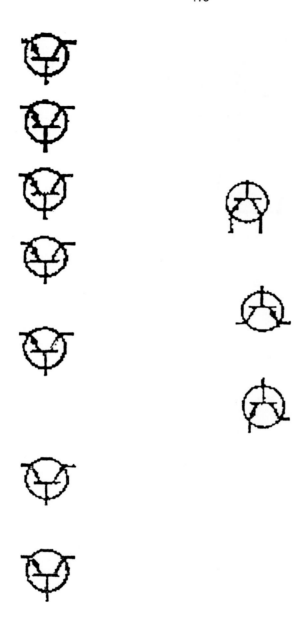

FIGURE 10(b)
Transistors Lifted from (a)

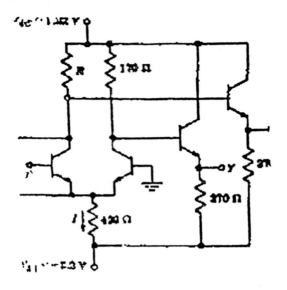

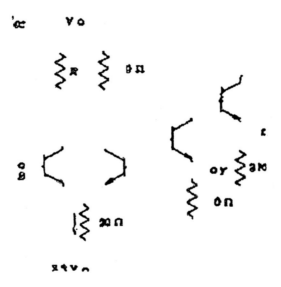

Figure 10(c)
A Circuit Diagram and its Segmented Components

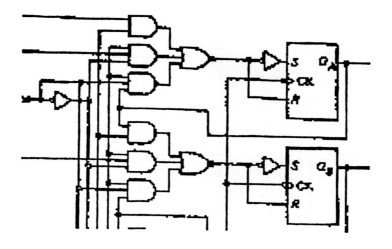

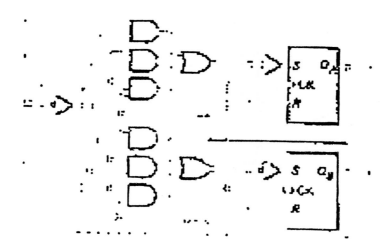

Figure 10(d)

A Logic Circuit Diagram and Its
Partially Segmented Components

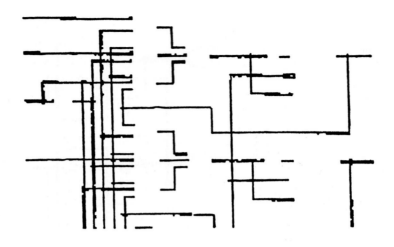

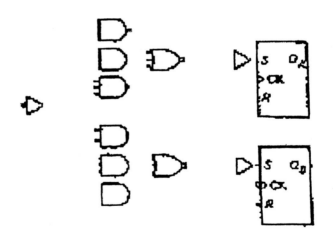

Figure 10(e)

Connection Lines and Components Lifted from (d)

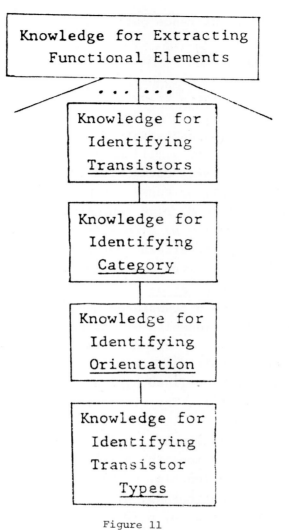

Figure 11

Knowledge Hierarchy

vectors serve as the reference patterns for the transistor categories.

The identification of transistor category is accomplished on the basis of pattern matching. Let $U = [u_1, u_2, u_3]$ be a pattern vector describing a transistor whose category is to be determined, where u_1, u_2 and u_3 denote the number of vertical lines, horizontal lines, and slant lines, respectively. We also define $V_i = [v_1(i), v_2(i), v_3(i)]$ as a standard pattern vector describing one of the three categories of transistors, where i denotes a Bipolar, JFET, or FOSFET transistor. We introduce a similarity measure given by $D_i = |u_1 - v_1(i)| + |u_2 - v_2(i)| + 3|u_3 - v_3(i)|$ where i = b, j, m. If D_i is minimum for i = b, then the transistor with pattern vector U is a bipolar transistor. This rule also applies to the other two categories.

(B) Orientation Identification

The orientation of a transistor may be characterized by the basic geometrical relationships of its primitives. We describe the orientation by an orientation vector which consists of six elements: parallel, colinear, left of, right of, above, and below. Each primitive is defined by its starting point (x_s, y_s) and its ending point (x_e, y_e). The six basic geometrical relationships are defined as follows:

(1) PARALLEL relationship, $p_1(k)$
(2) COLINEAR relationship, $p_2(k)$
(3) LEFT OF relationship, $p_3(k)$
(4) RIGHT OF relationship, $p_4(k)$
(5) ABOVE relationship, $p_5(k)$
(6) BELOW relationship, $p_6(k)$

We define the transistor orientation vector as

$$P_{jk} = \left[p_1(k), p_2(k), p_3(k), p_4(k), p_5(k), p_6(k) \right]$$

where index j denotes transistor categories and index k is the orientation index. The elements of the vector P_{jk} are the six basic geometrical relationships. The geometrical properties for each category-orientation are summarized in an orientation relation table shown on the following page. To derive the orientation relation table, we impose the following constraints:

(1) The relationship is unidirectional.

(2) The relationship among external primitives are not considered.
(3) In establishing the relationship between an internal primitive and an external primitive, the internal primitive is considered as the reference.

CATEGORY	ORIENTATION	$P_1(k)$	$P_2(k)$	$P_3(k)$	$P_4(k)$	$P_5(k)$	$P_6(k)$
I	1	0	0	0	2	0	0
	2	0	0	0	0	0	2
	3	0	0	2	0	0	0
	4	0	0	0	0	2	0
II	1	2	0	0	2	1	1
	2	2	0	1	1	0	2
	3	2	0	2	0	1	1
	4	2	0	1	1	2	0
III	1	3	3	0	6	0	0
	2	3	3	0	0	0	6
	3	3	3	6	0	0	0
	4	3	3	0	0	6	0

The internal primitives are not directly connected to the outside circuit, while the external primitives are directly connected to other functional elements of the circuit diagram. For a JFET transistor, there are three internal primitives and three external primitives. These constraints are introduced to simplify the establishment of the orientation relation table with no ambiguity. Let $Q = [q_1, q_2, q_3, q_4, q_5, q_6]$ be the measured orientation vector of a transistor whose orientation is to be determined. We introduce a similarity measure given by

$$S_k = P_{jk}^T Q$$

Then the orientation is determined by the orientation index k which makes S_k a maximum.

(C) Type Identification

Having identified the transistor category and its orientation in an electronic circuit diagram, the next problem in circuit diagram understanding is to identify the type of the transistor. The arrowheads in a transistor symbol generally provide necessary information for distinguishing between PNP type and NPN type, P-channel and N-channel type, and P-MOS and N-MOS type. To identify an arrowhead, we may apply template matching techniques to the terminal lines. If template matching results in only one significant correlation, this location is recognized as the presence of an arrowhead. If template matching does not yield a single significant correlation, the situation is ambiguous and the system has to make use of transistor circuitry knowledge in order to identify the transistor type. Such expert knowledge on transistor circuitry is stored in the knowledge base of the AUTORED system[6].

In the case of bipolar transistors, the arrowhead patterns may belong to one of five classes:

Class 1 -- One arrowhead only. There is no ambiguity.

Class 2 -- One arrowhead on each slant line. In this situation, the identification of the emitter and the collector is needed.

Class 3 -- Two arrowheads on one of the slant lines. In this situation, the identification of the direction of current flow is required.

Class 4 -- Two arrowheads on one of the slant lines and one arrowhead on the other. In this situation, the collector/emitter line is to be identified first. If the emitter line has one arrowhead, the ambiguity is resolved. If the emitter line has two arrowheads, the direction of current flow should be determined.

Class 5 -- No arrowheads or four arrowheads. In this situation, both the collector/emitter line and the current flow should be identified.

To facilitate the design of a knowledge-based system for interpretation, we divide the knowledge into judgmental knowledge and factual knowledge. The former is used to discriminate ambiguity, and the latter provides rules for solving the problem. Let C_i be the correlation for the ith arrowhead with a template. We may express the judgmental knowledge in terms of a set of judgmental rules corresponding to various arrowhead pattern classes:

Case 1 -- If $C_1 > \theta$, $C_2 < \theta$, $C_3 < \theta$ and $C_4 < \theta$,
or $C_4 > \theta$, $C_1 < \theta$, $C_2 < \theta$ and $C_3 < \theta$,
THEN the arrowhead is at position 1 or 4, and
this element is recognized as an NPN transistor.

Case 2 -- IF $C_2 > \theta$, $C_1 < \theta$, $C_3 < \theta$ and $C_4 < \theta$,
or $C_3 > \theta$, $C_1 < \theta$, $C_2 < \theta$ and $C_4 < \theta$,
THEN the arrowhead is at position 2 or 3, and
this element is recognized as a PNP transistor.

Case 3 -- IF $C_1 > \theta$, $C_3 > \theta$, $C_2 < \theta$ and $C_4 < \theta$,
or $C_1 > \theta$, $C_4 > \theta$, $C_2 < \theta$ and $C_3 < \theta$,
or $C_2 > \theta$, $C_3 > \theta$, $C_1 < \theta$ and $C_4 < \theta$,
or $C_2 > \theta$, $C_4 > \theta$, $C_1 < \theta$ and $C_3 < \theta$,
THEN the system generates data set 1 and infers
rule set 1.

Case 4 -- IF $C_1 > \theta$, $C_2 > \theta$, $C_3 < \theta$ and $C_4 < \theta$,
or $C_3 > \theta$, $C_4 > \theta$, $C_1 < \theta$ and $C_2 < \theta$,
THEN the system generates data set 2 and infers
rule set 2.

Case 5 -- IF $C_1 > \theta$, $C_2 > \theta$, $C_3 > \theta$ and $C_4 < \theta$,
or $C_1 > \theta$, $C_2 > \theta$, $C_4 > \theta$ and $C_3 < \theta$,
THEN the system generates data set 1 and infers
rule set 1.
IF slant line 2 is emitter line,
THEN EXIT; else generate data set 2 and infer rule
set 2.

Case 6 -- IF $C_2 > \theta$, $C_3 > \theta$, $C_4 > \theta$ and $C_1 < \theta$,
or $C_1 > \theta$, $C_3 > \theta$, $C_4 > \theta$ and $C_2 < \theta$,
THEN the system generates data set 1 and infers
rule set 1.
IF slant line 1 is emitter line,
THEN EXIT; else generate data set 2 and infer rule
set 2.

Case 7 -- IF $C_1 < \theta$, $C_2 < \theta$, $C_3 < \theta$ and $C_4 < \theta$,
or $C_1 > \theta$, $C_2 > \theta$, $C_3 > \theta$ and $C_4 > \theta$,
THEN the system generates data sets 1 and 2 and in-
fers rule sets 1 and 2.

The facutal knowledge is expressed in terms of four rule sets.
Rule set 1 contains the knowledge required to identify the collec-

tor/emitter terminal. Rule set 2 contains the knowledge for identifying the types of bipolar transistors. Rule set 3 and rule set 4 contain the necessary knowledge for recognizing the types of junction FET and MOSFET, respectively. Although most of the rule sets are pre-stored in the rule base, new rules may be generated via inference and added to the rule base.

The data sets associated with the four rule sets form the contents of the database. Although most of the data sets are pre-stored in the database, new data may be generated via inference. Data set 1 is used for checking the connections of terminals 2 and 3 to a voltage source, ground, a resistor, or a bipolar transistor. Data set 2 is used for checking the voltage value. Data set 3 is used for checking terminal connections for junction FET transistors. Data set 4 is used for checking terminal connections for MOSFET transistors.

Some experimental results illustrating the recognition and interpretation of functional elements are shown in Figure 12. The AUTORED system generates the segmented elements and identifies the transistors. AUTORED extracts features from the binary images, which are used for the identification of transistor categories and transistor orientation. To complete the interpretation process, AUTORED makes use of electronic circuit designer's knowledge and expertise which are stored in the knowledge base. Information on transistor category, transistor orientation, and primitive locations are the input data to the knowledge-based transistor type identification system. The knowledge-based system completes the interpretation.

CONCLUSIONS

An intelligent robot is made up of four fundamental components: machineware, hardware, software and brainware. An approach to the design of robotic brainware is to make use of knowledge-based systems. The incorporation of knowledge bases provides an effective design of intelligent robots and enables the robot to get access to desired knowledge, to do reasoning tasks, and to accumulate past experience which is needed for improving the performance.

Illustration of Electronic Circuit Diagram Interpretation by AUTORED

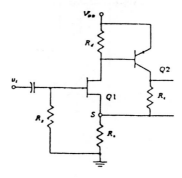

IDENTIFIED CATEGORY

	BT	JFET	MOSFET
COMP1	1	0	0
COMP2	0	1	0

IDENTIFIED ORIENTATION

	WE	NS	EW	SN
COMP1	1	0	0	0
COMP2	1	0	0	0

COMPONENT TYPES ARE IDENTIFIED AS FOLLOWS

COMP	BT NPN	BT PNP	JFET P	JFET N	MOS P	MOS N
1	0	1	0	0	0	0
2	0	0	0	1	0	0

Figure 12(a)

Circuit Diagram with Bipolar Transistor and Junction FET

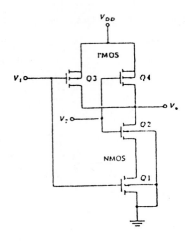

IDENTIFIED CATEGORY

	BT	JFET	MOSFET
COMP1	0	0	1
COMP2	0	0	1
COMP3	0	0	1
COMP4	0	0	1

IDENTIFIED ORIENTATION

	WE	NS	EW	SN
COMP1	1	0	0	0
COMP2	1	0	0	0
COMP3	1	0	0	0
COMP4	1	0	0	0

COMPONENT TYPES ARE IDENTIFIED AS FOLLOWS

COMP	BT		JFET		MOS	
	NPN	PNP	P	N	P	N
1	0	0	0	0	1	0
2	0	0	0	0	1	0
3	0	0	0	0	0	1
4	0	0	0	0	0	1

Figure 12(b)

Circuit Diagram with MOS

In this paper we have presented the design of the brainware for several robotic applications. The specific applications discussed are mobility improvement for roving robots, automated parts handling by robots, and automated trouble-shooting by robots. To improve the mobility of roving robots, we have designed a computer vision system which makes use of knowledge-based systems for directional guidance, obstacle avoidance, orientation determination, range finding, and object identification.

A fundamental problem in the design of automated parts handling by robot is to enable the robot to read labels and to recognize 3-D objects. We have designed the VIREC system to solve this problem. The VIREC is a knowledge-based system which conducts two modes of operation: Under Mode I operation, the VIREC enables roving robots to fetch parts from storage bins which are labeled with pictorial descriptions or to handle boxes of finished products for receiving, distribution and shipping. Under Mode II operation, the VIREC enables robots to visually recognize industrial parts for handling or for automated assembly. Mode I operation makes use of topological transformation techniques and Mode II operation is based upon the concepts of orthographic projection transformation.

In the design of robots to do independent trouble-shooting of electrical and electronic systems, a fundamental task is to enable the robot to read electronic schematic diagrams and drawings. The AUTORED system may be used for this type of robotic application. AUTORED is a knowledge-based system which automatically reads electronic circuit diagrams and schematics. The fundamental procedures in interpretation and understanding of circuit diagrams are the recognition of functional elements, the reading of denotations and labels, the tracing of connection line elements, and the determination of interconnecting relationships. The AUTORED system has been designed on the basis of the <u>multiple-pass pattern extraction technique</u>. Each pass of pattern extraction generates a page of information of one kind. The electronic circuit diagram is segmented according to the nature of the elements. The AUTORED system makes use of electronic and logic circuit designers' knowledge and expertise in conducting the interpretation and understanding process.

We hope the fundamental work presented here will provide the foundation for the design of the brainware of intelligent robots for factory floor materials handling, shipping and receiving, automated parts assembly, and autonomous trouble-shooting.

REFERENCES

1. J. T. Solberg, et al, "The Factory of the Future: A Framework for Research", *Proceedings of the NSF Conference on Computer-based Factory Automation*, 1984.

2. J. T. Tou, "Knowledge Engineering Revisited", *International Journal of Computer and Information Sciences*, Volume 14, Number 3, June 1985.

3. J. T. Tou, "Software Architecture of Machine Vision for Roving Robots", *Optical Engineering*, Volume 25, Number 3, March 1986.

4. J. T. Tou, "VIREC - A Visual Recognition Machine for Industrial parts", *Proceedings of Artificial Intelligence in Manufacturing Conference*, 1986.

5. J. T. Tou and J. M. Cheng, "Automatic Generation of Knowledge Base from Electronic Diagrams", in *Computer-based Automation*, Plenum Publishing Corporation, 1985.

6. J. T. Tou, et al, "Design of A Knowledge-based System for Understanding Electronic Circuit Diagrams", *Proceedings of the IEEE Conference on Artificial Intelligence Applications*, 1984.

7. R. W. DePree, et al, "A Relational Model for Knowledge Transfer", *Proceedings of 1980 International Computer Symposium*, 1980.

8. E. H. Shortlife, *Computer-based Medical Consultation: MYCIN*, Elsevier, New York, NY, 1976.

9. J. T. Tou, "MEDIKS - A Medical Knowledge System", *Proceedings of the 31st Annual Conference on Engineering in Medicine and Biology*, 1978.

10. J. T. Tou, "Design of A Medical Knowledge System for Diagnostic Consultation and Clincal Decision Making", *Proceedings of the 1978 International Computer Symposium*, 1978.

11. L. C. Chang and J. T. Tou, "MEDIKS - A Medical Knowledge System", *IEEE Transactions on Systems, Man and Cybernetics*, September 1984.

12. J. T. Tou and R. W. DePree, "UNIPS - A Universal Image Processing System", *Proceedings of the 5th International Conference on Pattern Recognition*, 1980.

13. J.T. Tou, "An Approach to Understanding Geometrical Configurations by Computer", *International Journal of Computer and Information Sciences*, Volume 9, Number 1, 1980.

14. J.T. Tou, *Lecture Notes on Computer Vision and Image Understanding*, Department of Computer and Information Science, University of Florida, 1984.

15. J.T. Tou and J.M. Cheng, "AUTORED: An Automated Electronic Diagram Reading Machine", *Proceedings of the IEEE International Conference on CAD*, 1983.

16. R.C. Gonzalez and P.A. Wintz, *Digital Image Processing*, Addison Wesley Publishing Company, Reading, MA, 1977.

17. C.L. Huang and J.T. Tou, "Knowledge-based Functional-symbol Understanding in Electronic Diagram Interpretation", *Proceedings of the SPIE Conference on Image Processing*, Orlando, FL, 1986.

ACKNOWLEDGEMENT -- this work was supported in part by NSF grant IST-8305815.

IMAGE UNDERSTANDING FOR ROBOTIC APPLICATIONS

Ramesh Jain

Computer Vision Research Laboratory
Electrical Engineering and Computer Science
The University of Michigan
Ann Arbor, MI 48109

ABSTRACT

Current generation robots work in a constrained environment. In most robotic applications, the environment is known and several aspects of it may be controlled. The knowledge about the environment may be used by image understanding algorithms to facilitate the recovery of information from images. Some difficult problems faced by general image understanding systems are simplified by developing techniques that exploit the available knowledge about the environment. We demonstrate the role of such knowledge in robotic applications for recovering information in dynamic scenes. In the first application, the knowledge of ego-motion parameters of a mobile robot is used for segmentation of a scene and recovery of depth information. In the second application, a hypothesize and test approach is used to find road edges in real scenes for an autonomous vehicle.

INTRODUCTION

An image understanding system recovers useful information from one or more images of a scene. The useful information is application dependent. Systems designed for specific applications may use knowledge about the applications, the environment in which the system works, and the specific sensors used. If the knowledge about the application is embedded in the system, the maintenance and modifications to cope with slight variations in the application may cause serious problems for the system. It is desirable to use only general knowledge about the application domain in a system. If this general knowledge is used explicitly in the system, then the system will have enough flexibility to cope with normal variations encountered in the application. The knowledge used by an image understanding system does not, necessarily, have to be high-level knowledge about the objects appearance in the scene.

It is well recognized that information about the 3-D world can be recovered from images using constraints. A constraint explicitly states an assumption (or knowledge) that plays a vital role in the recovery process. The last decade has seen emergency of many powerful techniques that exploit these constraints to recover the 3-D shape of surfaces. Most research in discovering appropriate constraints has addressed general purpose vision systems. This research has resulted in identifying powerful constraints about the nature of surfaces, illumination, and rigidity of objects. Since many researchers are interested in discovering constraints that are useful in the human visual system also, the constraints related to camera parameters and camera motion did not attract much attention. In robot vision however, known camera parameters will play a primary role in the recovery of information.

In many robotic applications, the camera is mounted at a fixed and known location with respect to the workplace. If the objects are on a known surface, it is possible to use the camera parameters and image position to find information about the location of the object. Illumination in the scene may also be controlled to provide good contrast. Many such techniques are commonly used in industrial applications.

With progress in image understanding and robotics, the applications are becoming complex. One of the fastest growing areas is mobile robots. Clearly, mobile robots will open several new areas for the application of robotics. A problem involved in applications of mobile robots is the fact that unlike earlier applications, the environment cannot be controlled as much. The work area of mobile robots will be much bigger and more unstructured. In some cases, mobile robots may even work in an outdoor environment. If such robots are expected to navigate unknown terrains, the vision task may become quite complex. There may be moving obstacles and the robot should avoid them.

It may be difficult to control the environment for mobile robots, but it is not difficult to obtain camera parameters. Moreover, it is possible to get motion parameters of the robot, and hence of the camera mounted on the robot. These parameters may be used to recover information.

In this paper, we discuss two applications of knowledge of camera parameters for mobile robots. In the first application, we discuss Ego-Motion Complex Logarithmic Mapping (ECLM) for segmenting dynamic scenes acquired by a mobile robot and show that this transformation may be useful in recovering depth of objects. The second application discusses a hypothesize-and-test approach for finding road edges.

USE OF EGO-MOTION PARAMETERS

Dynamic scene analysis is receiving increasing attention from researchers due to applications of vision in many complex situations. Most early dynamic scene analysis systems considered an image sequence acquired using a stationary camera [Jai85]; the last few years have seen increased interest in image sequences acquired using a mobile camera. As in static vision, one of the most important early processing steps is segmentation. Approaches have been developed for identifying moving objects using changes in intensity characteristics at the corresponding locations in the image sequence. When the camera remains stationary, intensity characteristics, assuming no illumination changes, will change only due to motion of objects in the scene. By detecting such changes, one may initiate processes that lead to recovering images of moving objects.

Identifying moving objects in image sequences acquired using a mobile camera is more difficult. The camera motion assigns a motion component to each point in the image. The relative motion of all objects in the scene makes the application of processes that may be used in the stationary camera case difficult. By obtaining the relative velocity of each object in a scene, it is possible to segment an image into different objects. Optical flow has been proposed as a tool to be used in scene segmentation. Optical flow is the field of velocity vectors at each point of an image. By determining optical flow, it may be possible to segment a scene [Sch85].

It has been shown that optical flow carries information about the structure of the environment and the motion of the observer [Clo80, Gib79, Lee80, Pra80]. When an observer moves through a scene, all points in the scene are in motion relative to him. Points which are close to the observer appear to move relatively faster than points which are further away. The flow vectors due to the stationary components of a scene intersect at a point, as shown in Figure 1. This point is called the Focus of Expansion (FOE). It has been shown that the FOE plays a vital role in the recovery of information from the optical flow field [Lee80,Pra80].

The last few years have seen increasing efforts to use optical flow in the analysis of dynamic scenes. Many approaches have been proposed for the computation of optical flow from images of a dynamic scene. Most of these proposed approaches consider two frames of a scene to compute the flow field. Based on the research reported in the literature, it appears that the computation of acceptable quality optical flow for real world scenes is a very difficult problem. Moreover, the methods for the recovery of the information are sensitive to the noise in the optical flow. Thus, in most realistic applications, the information obtained from the computed flow fields may not be reliable.

A problem in the computation of optical flow is that the smoothness constraint used for the determination of optical flow, implies that the information for one surface should not be propagated to other surfaces. Since optical flow is expected to give the segmentation, before optical flow is determined there is no knowledge about the surfaces in an image. Thus, a deadlock is created: without knowing surfaces, one cannot determine reliable optical flow; and without determining optical flow, one cannot find surfaces.

Ego-Motion Polar Mapping

Given that the camera motion parameters are known or can be obtained, this knowledge about the ego-motion parameters can be used to break the deadlock mentioned above. The idea is to try to simplify the problem by using the information that can be easily obtained. As shown in this section, in the case of known translational motion of the camera, it is possible to use characteristics of optical flow without actually computing optical flow.

If the camera displacement between two frames is dX, dY, dZ, then the FOE for these frames is

$$F_x = \frac{dX}{dZ} \tag{1}$$

$$F_y = \frac{dY}{dZ} \tag{2}$$

The information about the camera motion parameters, and hence about the FOE, may be used in the recovery of information. Jain tried to exploit characteristics of optical flow without actually computing the optical flow field Jai83,Jai84 . He used a transformation on images acquired using a moving camera to segment a dynamic scene into its stationary and nonstationary components. This transformation, called the Ego-motion Polar transform (EMP) is centered around the FOE and converts the original image $I(x,y)$ into an image $I^*(r,\theta)$ using

$$I^*(r,\theta) = I(x,y) \tag{3}$$

where

$$r = \sqrt{(x-F_x)^2 + (y-F_y)^2} \tag{4}$$

and

$$\theta = \arctan \frac{(y-F_y)}{(x-F_x)} \tag{5}$$

This transformation is shown in Figure 2.

It is shown in [Jai84], that for a moving observer, all stationary points in a scene will show only horizontal displacement in the EMP transformed image. This fact can be used to determine whether an object is moving or not. As shown in Figure 3, the apparent motion of stationary points is converted from assorted directions, depending on their locations in the image plane, to unidirectional motion in the EMP space. The stationarity of an object is judged by the absence of a θ component for the region corresponding to

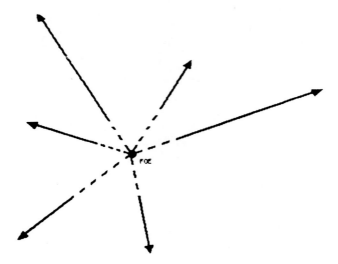

Figure 1 - The optical flow field for a moving observer in a stationary environment. The vectors intersect at the focus of expansion

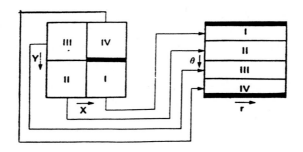

Figure 2 - An image I(x,y) is converted to another image I(r,θ) using the EMP transform

the object in the EMP space. An algorithm was developed to implement this scheme. The results for real world scenes are reported in [Jai84].

Complex Logarithmic Mapping

Schwartz showed that the retino-striate mapping can be approximated using a Complex Logarithmic Mapping (CLM). Retino-striate mapping, a common feature of vertebrate sensory information processing, is a spatial mapping of the peripheral sensory receptive

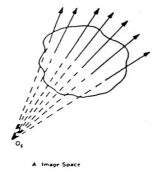

A Image Space

B EMP Space

Figure 3 - The assorted directions of the velocity vectors for a stationary object are transformed to one direction in the transformed image.

surfaces onto corresponding parts of the central nervous system [Sch80, Sch81]. In our own human vision system, as well as those of lower animals, it has been found that the excitement of the striate cortex can be approximated by a Complex Logarithmic Mapping (CLM) of the eye's retinal image. In other words, what we see as the real world and what is focused on the retinas of our eyes, is reconfigured onto the striate cortex by a process similar to complex logarithmic mapping [Sch80,Sch81] before it is examined or interpreted in our brain. Schwartz further argued that this mapping is responsible for the scale, rotation and projection invariances in the human visual system. As is well known, these invariances play a vital role in human visual perception. Cavanaugh [Cav78,Cav81], however, showed that Schwartz's claims about the CLM resulting in the invariances are correct only under certain conditions. The rotation and scale invariances are obtained if the object is in the center of the image and the rotation and scale changes are with respect to the origin. The projection invariance is obtained only if the direction of the observer's gaze and motion are the same.

Let us look at the mathematical definition of CLM. Complex log mapping may be written mathematically as

$$w = \log z \tag{6}$$

where w and z are complex variables:

$$z = x + iy = r(\cos\theta + i\sin\theta) = re^{i\theta} \tag{7}$$

and

$$w = u(z) + iv(z) \tag{8}$$

In this way, a function or image in z-space with coordinates x and y is mapped to w-space with coordinates u and v. The mapping is obtained from the simplified equations:

$$u(r,\theta) = \log r \tag{9}$$

$$v(r,\theta) = \theta \tag{10}$$

There are many attractive features of this mapping ChW79, BGT79, SaT80. From the psychological viewpoint, it is the only analytic function which maps a circular region, such as an image on the retina, into a rectangular region. This is a desirable feature for the study and modelling of the human visual system. The mapping of two regular patterns are shown in Figure 4 to result in similarly regular patterns. It is seen in Figure 4a that concentric circles in an image or the z-plane become vertical lines in the mapped w-plane. This becomes obvious when one examines the CLM definition above. A single circle maps to a single vertical line since the constant radius, r, at all angles, θ, of the circle gives a constant u coordinate for all v coordinates in the mapped space. Similarly in Figure 4b, an image of radial lines which have constant angle but variable radii, result in a map of horizontal lines.

Through these mappings, we can demonstrate some of the invariances of CLM that may be helpful in image understanding. The first such invariance is that of rotation. In Figure 4a, we saw that for a circle, all possible angular orientations of a point at the given radius will map to the same vertical line. Thus, if an object is rotated between successive images, this will result in only a vertical displacement of the mapped image. This same result can be seen in Figure 4b. As a radial line rotates

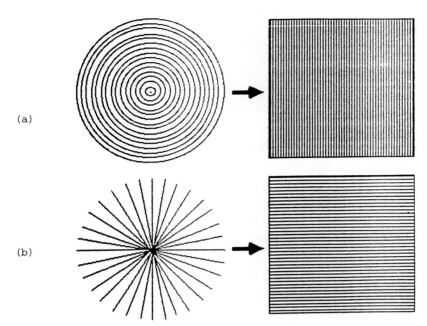

Figure 4 - The CLM results in the transformation of certain regular patterns in the z-plane into another regular pattern in the w-plane

about the origin, its entire horizontal line mapping moves only vertically.

Another characteristic of CLM is the size invariance. This also can be seen in Figure 4. As a point moves out from the origin along a radial line in Figure 4b, its mapping stays on the same horizontal line moving only from left to right. The mappings of the concentric circles of Figure 4a remain vertical lines and only move horizontally as the circles change in size.

A third important invariance is that of projection. When an observer translates in space, the images of objects appear to remain unchanged. Thus, though the images of stationary objects do change on the retina, the object perceived on the striate cortex does not change. This is due to the fact that in the CLM space, translation of the observer only causes the object image to be displaced in the horizontal direction; the size and shape of the object image remain unchanged.

Ego-Motion Complex Log Mapping

To achieve the invariances, which are so important, the images must be obtained under certain constraints. The scale and rotation invariances are present only if the object is centered in the image, and the scale and rotation changes are with respect to the origin. In other cases, these invariances are not obtained. The projection invariance is only obtained by a camera translating along its optical axis. In this case the direction of the gaze and the direction of the motion are the same. This is a serious constraint. Indeed, in this case the FOE is (0,0) and hence the projection invariance really is the same as the scale invariance. If the observer motion is translational and is known, then the FOE is also known. The CLM is then taken so that all radii, r, and angles, θ, are in reference to this calculated FOE. This transformation is called Ego-Motion Complex Logarithmic Mapping (ECLM), since the mapping is performed with regard to the motion of the camera/observer. Let us consider this transformation for a point in the 3-D space.

When the observer moves in the direction of his gaze, the (X, Y, Z) coordinates of objects which are stationary relative to the observer, change only in the Z coordinate. With the perspective projection, the invariance resulting from the ECLM gives only a horizontal displacement between images for cooresponding points.

For a stationary point in the environment, with real world coordinates (X,Y,Z) relative to the observer at a time instant, the perspective projection (x,y) of this point onto the image plane is given by

$$x = \frac{X}{Z} \tag{11}$$

$$y = \frac{Y}{Z} \tag{12}$$

assuming that the projection plane is parallel to the XY plane at Z=1. For a translational motion along the direction of the gaze of the observer, the relationship between the distance, r, of the projection of the point from the FOE, and the distance, Z, of the point from the observer is

$$\frac{dr}{dZ} = \frac{d\sqrt{x^2+y^2}}{dZ} = -\frac{r}{Z} \qquad (13)$$

By the chain rule

$$\frac{du}{dZ} = \frac{du}{dr} * \frac{dr}{dZ} \qquad (14)$$

and from the equation (9),

$$\frac{du}{dr} = \frac{1}{r} \qquad (15)$$

Therefore, we have

$$\frac{du}{dZ} = \frac{1}{Z} \qquad (16)$$

Similarly, to find $\frac{dv}{dZ}$,

$$\frac{d\theta}{dZ} = \frac{d(\tan^{-1}\frac{y}{x})}{dZ} = 0 \qquad (17)$$

and

$$\frac{dv}{dZ} = \frac{dv}{d\theta} * \frac{d\theta}{dZ} = 0 \qquad (18)$$

In equation (16) we see that the depth, Z, of a point can be determined from the horizontal displacement, du, in the ECLM for that point, and from the velocity, dZ, of the observer. Furthermore, the axial movement of the observer will result in only a horizontal change in the mapping of the image points since dv/dZ=0. There will be no vertical movement of the mapped points and thus correspondence of points between the two stereo pictures will become easier. Note that this is similar to the epi-polar constraint used in the lateral stereo. Now assuming that there is sufficient control of the camera to be able to determine the amount of its movement, both variables necessary to determine image depths are readily available. Thus, it is possible to recover depth, in principle, if the camera motion is along its optical axis.

What is more interesting is that the depth can be recovered using the above technique even if the camera motion is not along its optical axis. It is shown in [JB086] that even for arbitrary translation motion of an object, if the mapping is obtained with respect to the FOE, then the displacement in the u direction depends only on the Z coordinate of the point.

Another interesting feature of this stereo approach is that, if required, we can obtain many frames for solving ambiguities that cannot be resolved based only on two frames. Moravec [Mor81] developed a technique for interpolating over nine frames which he used with common stereo. This technique may be even more applicable to motion stereo, because the series of frames can be naturally extended each time the observer moves. The frame sequence can be constantly updated by merely pushing back the current series by one time instant and adding a new frame to the front of the sequence.

Experiments Using ECLM

To study the efficacy of the proposed research for real scenes, we performed several experiments in our laboratory. We mounted a camera on a PUMA robot. This set up allows us to move the camera in a desired direction by a desired amount.

The objects used were wooden blocks with dimensions less than 6 inches. We were forced to place objects within a small depth of each other due to the limited depth of field of the camera. By placing objects far away, we could do some more experiments, but our laboratory set-up did not allow us to perform these experiments. Moreover, we were more interested in studying the efficacy of the proposed approach, which could be done by considering a limited set of scenes. In the first image the lower right corner of the farthest block was placed in the center of the image. A paper triangle was mounted on a block and positioned so an acute angle lined up with the block corner. The position of the triangle was marked and experiments were done to guarantee the reproduceability of its position. The image was obtained with the triangle removed, then the camera moved 6 inches forward. The triangle was replaced and the camera location adjusted so the acute angle and the corner of the block lined up. This process was repeated until 5 images had been obtained. The 512 x 512 images were shrunk to 128 x 128. Three frames of the sequence obtained are shown in Figure 5.

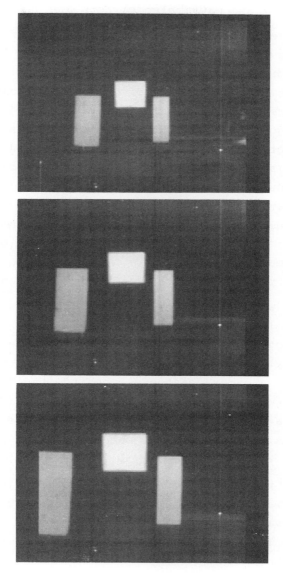

Figure 5 - Three frames of the laboratory sequence used in our experiments

Corners were found and the coordinates in complex log space of these points were calculated. The corner detector used simple masks to find corners, so the corner images produced were noisy. No effort was made to remedy this since it served to make the problem more challenging. The corners obtained using our algorithms are shown in Figure 6. Note that the location of the corners is noisy and many false corners are detected.

The corners are mapped to the ECLM space. Figure 7 shows the corners of Figure 6 mapped in a composite frame in the ECLM space. The correspondence between corners was established in the ECLM space. For each pair of images, the coordinates in the ECLM space were calculated for all the "corners" and saved in a list. The algorithm knows the number of objects and the corners for each object. The number of corners detected in different images varies significantly, however. Theoretically, two matching points should have the same θ value (v coordinate). Due to digitization error, blurring in the images from shrinking, and errors in corner detection, matching points do not have the same θ value. So a threshold for dv (the difference between the θ values) is used. Some heuristics were used to establish correspondence for points that are within the selected threshold.

The algorithm was run on every pair of images using three different thresholds for dv: 0.01, 0.02, and 0.05. A match is considered correct if the points belong to the same object. By using two simple rules, all errors are detected by the system. The rules are: a camera cannot see behind itself and if you pass an object, it won't be in the second image. This eliminates all matches with a distance less than the amount the camera moved. Figure 8 shows some examples of matches and the results of applying the rules to the matches.

The effect on the average depth determination with a tighter bound on allowable depths for individual pairs was also studied. An arbitrary value larger than the distance the camera moved was chosen as the lower bound. An arbitrary value of 100 was chosen as the upper bound.

The depth of an object was obtained by averaging the depth obtained for its corners. As the threshold increased for dv, the number of pairs that "match" increases as does the match error rate. However, since the system is so good at finding the erroneous matches, the extra information from the larger number of points makes the distance deterioration more accurate, overall.

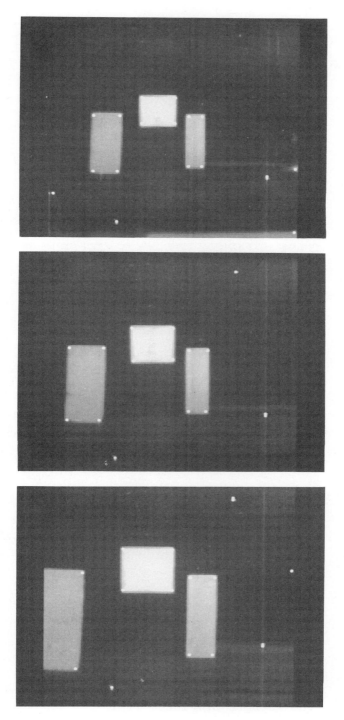

Figure 6 - Corners detected in the frames. Note the poor quality of the corners.

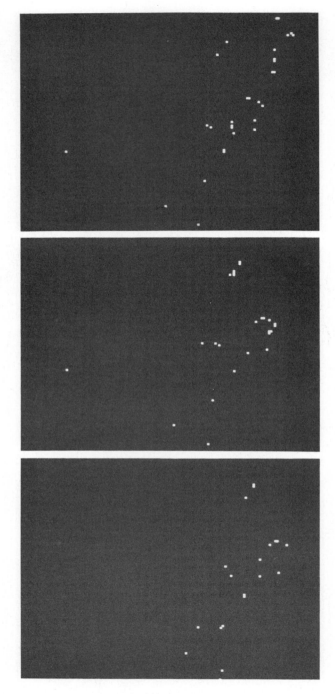

Figure 7 - The corners of Figure 11 in the ECLM space

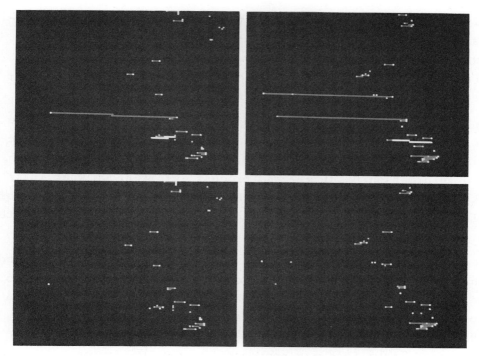

Figure 8 - Match found by the algorithm. Images on the left are for frames 1 and 3. Images on the right are for frames 2 and 4. The top images show all the matches found. The bottom images show the corrected matches.

The calculated depths tended to be lower than the real depths. We did not use the focal length of the camera in our depth computation, since for the camera we used, the focal length is not known. Without the focal length, the depth values should indicate the relative depth values, rather than the absolute depth. Depth determination from images where the camera was closer to the objects was more accurate. Larger camera movement gave better results when the camera was far from the objects, but not when the camera was closer. We ran the experiments for several frame pairs. In some cases, the results indicated wrong depth order for the objects.

HYPOTHESIZE-AND-TEST APPROACH

Often a complex problem can be simplified by formulating it a hypothesize-and-test approach. The hypothesize-and-test method is a form of goal-directed, or top-down, approach. Instead of analyzing an image to recover certain information, in the hypothesize-and-test approach one assumes that the information is known, and then analyzes the image to verify the assumption. The advantages and disadvantages of top-down and bottom-up approaches are well known, and will not be discussed here.

The hypothesize-and-test approach is very attractive in robotic applications. Particularly, if a robot is working in a dynamic environment, then its actions are purposive. It does not receive information passively and process it; it knows what it is looking for and acquires only relevant information. The role of a vision system in this application is to provide the robot with the relevant information.

Here we discuss the hypothesize-and-test approach to detect road boundaries for an autonomous vehicle. Of course the goal is to have the vehicle remain on the road. Many approaches have been proposed to detect road boundaries [Dak85,GSC79,IMB84,WMS85, WST85,YIT83]. Most approaches apply an edge detector and then try to find lines that may approximate road boundaries.

By simplifying the problem using some ideas of the Vanishing Point (VP) analysis, we propose an algorithm called the Vanishing-Point Road-Edge Detector to extract the roadway boundaries. In order to avoid using any data-driven line fitting or approximation technique in the algorithm, a set of convergent lines is used to speed up the process. Essentially, for a hypothesized vanishing point we collect evidence supporting it from the image. The supporting evidence comes from the strength and direction of edge points and length of line segments. We define functions to measure the support of each evidence and combine them in the performance measure that gives confidence in the VP. The VP having highest confidence is selected and the corresponding convergent lines are considered as road boundaries.

Camera Model

A camera mounted on the front of the vehicle acquires the image. Use of this image to find the roadway or measure distance requires an explicit camera model. The camera models have been extensively discussed in [DuH73, NeS79]. The version we use in our paper is based on [DuH73], with a change in coordinate axes. The camera is translated from the origin, panned through an angle θ, and tilted through an angle ϕ. Also, we use the two coordinate systems -- global coordinate system formed by (X,Y,Z) and picture coordinate system represented by (X',Y') -- to rectify the camera centric situation. In order to use a linear matrix operator to represent a nonlinear projection, the homogeneous coordinates are commonly used to represent both object and picture points. For convenience, we write the offset l, the vector from gimbal center to the lens center, as $(l_1, l_2+f, l_s)^t$.

Applying the system geometry stated above to the direct transformation equations presented in [DuH73] gives the coordinates of a point (x',y',z) in an image as

$$x' = f \frac{(x-x_o)\cos\theta + (z-z_o)\sin\theta - l_1}{-(x-x_o)\cos\phi\sin\phi + (z-z_o)\cos\phi\cos\theta + (y-y_o)\sin\phi - l_2} \qquad (19)$$

Road Model

We assume that the road segments can be approximated by straight lines. Thus, segments of road boundaries can be modelled as:

$$y = mx + b \qquad (20)$$

where m is the slope of roadway boundaries and b is a constant.

Note that for controlling the vehicle, the boundaries near the vehicle are more important. In most cases, the assumption of linear boundaries is satisfied reasonably well for the boundaries near the vehicle.

Vanishing Point Analysis

The perspective projection of any set of parallel lines which are not parallel to the image plane converges to a vanishing point. The location of the vanishing point depends on the orientation of the lines. Thus, the location allows us to constrain the possible positions of roadway boundaries and other known structures in an image. Determination of the vanishing points has, therefore, attracted the attention of many researchers [Bar82, MaA84].

Problem Requirements and Analysis

To find edges, one approach is to detect all edge points in a frame and then interpret these points. Conventional edge detectors do not provide accurate solutions to the roadway boundary location problem without some kind of line fitting operation after edge detection. Since most techniques for line fitting are slow, the task of roadway boundary location in real time is usually difficult.

To simplify the problem, low curvature roadway boundaries and locally level ground driving are assumed. In that case, the tangent direction to the roadway edges, at any time instant, can provide sufficient information for vehicle guidance. Let us define the road vanishing point at time t, denoted by RVP_t, to be a vanishing point at which the tangents to the current roadway boundaries at time t will converge.

Now, since the location of the vanishing point does not change much between two contiguous frames, we can assume that its location in the current frame will be about the same as in the previous frame. This fact allows combination of road edge detection and vanishing point determination into one step. By assuming a search window around the previous-frame position of the vanishing point, we may find the best vanishing point in the current frame by using a measure based on the edges and their characteristics in the frame.

If the exact tilt angle is detected by some physical instruments as δ, the search space of RVP at time t is easily determined as (see Figure 9)

$$S_t = \left\{(x,y) \mid |x - RVP\_X_{t-1}| \leq w \text{ and } y = \frac{-f\sin\delta}{\cos\delta}\right\} \quad (21)$$

where w is a constant and RVP_X is the x coordinate of RVP.

$$y = \frac{-f\sin\delta}{\cos\delta} \quad (22)$$

is the so-called vanishing line [SKK82].

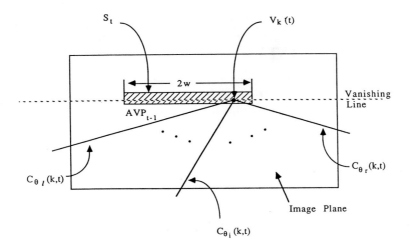

Figure 9 - Convergent Lines

Now, if the location of a vanishing point for a road is given, then we may locate the road boundaries by considering all possible convergent lines to this point. We may use the information about the width of the road and the camera model to select the appropriate pair of lines as the correct road boundaries. These road boundaries may be verified against the ground truth, i.e. the image. This verification will involve studying the image to find support for the presence of the road boundaries.

In the proposed approach, a window containing the correct VP is given. To find the best VP, we use a maximum approach. The point that has maximum support in the image is selected as the RVP.

Since the RVP is the point where the extensions of the two most obvious straight edges, i.e. the left and right roadway boundaries, meet, our performance measure should try to detect the two best converging straight lines. Due to noise and other problems, parts of lines may not be present. The factors influencing the performance measure should be based on:

1) length of the line,
2) average gradient magnitude of all the points on the line, and
3) consistency of the direction of each point on the line

that converge to the hypothesized vanishing point.

Besides, the position of the line also plays a vital role in determining road edges. We decided to use the following four factors which dominate the RVP determination problem:

(a) the length of the detected line formed by continuous line segments associated with a vanishing point.
(b) the positions of these detected lines.
(c) the average gradient magnitude of the points on the detected lines.
(d) the consistency of those line segments on a detected line.

Unlike most approaches to controlling the vehicle by finding the road center, the proposed technique may even simplify this computation. The turning angle can be obtained by a simple lookup table, in which each entry indicates the value we got from $RVP_t - RVP_{t-1}$. Since this paper is only concerned with the solution to the roadway location problem, we will not discuss the control problem in more detail.

Proposed Vanishing-Point Road-Edge Detector

The proposed algorithm can be implemented on multiprocessors such that each processor essentially works using a hypothesized vanishing point.

According to Eq. (21) and within a preset tolerance a, the coordinates of all possible vanishing points can be calculated as:

$$x_k = RVP\_X_{t-1} - w + Ka \qquad (23)$$
$$y_k = \frac{-f \sin \delta}{\cos \delta} \quad \forall \ k = 0, \frac{2w}{a}$$

Let us denote that kth vanishing point at time t as $V_k(t)$ and define the Convergent Line with an angle s, represented by $C_s(k,t)$ to be a line segment that crosses the line plane and its extension passes the vanishing point $V_k(t)$.

If we assume that both roadway boundaries are located in the range from θ_l to θ_r, the number of convergent lines associated with any vanishing point, for the given resolution θ_a, is

$$\frac{\theta_l - \theta_r}{\theta_a} + 1$$

And the coordinates of the points which are located on these different convergent lines can also be calculated before the vehicle begins moving.

Let us define Pixel Element $(d,m)_{i,j}$ to be a pair of data derived from the estimate of the gradient magnitude m and edge orientation d at pixel location (i.j) and let ι_r be the adjustment ratio used for the safety purpose.

According to Eq. (23), for the VPRE detector algorithm, $\frac{2w}{a} + 1$ processors are used to achieve the parallel performance. This algorithm is stated as

1. For processor k, $(k=0, \frac{2w}{a})$, we start from $V_k(t)$ at time t
   * produce the corresponding pixel elements
   * scan the image and get the next convergent line $C_s(k,t)$ as illustrated in Figure 2
   * compute the performance measure of each convergent line
   * divide all the line sets into two categories. One is the set of all lines with positive slopes, and the other is the one with negative slopes
   * for each category,
     * find the line set i with the maximum performance measure m, of all line sets
     * if there exists a line set j with performance measure m_j such that
     $$m\_set\_found^{-m}j| \leq \iota_r * m_{set\_found} \qquad (24)$$
     Set set_found to i, value_found to m_i
     * Line set set_found is the roadway boundary in this category
   * Set the performance measure of $V_k(t)$ to be the minimun of the performance measures in these two categories
2. Select the vanishing point which has the maximum performance measure as the RVP and its corresponding detected lines in each category as the road edges.

Implementation Details

Due to good computational efficiency and reasonable performance, the Sobel edge finder is used to compute the direction and magnitude of the gradient at a point i.j, where the intensity is $f_{i,j}$. Thus, we define the partial derivative estimates

$$S_x(i,j) = f_{i+1} + 2f_{i+1,j} + f_{i+1,j-1} - f_{i-1,j+1} - 2f_{i-1,j} - f_{i-1,j-1} \qquad (25)$$

$$S_y(i,j) = f_{i-1,j+1} + 2f_{i,j+1} + f_{i+1,j+1} - f_{i-1,j-1} - 2f_{i,j-1} - f_{i+1,j-1} \qquad (26)$$

The S_x and S_y partial derivative estimates are then combined to form an estimate of the gradient magnitude m, and direction d by

$$m = |S_x| + |S_y|$$
$$d = \tan^{-1}\left[\frac{S_y}{S_x}\right] \qquad (27)$$

The magnitude and direction maps are both quantized to eight bits to form the pixel element.

Now let us define the Edge Point to be a point whose pixel element is (d,m) such that for a convergent line $C_s(k,t)$

$$|d-s| \leq t_8 \text{ and } m \geq t_s \qquad (28)$$

where t_8 and t_s are both threshold values.

Note that we are interested only in those edge points that are on convergent lines. Other points will not enter into considerations irrespective of their strength. Next, we form line segments out of these edge points. We use t_d to represent the allowed maximum length of continuous non-edge points, and t_1 to mean the minimum length of continuous edge points to form a line segment.

From all points along $C_s(k,t)$ for some s and $V_k(t)$, we extract all the qualified line segments. Finally, a decision is made over all convergent lines by a measure of the system's confidence is how well the lines is formed.

For finding good lines, we used the following performance measure functions with normalized value in a range from 0 to 1.

A. The length of the detected line formed by continuous line segments associated with a vanishing point

$$\text{Leng}(x,k) = \frac{D_{xk}}{D_{Max}} \quad (29)$$

Where D_{xk} = total length of all line segments detected along the convergent line, $C_x(k,t)$, and D_{Max} is the maximum length value for all possibilities.

B. The position of these detected lines

$$\text{Close}(x,k) = \frac{1}{1+q_y} \quad (30)$$

where q_y is the y coordinate of the points which is nearest to the X axis and also located on the detected lines.

C. The average gradient magnitude of the points on the detected lines

$$\text{Mag}(x,k) = \frac{\sum_{y \in E_x} S_y}{255 N(E_x)} \quad (31)$$

where S_y means the gradient magnitude component of the pixel element $(d,s)_y$ and $N(E_x)$ means the amount of all edge points found along the convergent line $C_x(k,t)$.

D. The consistency of those line segments on a detected line whose direction is θ_x

$$\text{Cons}(x,k) = \frac{\sum_{y \in E_x} \frac{\pi - 2|d_y - \theta_x|}{\pi}}{N(E_x)} \quad (32)$$

where d_y means the direction component of the pixel element $(d,m)_y$

The performance measure $\lambda(x,k)$ is then defined as

$$\lambda(x,k) = w_1 \cdot \text{Leng}(x,k) + w_w \cdot \text{Close}(x,k) + w_s \cdot \text{Mag}(x,k) + w_4 \cdot \text{Cons}(x,k) \quad (33)$$

where W_i, i=1..4 are the weights and the value of x and k are associated with a convergent line $C_x(k,t)$ at time instant t.

And, by using λ+ and λ- to represent the measure in the positive-slope and negative-slope categories separately, the purpose is to find x and y such that

$$\lambda+(x,k) = MAX \{\lambda+(i,k)\} \quad \forall\ i \in A$$
$$\lambda-(y,k) = MAX \{\lambda-(i,k)\} \quad \forall\ i \in B \tag{34}$$

Where A and B indicate the set of convergent lines in the positive-slope and negative-slope categories separately.

and define x as

$$x(k) = MIN \{\lambda+(x,k), \lambda-(y,k)\} \tag{35}$$

By combining the information from RVP_{t-1} as well as x, the process will make the final decision on the RVP_t position. The decision is made by selecting the vanishing point $V_1(t)$ such that

$$x(1) = MAX \{x(k)\} \quad \forall\ V_k(t) \in S_t \tag{36}$$

Experimental Results

The performance of the Vanishing-Point Road-Edge Detector was tested using several real-world road image sequences.

Figures 10-12 show the original pictures and the road edges with the vanishing points found by our approach. In each Figure, (a) is the original picture. (b) and (c) are the direction and magnitude maps derived from the Sobel edge detector. The output picture with the lines showing the roadway boundaries and the box illustrating the search space is presented in (d). We found that the resulting edges are very good even for the poor contrast pictures.

CONCLUSIONS

Robot vision systems allow the use of camera parameters, in addition to other forms of knowledge, for recovering information. In this paper, we have discussed two applications of known camera parameters to vision problems. We showed that knowledge of ego-motion parameters can simplify the task of segmentation, as

Figure 10 - The first frame is Picture Sequence 1
(a) Original Picture, (b) Direction Map, (c) Magnitude Map, (d) Output Picture

Figure 11 - The first frame in Picture Sequence 2
(a) Original Picture, (b) Direction Map, (c) Magnitude Map, (d) Output Picture

Figure 12 - The first frame in Picture Sequence 3
(a) Original Picture, (b) Direction Map, (c) Magnitude Map, (d) Output Picture

well as provide depth information. We also showed that in a road navigation problem, knowledge of the camera tilt angle and the vanishing point (which should be determined in initial processing stages) will ease the determination of the vanishing point and the road edges in subsequent stages. The idea of using knowledge such as camera perameters and other information is very attractive in industrial applications of image understanding systems.

ACKNOWLEDGEMENT

Nancy O'Brien, Sandy Bartlett, and Shih-Ping Lou developed systems reported in this paper. They contributed many ideas in these projects. Moreover, it is a pleasure to have such nice friendly co-researchers. I am thankful to them.

This work was partially supported by NSF Grant No. DCR-8517251.

REFERENCES

[Bar82] Barnard, S.T., "Methods for interpreting perspective images", Proc. of the Image Understanding Workshop, Stanford University, Palo Alto, Calif., 1982, 193-203.

[Cav78]   Cavanaugh, P., "Size and position invariance in the visual system", Perception, vol. 7, pp. 167-177, 1978.

[Cav81]   Cavanaugh, P., "Size invariance: reply to Schwartz", Perception, col. 10, pp. 469-474, 1981.

[ChW79]   Chaikin, G. and C. Weiman, "Log spiral grids in computer pattern recognition", Computer Graphics and Pattern Recognition, vol. 4, pp. 197-226, 1979.

[Clo80]   Clocksin, W.F., "Perception of surface slant and edge labels from optical flow: A computational approach", Perception, vol. 9, 1980, pp. 253-269.

[DaK85]   Davis, L.S., and T.R. Kushner, "Road Boundary Detection for Autonomous Vehicle Navigation", CS-TR-1538, Center for Automation Research, University of Maryland, July 1985.

[Gib79]   Gibson, J.J., The ecological approach to visual perception, Houghton Mifflen, Boston, 1979.

[GSC79]   Giralt, G., R. Sobek, and A. Chatila, "A multi-level planning and navigation system for a mobile robot", Proc. 6th IJCAI, Tokyo, 1979, 355-337.

[IMB84]   Inigo, R.M., E.S. McVey, B.J. Berger, and M.J. Wirtz, "Machine vision applied to vehicle guidance", IEEE Trans. on Pattern Analysis and Machine Intelligence, 6, No. 6, 1984, 820-826.

[Jai84]   Jain, R., "Segmentation of frame sequences obtained by a moving observer", IEEE Trans. on PAMI, pp. 624-629, Sept. 1984.

[Jai85]   Jain, R., "Dynamic scene analysis", in Progress in Pattern Recognition, Vol, 2, Ed. L. Kanal and A. Rosenfeld, North Holland, 1985.

[Ja085]   Jain, R. and N. O'Brien, "Ego-Motion Complex Logarithmic Mapping", SPIE, Nov. 1984.

[JBP86]   Jain, R., S. Bartlett, and N. O'Brien, "Motion Stereo Using Ego-Motion Complex Logarithmic Mapping", Technical Report, RSD-TR-3-86, The University of Michigan, Feb. 1986.

[JeJ84]   Jerian, C. and R. Jain, "Determining motion parameters for scene with translation and rotation", IEEE Trans. on PAMI, Vol. 6, No. 4, July 1984, 523-530.

[Lee80]   Lee, D.N., "The optic flow field: The foundation of vision", Phil. Trans. Royal Society of London, vol. B290, 1980, pp. 169-179.

[MaA84]   Magee, M. J. and J.K. Aggarwal, "Determining vanishing points from perspective images" Computer Vision, Graphics and Image Processing, 26, 1984, 256-267.

[Mor81]   Moravec, H.P., Robot Rover Visual Navigation, UMI Research Press, Ann Arbor, 1981.

[MST85]   Massone, L. and G. Sandini and V. Tagliasco,"'Form-Invariant' Topical Mapping Strategy for 2D Shape Recognition", Computer Vision, Graphics and Image Processing, Vol. 30, pp. 169-188, 1985.

[ObJ84]   O'Brien, N. and R. Jain, "Axial Motion Stereo", Proc. of Workshop on Computer Vision, pp. 88-92, Annapolis, Maryland, April 1984.

[Pra80]   Prazdny, K., "Egomotion and relative depth map from optical flow", Biological Cybernetics, vol. 36, 1980, pp. 87-102.

[Sch80]   Schwartz, E.L., "Computational anatomy and functional architecture of striate cortex: a spatial mapping approach to coding", Vision Research, 20, 1980, pp. 645-669.

[Sch81]   Schwartz, E.L., "Cortical anatomy, size invariance, and spatial frequency analysis", Perception, vol. 10, pp. 455-468, 1981.

[Sch85]   Schunck, B.G., "Image Flow: Fundamentals and Future Research", Proc. of IEEE Conf. on Computer Vision and Pattern Recognition, San Francisco, June 1985.

[WMS85]   Waxman, A. M., J. Le Moirne, and B. Srinivasan, "Visual navigation of roadways", Proc. 1985 IEEE International Conf. on Robotics and Automation, St. Louis, March 1985, 862-867.

[WST85]   Wallace, R., A. Stentz, C. Thorpe, H. Moravec, W. Whittaker and T. Kanade, "First Results in Robot Road-Following", Proc. of 9th IJCAI, Los Angeles, CA, Aug. 18-23, 1985, 1089-1095.

[YIT83]   Yachida, M., T. Ichinose, and S. Tsuji, "Model-guided monitoring of a building environment by a mobile robot", Proc. 8th IJCAI, Munich, 1983, 1125-1127.

MACHINE-INTELLIGENT ROBOTS: A HIERARCHICAL CONTROL APPROACH

George N. Saridis
ECSE Department
Rensselaer Polytechnic Institute
Troy, New York 12180

1. INTRODUCTION

Intelligent Machines capable of performing autonomously in uncertain environments, have imposed new design requirements for modern engineers. New concepts, drawn from areas like Artificial Intelligence, Operations Research and Control Theory, are required in order to implement anthropomorphic tasks with minimum intervention of an operator. This work deals with the definition of Hierarchically Intelligent Control and the Principle of Decreasing Precision with Increasing Intelligence. A three level structure representing Organization, Coordination and Execution will be developed as a probabilistic model of such a system and the approaches necessary to implement each one of them will be discussed. Finally, Entropy will be proposed as a common measure of all three levels and the problem of Intelligent Control will be cast as the mathematical programming solution that minimizes the total Entropy.

2. INTELLIGENT CONTROL: A THEORETICAL REVIEW

Intelligent Control systems were defined by Fu in 1971, as the area beyond adaptive, learning and self-organizing control systems, where artificial intelligence automatic control and operations research meet. In the 1970's adaptive, learning and self-organizing controls used mathematical methods borrowed from behavioral sciences to control complex engineering systems (Saridis, 1977).

However, self-organizing controls were not sufficiently powerful to drive systems to perform anthropomorphic tasks in structured or unstructured but unfamiliar environments with minimum interaction with a human operator. New methodologies proposed during the 1970's to handle advanced machine intelligence and decision making, are based on the computational capacity of modern computers in conjunction with sophisticated execution hardware, and come under the name of <u>Intelligent Control Systems</u> (Saridis, 1979, 1980).

Several researchers have proposed heuristic techniques based on the ability of the computer to reason and function in a logical manner. Such methodologies come under the general category of <u>Artificial Intelligence</u> (Albus, 1975) (Rosen & Nilsson, 1967), (Winston, 1977). Other researchers have been trying to apply system theoretic techniques to implement anthropomorphic theoretic techniques to implement anthropomorphic functions. Such techniques come under the category of <u>General Systems Theory</u> and require a very sophisticated mathematical formulation of the problem (Klir, 1976), (Mesarovic, 1970).

The approach proposed by Saridis is a different one and applies to intelligent robots as well as any Advanced Automation process where minimal interaction with a human operator is required (Saridis, 1983). It involves a mathematical formulation of the problem by combining the methodologies of artificial intelligence, operations research and mathematical system theory (Saridis, 1977, 1979, 1980). A precise definition is given in the sequel.

Intelligent Control Theory utilizes the powerful high-level decision making of the digital computer with advanced mathematical modelling and synthesis techniques of system theory to produce a unified approach suitable for the engineering needs of the future. They may be thought of as the result of intersection of three major disciplines: <u>Artificial Intelligence</u>, <u>Operations Research</u> and <u>Control Theory</u>, as shown in Figure 1.

One of the most demanding applications of intelligent control theory is in manipulative and guidance systems. These systems may involve the control of a general-purpose manipulator for battlefield support services like surveillance and maintenance for space exploration, like the Mars rover, or a hazardous-environment robot for operation in a nuclear containment, or a hospital-aid manipulator, or an electrically-driven prosthetic limb to replace an amputated arm. Such devices impose special conditions on the design process such as light weight, small physical dimensions, real time performance, human limb appearance and functionality, and most restrictive, a small number of non-interactive command sources, such as a command vocabulary and a small number of sensors. The above constraints exclude computationally complex algorithms or long computation time. Hence, such systems must maximize flexibility of performance subject to a minimal input dictionary and minimal computational complexity. A particular theory in intelligent controls is presented in the sequel.

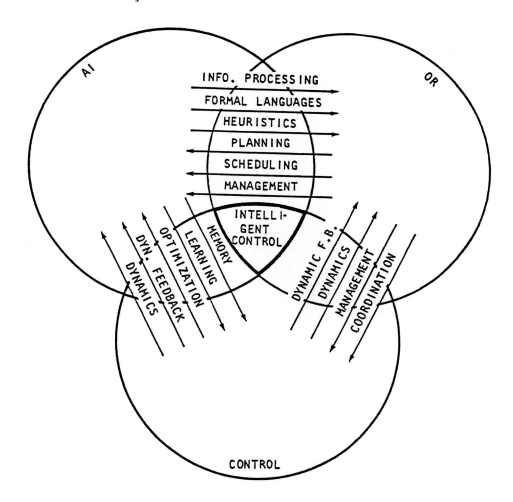

Fig. 1. Intersection of Artificial Intelligence Operations Research, and Control Theory and the Resulting Intelligent Control.

3. HIERARCHICALLY INTELLIGENT CONTROL: A REALIZATION OF INTELLIGENT MACHINES

In order to solve the modern technological problems that require control systems with intelligent functions such as simultaneous utilization of a memory, learning or multilevel decision making in response to "fuzzy" or qualitative commands, <u>Intelligent Controls</u> have been developed by Saridis (1977, 1979). They utilize the results of cognitive systems research effectively with various mathematical programming control techniques (Birk & Kelley, 1981).

Cognitive systems have been traditionally developed as part of the field of artificial intelligence to implement, on a computer, functions similar to the ones encountered in human behavior (Albus, 1975, Minsky, 1972, Winston, 1977, Nilsson, 1969). Such functions as speech recognition and analysis, image and scene analysis, data base organization and dissemination, learning and high-level decision making, have been based on methodologies emanating from a simple logic operation to advanced reasoning as in pattern recognition, linguistic and fuzzy set theory approaches. The results have been well documented in the literature.

Various pattern recognition, linguistic or even heuristic methods have been used to analyze and classify speech, images or other information coming in through sensory devices as part of the cognitive system (Kelley, 1979). Decision making and motion control were performed by a dedicated digital computer using either kinematic methods, like trajectory tracking, or dynamic methods based on compliance, dynamic programming or even approximately optimal control (Saridis and Lee, 1979).

A <u>Hierarchically Intelligent Control</u> approach has been proposed by Saridis (1979), as a unified theoretic approach of cognitive and control systems methodologies. The control intelligence is hierarchically distributed according to the <u>Principle of Decreasing Precision with Increasing Intelligence</u>, evident in all hierarchical management systems. They are composed of three basic levels of controls even though each level may contain more than one layer of tree-structured functions (Figure 2):

1. The organization level.
2. The coordination level.
3. The execution level.

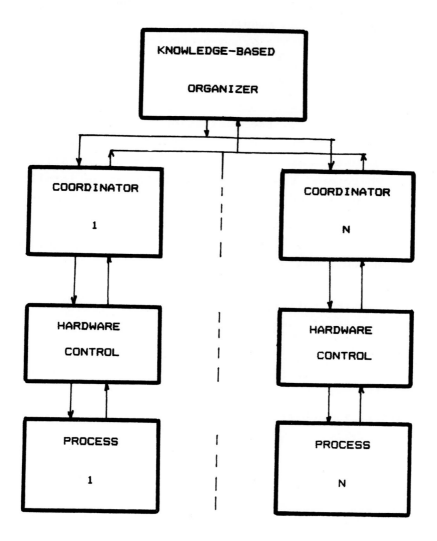

Fig. 2. Hierarchical Intelligent Control System.

The Organization level, which should perform such operations as planning, high level decision from long term memories. It may require high level information processing like the knowledge based systems encountered in Artificial Intelligence (Albus, 1975, Winston, 1977) which requires large quantities of knowledge processing with little or no precision.

The functions involved in the upper levels of an intelligent machine are imitating functions of human behavior and may be treated as elements of knowledge-based systems, as in Hayes-Roth, et. al. (1983). Actually, the activities of planning, decision making, learning, data storage and retrieval, task coordination, etc. may be thought of as knowledge (information) handling and management. Therefore, the flow of knowledge in an intelligent machine may be considered as the key variable of such a system.

Knowledge is the function of removing the ignorance or uncertainty in the operation of an intelligent machine and may be measured by entropy, which is a measure of uncertainty.

The Rate of Knowledge F is related to the flow of knowledge in the machine, has the ability to reduce uncertainty and it is a measure of its Intelligence.

The last one is measured in terms of entropy rates and must satisfy the generalized partition law of information rates, (Conant, 1976),

$$F = F_T(x,y) + F_B(x,y) + F_C(x,y) + F_D(x,y) + F_N(x,y) \quad (1)$$

where x are the inputs and y are the states of the machine and

$\quad$ F symbolizes the total rate and

$\quad F_T$ is Throughput Rate,

$\quad F_B$ is Blockage Rate,

$\quad F_C$ is Coordination Rate,

$\quad F_D$ is Internal Decision Rate,

$\quad F_N$ is Noise Rate.

Knowledge flow in a knowledge-based system is composed of

1. Knowledge Representation
2. Reasoning

3. Cognition

4. Languages

In an intelligent machine's organizational level, knowledge flow represents respectively,

1. Data Handling and Management performed through the computer memory.
2. Planning and Decision performed by the CPU.
3. Sensing and Data Acquisition obtained through I/O's.
4. Formal Languages which define the software.

Subjective probalistic models are assigned to the individual function and their <u>entropies</u> may be evaluated for every task executed, thus providing an analytic measure of the total activity.

The equivalence of knowledge representation is given in Figure 3.

Artificial Intelligence methods also applicable for the processing of knowledge and knowledge rates of the organization level of an intelligent machine have been developed by Meystel and his colleagues at Drexel (1985), and are discussed in a separate part of this paper.

<u>The coordination level</u> is an intermediate structure serving as an interface between the organization and execution level.

It is involved with coordination, decision making and learning on a short term memory, e.g. a buffer. It may utilize <u>linguistic decision schemata</u> with learning capabilities defined in Saridis and Graham (1984), and assign subjective probabilities for each action. The respective entropies may be obtained directly from these subjective probabilities.

A decision schemata is a software device that maps a string from an input language $L(G_i)$ to each possible string belonging to one or more output languages $L(G_{oj})$

$$L(G_i) = \left\{\begin{matrix} x_1 \\ \vdots \\ x_n \end{matrix}\right\} \rightarrow P\left(\bigcup_{j=1}^{m} L(G_{oj})\right) = \left\{\begin{matrix} y_{11} \\ \vdots \\ y_{1m} \end{matrix} \cdots \begin{matrix} y_{n1} \\ \vdots \\ y_{nm} \end{matrix}\right\} \quad (2)$$

Fig. 3. Equivalence of Knowledge Representation Techniques

	KNOWLEDGE			
KNOWLEDGE-BASED EXPERT SYSTEMS	KNOWLEDGE REPRESENTATION	REASONING	NATURAL LANGUAGES	COGNITION
INTELLIGENT MACHINES	DATA MANAGEMENT	INFERENCE	FORMAL LANGUAGES	SENSING
DIGITAL COMPUTER	MEMORY	CPU	OPERATING SYSTEM	I/O DEVICES

To each mapping there is associated a subjective probability $P\{P_{ijk}, i=1,\ldots,n, j=1,\ldots l, k=1,\ldots m\}$ which may be used to select the paper output string related to the particular input string.

Given the ith input (command) the entropy associated with the output of the schemata is:

$$H_i = \sum_j \sum_k P_{ijk} - \sum_j \sum_k P_{ijk} \ln P_{ijk} \qquad (3)$$

Several special decision schemata and associated decision codes have been proposed by Saridis and Graham (1984). However, the easiest to implement is the so-called <u>Vocabulary Optimal Decision Schema</u> in which the input and output languages retain the same syntax, e.g. structural form, but there are several output terminal vocabularies to be selected for different tasks. They can all be implemented with a transducer automaton and require only short term memory to implement the function of learning. The coordination level is essential for dispatching organizational information to the next level, the execution level.

<u>The hardware execution level</u>, is the lowest one in the hierarchy and is evaluated by the cost of executing the appropriate control functions. This performance measure will be expressed as an entropy, thus completing the functions of an "intelligent machine" to be evaluated by entropies. Precision expressed in terms of certainty in execution comes nicely into the picture.

Optimal control theory utilizes a non-negative functional of the states of the system $x(t) \epsilon \Omega_x$ the states space, and a specific control $u(x,t) \epsilon \Omega_u$ the set of all admissible controls, to define the performance measure for some initial conditions $V(x(t),t)$, representing a generalized energy function, of the form,

$$V(x,t) = \int_t^{t_f} L(x,u(x,t),t)\, dt \qquad (4)$$

where $L(x,u(x,t),t) > 0$, subject to differential constraints dictated by the underlying process

$$\dot{x} = f(x,u(x,t),t); \quad x(t_o) = x_s, \quad x(t_f) \epsilon M_f \qquad (5)$$

with M_f a manifold in Ω_x. The trajectories of the system (3) are defined for a fixed but arbitrarily selected control $u(x,t)$ from the set of admissible controls Ω_u.

For an appropriate density function $p(x,u(x,t,t)$ satisfying Jaynes' Maximum entropy principle was shown by Saridis (1984) that the expression H_u representing the entropy for a particular control action $u(x,t)$ is given by

$$H_u = E_x [V(x,u(x,t),t] \tag{6}$$

This result implies that the average, with respect to initial state performing measure of feedback control problem corresponding to a specifically selected control, is an entropy function. The optimal control u^* that minimizes $V(x,u(x,t),t)$, maximizes $p(x,u(x,t),t)$ and consequently minimizes the entropy H_u:

$$u^* : E_x[V(x,u^*(x,t),t)] = \underset{u}{\text{Min}} \int \Omega_x V(x,u(x,t),t) \cdot p(x,u(x,t),t) \, dx \tag{7}$$

This statement establishes equivalent measures between information theoretic and optimal control problems and provides the information and feedback control theories with a common measure of performance. Entropy satisfies the additive property and any system composed of a combination of such subsystems will be optimal by minimizing its total entropy.

It can be that all levels of a hierarchical intelligent control can be measured by entropies and their rates. Then the optimal operation of an "intelligent machine" can be obtained through the solution of the following mathematical problems.

The theory of intelligent machines may be postulated as the mathematical problem by finding the right sequence of decisions and controls for a system structured according to the principle of increasing precision with decreasing intelligence (constraint) such that it minimizes its total entropy.

The above analytic formulation of the "intelligent machine problem" as a hierarchically intelligent control problem was based on the use of entropy as a measure of performance at all the levels of the hierarchy. It has many advantages because of the tree-like structure of the decision making process, and brings together functions that belong to a variety of disciplines. Figure 3 depicts such a system designed for a Robotic Arm.

4. APPLICATION TO ROBOTS

The previous sections discussed the details of the generic structures of a hierarchically intelligent machine and the analytic methodologies involved in controlling them to perform a series of tasks with minimum interaction with a human operator. However, the complete structure of such a machine, cannot be defined in general, but must be specified for each particular case under consideration (Saridis, 1980, 1983), (Saridis and Lee, 1979).

Even at the present time there is a large variety of applications for intelligent machines. Automated material handling and assembly in an automated factory, automated inspection, sentries in a nuclear containment are some of the areas where intelligent machines have and will find a great use. One of the most important applications though is the unmanned space exploration where, because of the distances involved, autonomous anthropomorphic tasks must be executed and only general commands and reports of executions may be communicated.

Such tasks are suitable for "intelligent robots" a type of intelligent machine capable of executing anthropomorphic tasks in unstructured uncertain environments. They are structured uncertain environments. They are structured usually in a human like shape and are equipped with vision and other tactile sensors to sense the environment, two areas to execute tasks and locomotion for appropriate mobility in the unstructured environment. The controls of such a machine are performed according to the theory of Intelligent Machines previously discussed, Saridis and Stephanou (1977), (Saridis, 1983, 1984). The three levels of controls, obeying the Principal of Decreasing Intelligence and Increasing Precision, are implemented with appropriately selected feedback as shown in Figure 4.

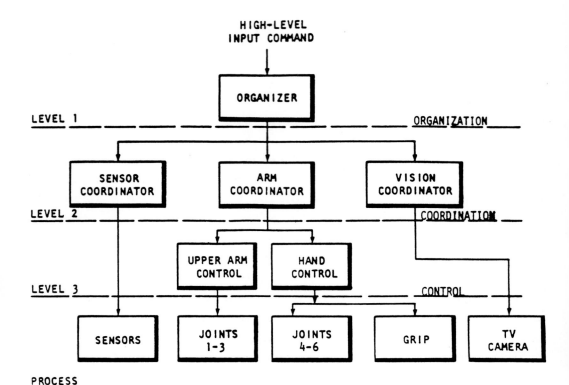

Fig. 4: Hierarchically Intelligent Control for a Manipulator with Visual Feedback

5. CONCLUSIONS

A mathematical theory for intelligent machines was proposed and traced back to its origins. The methodology was developed to formulate the "intelligent machine" as a mathematical programming problem as using the aggregated entropy of the system as its performance measure. The higher levels of the machine structured according to the Principle of Increasing Precision with Decreasing Intelligence can adopt performance measures easily expressed as entropies. This work is based to a major extent on the equivalence between the control performance and the entropy of selecting an appropriate control, thus integrating the execution level of the machine into the overall mathematical programming problem. Optimal solutions of the "intelligent machine" problem can be obtained by minimizing the overall entropy of the system. The entropy formulation presents a tree-like structure for this decision problem very appealing for real-time computational solutions.

ACKNOWLEDGEMENT

This work was supported by NSF Grant ENG ECS - 831279.

REFERENCES

Albus, J.S., (1975), "A New Approach to Manipulation Control: The Cerebellar Model Articulation Controller", *Trans. of ASME, Journal of 97*, pp. 220-227.

Birk, J.R. and Kelley, R.B., (1981), "An Overview of the Basic Research Needed to Advance the State of Knowledge in Robotics", *IEEE Trans. on SMC,* SMC-11, No. 8, pp. 575-579.

Conant, R.C. (1976), "Laws of Information Which Governs Systems", *IEEE Trans. on SMC*, Vol. SMC-6, No. 4, pp. 240-255.

Fu, K.S., (1971), "Learning Control Systems and Intelligent Control Systems: An Intersection of Artificial Intelligence and Automatic Control", *IEEE Trans. AC,* Vol. AC-16, No. 1, pp. 70-72.

Hayes-Roth, et al., (1983), *Building Expert Systems*, Addison-Wesley, N.Y.

Klir, G., (1975), International Journal of Intelligent Control, TR-EE 80-34, General Systems, Vol. 12, p. 149.

Mesarovic, M., D. Macko and Takahara, (1970), *Theory of Hierarchical Multilevel Systems*, Academic Press, N.Y.

Meystel, A., (1985), "Proc. of IEEE Workshop on Intelligent Control", Computer Society Press.

Minsky, M.L., (1972), *Artificial Intelligence*, McGraw-Hill, N.Y.

Nilsson, N.J., (1969), "A Mobile Automaton: An Application of Artificial Intelligence Techniques", *Proc. of International Joint Conference on Artificial Intelligence,* Washington, D.C.

Rosen, C.A. and N.J. Nilssen (1967), "An Intelligent Automaton", *1967 IEEE International Conv. Record.*, Part 9, N.Y.

Saridis, G.N. (1977), *Self-Organizing Control of Stochastic Systems*, Marcel Dekker, N.Y.

Saridis, G.N. (1979), "Toward the Realization of Intelligent Controls", *IEEE Proceedings*, Vol. 67, No. 8.

Saridis, G.N. (1980), "Intelligent Controls for Advanced Automated Process", *Proc. of Conf. on Automated Decision Making and Problem Solving*, NASA, Langley.

Saridis, G.N. (1983), "Intelligent Robotic Control", *IEEE Trans. on AC,* Vol. AC-28, No. 5, pp. 547-557.

Saridis, G.N. and J.H. Graham (1984), "Linguistic Decision Schemata for Intelligent Robots", *Automatica*, Vol. 20, No. 1, pp. 121-126.

Saridis, G.N. (1984) "Control Performance as an Entropy: An Integrated Theory of Intelligent Machines", *Proc. of International Conference on Robotics,* Atlanta, GA.

Saridis, G.N. and C.S.G. Lee (1979), "Approximation of Optimal Control for Trainable Manipulators", *IEEE Trans. on SMC,* Vol. SMC-8, No. 3.

Saridis, G.N. and H.E. Stephanou (1977), "A Hierarchical Approach to the Control of a Prosthetic Arm", *IEEE Trans. on SMC,* Vol. SMC-7, No. 6, pp. 407-420.

Shannon, C. and W. Weaver (1963), *The Mathematical Theory of Communication*, Illini Books.

Winston, P. (1977), *Articial Intelligence*, Addison Wesley, Reading, MA.

ON THE APPLICATION OF INTELLIGENT PLANNING TECHNIQUES IN INDUSTRIAL ROBOTICS

M. H. Lee
Department of Computer Science
University College of Wales
Aberystwyth, Wales, U.K.

ABSTRACT

This paper explores some of the problems encountered when attempting to apply AI techniques within the domain of industrial robotics. The use of planning techniques is investigated through experiments with two classic AI paradigms. The limitations of these methods are explored by contrasting the knowledge contained in the case studies with that available in realistic industrial applications. A criticism of the blocks world model is used to highlight the important features of real robot tasks and recommend directions for future development.

INTRODUCTION

The aim of this paper is to explore some of the problems encountered when attempting to apply Artificial Intelligence (AI) techniques within the domain of industrial robotics. We are interested in how much AI machinery is needed in order to improve the intelligent performance of industrial robot tasks. To this end, we take some basic AI methods and examine their utility in a typical application. In particular, the use of planning techniques is investigated through experiments with two classic AI paradigms. This approach can be seen from a knowledge engineering viewpoint, that is, what is the knowledge content and quality that is necessary in an application in order to achieve a given performance. We show that the knowledge employed in AI case studies is often quite different from that available or desirable in industrial situations.

Consider the simple assembly task shown in Figure 1. A robot has been programmed to perform a well defined task in a repetitious cycle. Components are taken from feeders, assembled together and then transported out of the work cell. The diagram illustrates typical elements of an industrial assembly cell; a manipulator (an active, movable device), active feeders (stationary devices but not able to act on the world), jigs, fixtures and passive feeders (stationary and passive devices), and component parts that flow through the work cell (passive but movable items).

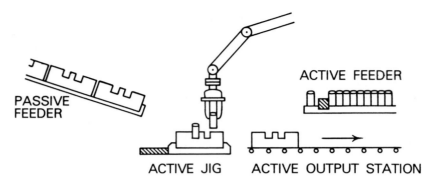
Figure 1

For normal error-free operation with fixed work-cycles AI techniques will not be needed. The system's "knowledge" of the task is embedded in the cell control program which executes in the robot controller or other supervisory computer. Significant task changes will usually be implemented by program modifications, and hence such systems embody only a very limited and implicit coding of the task specification.

However, with the emergence of flexible manufacturing systems much more autonomy will be desirable if such technology is to be cost effective. Examples of potential application areas include: automatic reconfiguration of the work cell for handling frequent batch variation, the completion of under-specified tasks due to environmental variability, and error recovery facilities to deal with minor irregular events. These facilities introduce major problems of both analysis and synthesis. Clearly, much sensing will be required to determine unknown variables and this will involve analytical reasoning from observed symptoms to their underlying meaning. Such reasoning is akin to that performed by expert systems which handle diagnosis and interpretation problems. In addition, there is a major synthesis requirement because such adaptive systems will need to plan and execute new actions which have not been anticipated in the control program. The automatic generation of action sequences is also a major goal of research into high level robot programming languages. In this paper we will consider only synthesis aspects and restrict this to the problem of planning small sequences of actions as might be required to compensate or correct faulty states during an assembly task.

An illustrative example is seen in Figure 2 where a set of components is stacked to form an assembly.

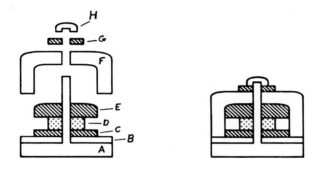

8 PIECE ASSEMBLY

Figure 2.

If component c has a malformed edge, this might remain undetected until component f fails to slide home. In order to recover from this state, components f, e, d and c must be removed in order, c must then be discarded and replaced, and then the assembly process should be restarted from c. Such a sequence of actions could be generated by a planner that understood the main parameters of the assembly task and had access to sufficient knowledge to facilitate the relevant reasoning processes. The fundamental question is: what sort of knowledge and reasoning is needed to allow the planning of new actions? Associated questions concern the quantity of knowledge required, and its availability in typical industrial applications. We turn to AI to look for solutions and guidance with these problems.

THE BLOCKS WORLD

Robotics research has always had intimate links with AI research. From the earliest days there have been AI programs that guide mobile robots, schedule tasks, coordinate dynamic manipulators and interpret sensory signals. Concerning our planning problem, the literature contains much material on action planning, with blocks world models being ubiquitous in robot planning examples. The blocks world is a simplified abstract model of a world in which standardised objects, "blocks", can be moved and configured by an idealised robot "hand". The model is usually implemented as a simulation of a severely restricted environment containing simple objects stacked in various ways on a table top. A simulated robot gripper is able to approach an object from above and grasp and move single objects. This model has been used since the late 1960's when Winograd developed the idea as a testbed domain for his language understanding program. The use of a restricted world was important in natural language research because it produced a controlled domain of discourse. Another development of the blocks idea can be seen in vision research in the 1970's where idealised polyhedral blocks were used in order to reduce the complexity of the visual scene. Many classic scene analysis programs of this period were founded on the assumptions of perfect geometric regularity and extreme object simplicity. Winston [1] has described the MIT project which used scene analysis to control a robot that could build a duplicate configuration of a viewed pile of blocks. Since this early work the blocks world model has become a classic framework in AI for the discussion of planning and control issues and is used widely by exponents of robot planning theory. Clearly we should carefully examine the blocks world model as a candidate methodology for our problem. The following sections describe a basic model and its use in experiments with two classic textbook techniques.

AN EXPERIMENTAL SYSTEM

In Winston's popular textbook on Artificial Intelligence [2] a blocks world planner called MOVER is described in chapter 2. In the companion text on Lisp [3], chapters 13, 14 and 15 deal with the programming and implementation issues behind MOVER. The author (together with help from his students) has created a simplified version of Winston's system for practical tutorial purposes. This consists of a two dimensional world with only one type of block. Despite its restrictions, this model closely corresponds to Winston's system and contains mappings of all of the main features. A Franz Lisp program, GOALPLAN, is available for experimentation and analysis. (All programs mentioned in this paper are available from the author.) A Prolog version was also used to cross-check the results.

Situations and actions in MOVER are mirrored by less complicated but equivalent effects in GOALPLAN; the main differences are in the geometry. Spatial positions in GOALPLAN are limited to discrete locations, thus a position is given by (x y) where x and y are positive integers. An origin is assumed at (0 0) so space can be labelled:-

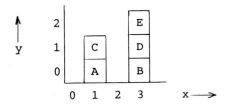

Only square, unit sized blocks are allowed and only one block can be directly supported by any other block. Blocks with consecutive integer values are adjacent. A robot gripper can move in the same coordinate system and can move a single block provided it is grasping the block. In order to grasp, the gripper's position must equal the block's position and the top must be free. The atoms A through Z represent blocks and have property lists containing PLACE, SUPPORTED-BY and DIRECTLY-SUPPORTS values. The atom HAND has properties for GRASPING and PLACE and the atom TABLE supports blocks with zero y coordinates. A list of existent blocks is given by BLOCKS, and WORLD contains a similar list plus the hand and table.

This model includes the use of spatial data and primitive geometric operations. For example, the function FIND-SPACE performs a linear scan across the table, starting at (0 0) and moving through increasing x until either a vacant space is found or (10 0) is reached. In MOVER no spatial locations or geometry are used. The differences between GOALPLAN and MOVER are:

- 2 dimensions, not 3 (smaller world)
- only one block type (no balls or pyramids)
- only one block size (one-to-one support relations)
- no block colour   (identification by name only)
- blocks have spatial locations (coordinates)

GOAL DIRECTED PLANNING

The methodology behind MOVER is based on goal directed procedures. The basic design premise dictates a large collection of short procedures that have access to a common data base. Each procedure is directed at achieving some given goal and can break up its task into a series of subgoals. Each procedure is self contained in that it might be called at any stage and should be able to encounter any state of the system without reacting in an erroneous or ill-conditioned manner. There is no overall executive control algorithm; each procedure calls others and the flow of control follows data directed paths through many levels of mutual recursion between procedures.

In order to operate GOALPLAN, the "top level" goal procedure PUT-ON is called which activates other procedures to satisfy various subgoals. While the system runs, the procedures GRASP, UNGRASP, MOVE-HAND and MOVE-OBJECT add entries to the list PLAN in order to generate the plan of actions.

Two tracing aids have been provided. PIC is a function that displays the status of all the objects in the World. PIC prints a table with the columns showing the block names, positions, supports and the blocks above. The other aid is an automatic trace function which is used just like DEFUN to define functions but adds an extra bit of code to each definition. This code controls an indentation counter and prints the names and arguments of functions when they are called. This provides a history feature. Example output showing the use of PIC and the structured trace are given in the appendix.

Some Failure Cases

When GOALPLAN is running, many operations can be examined in detail. The author has found these practical demonstrations extremely useful in teaching situations. While most operations perform as expected there are a number of planning failures. These are frequently trivial but there are some serious cases.

First, some trivial cases.
PUT-ON (A B) for the situation:-

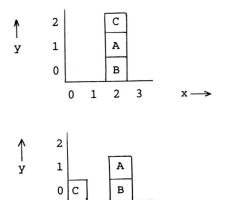

during which A is removed from B and then replaced.
Likewise PUT-ON (A TABLE) for:-

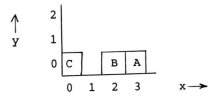

will result in:-

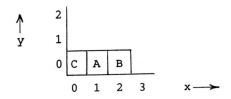

Both these occur because there is no check to see if the desired goal is <u>already</u> satisfied. In the first example, PUT-ON uses the function MAKE-SPACE to clear the top of B and then puts A back onto B. In the second example, function FIND-SPACE locates (0 1) as the first free place on the table to which A is then moved. The necessary modifications required are minor but these results illustrate how easily odd behaviour can be produced by goal directed systems.

In fact, we should add even more tests to make sure that the system doesn't try to put a block onto itself, try to put the table onto a block or try to move nonexistant blocks. (Winston originally had more of these tests in the earlier versions in the first editions of his textbooks. It is not clear why they were removed.) In the Prolog version, PUT-ON (TABLE A) caused considerable block movement, before failing, because the backtracking mechanism kept generating attempts to clear the top of the table!

Now for some serious cases.

First consider PUT-ON (B TABLE) for:-

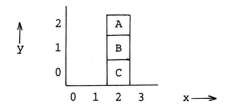

this results in two blocks "sharing" the same space:-

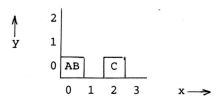

In order to place B on the table, function FIND-SPACE first locates (0 0) as a target for PUT-AT which then calls GRASP(B). GRASP then calls CLEAR-TOP(B) which calls GET-RID-OF(A). GET-RID-OF then uses FIND-SPACE and PUT-AT to place A at (0 0). GRASP then continues and B is moved to its target location, which has now already been used! The reason for this is found in the communication (or lack of it) between activated functions. When one function is invoked it might need to know the <u>intentions</u> or <u>partial results</u> of a whole range of functions which perhaps only relate to it through very indirect routes.

Another example illustrates this:-

Consider PUT-ON (C A) for:-

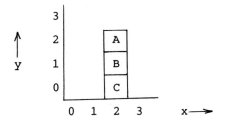

This results in a system failure. First FIND-SPACE(A) determines (2 3) as the target space for C (because it is free and can provide support). Then PUT-AT is called to move C to (2 3). This results in GET-RID-OF (via GRASP and CLEAR-TOP) removing A and B. Then when UNGRASP tries to release C at (2 3) it finds that the support has gone.

Clearly, we <u>might</u> get round these problems by redesigning the programs. We could try clearing the top of the grasped object <u>before</u> locating the target position. At first sight it seems this would satisfy the above cases. However, this would <u>also</u> fail in analogous tasks where clearing the target location alters the previously determined grasp site. There are two design options for PUT-ON; either the target location is cleared before the grasped block is cleared, or it is cleared afterwards. In <u>both</u> cases there are situations where some of the subgoals are able to disturb the conditions set up by their calling procedures or earlier subgoals. These examples expose serious difficulties in this approach to goal directed programming. The fundamental problem is that any subgoal might change the state of the world in a way that has not been anticipated by its superior goals.

The fact that such errors easily occur in this minimal problem space bodes ill for anything more complicated. Consider a higher level task such as moving a stack of blocks. Assume we require a function PUT-STACK (A D) which places A and any blocks above A on top of D.

Thus:-

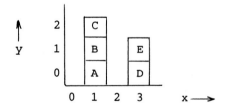

becomes:-

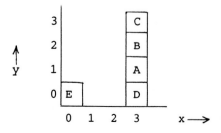

In order to put A onto D we must first remove C and B and <u>remember</u> to place them back onto A. However, it is not clear which procedures should be given the responsibility for maintaining this information or even which will need access to it. If several stacks are being moved it will be necessary to have distinct data for each stack. The dilemma is that if this information is held locally in "higher level" functions (i.e. good structured design) it will prevent occasional access by other functions, whereas if it is held at the lowest levels (globally) there may be conflicts in maintaining the data by different procedures.

To summarise, there are two main results from our experiments: (a) it can be difficult for programmers to anticipate all the necessary preconditions to prevent inappropriate subgoals being pursued, and (b) it is easy for one subgoal to undo or destroy the conditions set up by another.

Critique of GOALPLAN

Winston presents a method of planning based on the concept of distributed control. The planner consists of many, small, mutually recursive procedures and resembles a data driven network of procedural nodes. Winston argues that this style of programming provides the adaptability needed in such tasks, together with elements of good programming practice. He offers the MOVER system as a step towards a general model for distributed control.

From our investigation we make two observations:-

(a) There are significant shortcomings in the goal directed methodology. Unless the procedures can be mapped onto completely independent subgoals, there will be cases where their interaction causes effects not readily determined from their specifications.

(b) For similar reasons, we feel that, from a software engineering viewpoint, the claims for good programming practice will not hold for large practical examples.

It is important to state that we are not critical of many of the pedagogic features illustrated by the goal directed concept, e.g. data-driven procedures, question/answer reasoning and automatic plan generation. It is only the control aspects that we concentrate on; in particular, the three interlocking principles claimed by Winston [2] :-

- control should be distributed

- procedures should be goal oriented

- procedures should be self-contained

Our main argument is that after many years [4], a sound theoretical basis for building such systems is still missing and there appear to be serious conflicts between these three principles.

Protection Violations and the Frame Problems

If we return to Winston's arguments for a large collection of small procedures we see that the lack of control structure in such loose organisations may be the root of our problem. Winston's procedures are short, goal oriented and self-contained. They should be able to compute all relevant preconditions before invoking their subgoal procedures. While they execute they have no knowledge of the state of each other so that partial results are not accessible. This principle of total independence rules out any communication between partially completed procedures and assumes that the planning task can be decomposed into a series of independent, non-interacting sub-tasks. (In order for this to work not only must the task be decomposable but we must also be able to _find_ a suitable decomposition and map it onto a set of goal directed procedures). In fact, in most realistic planning domains this is not true and there are many interactions between subgoals. Consequently, the independence assumption must be dropped and some additional features must be incorporated into goal based planners. The most appropriate technique is the idea of protection mechanisms. These ensure that once a subgoal has been achieved it will be protected while other parts of the plan are completed. When goals are decomposed, special definitions can be given to specify the protected items and their duration. Waldinger [5] has looked at alternatives to the single minded, one-goal-at-a-time process by trying to achieve, and/or protect, several subgoals at once. His notion of "protected lists" can be used to hold locations which must remain empty or supports which are not to be moved. Also blocks which satisfy some partial goal could be designated "fixed". However, even ignoring the problems of managing protected goals (sometimes a protection violation is actually helpful in planning!), it is still not always possible to find all the sub-goals, or other items, that must be protected due to the complexity of their interactions. In our simple examples, protecting clear tops would solve many errors but, in general, it is not easy to decide which blocks, locations or subgoals need protection.

Our difficulties can also be seen to be related to the frame (of reference) problem [6]. In theory, any action can change any of the relations in a task and so there has to be some method of computing the way the world has changed after each action. Thus, a planner must have knowledge of all the consequences of each action, including possible side effects and details such as subgoal ordering sensitivities, so that it's model of the world can be correctly maintained. Rather than list all the relations that are not altered (i.e. the frame axioms), it is nowadays more usual to follow the STRIPS idea in which all relations are assumed unchanged unless explicitly mentioned in the specifications of the actions [7]. Hence, goal directed procedures may embrace the STRIPS solution to the frame problem provided they can arrange to compute all the consequences of their actions. This is fairly easily satisfied if the actions are reasonably independent and only change a small proportion of the total relations. Difficulties arise when actions become "influential" and produce complex indirect side effects. For example, inverting the assembly shown in Figure 2 might change many of the component relations. Notice that the nature of these changes will depend dramatically on whether certain components (such as, c and f, or a and h) are jammed together. In such cases, there are so many interactions that each individual procedure must contain elaborate checks on many of the world relations. As these interactions increase so each procedure becomes a large and cumbersome program. There will also be much duplication of effort across the procedures. It seems that even small realistic problems have far too many potential interactions for this style of programming to be effective.

One of the most advanced block world planning systems is BUILD by Fahlman [8]. In BUILD failures are handled by a choice and gripe mechanism. Gripe handlers have access to a range of information and deal with decision backtracking at many levels. Fahlman also used a system of primary and secondary relations, in which secondary changes were deduced from primary effects while primary changes follow directly from the actions. These additional mechanisms offer features to handle the problems mentioned above. Even so, Fahlman argues for future work on inter-goal cooperation i.e. ".. the protection of the accomplishments of one goal from premature destruction by later goals, and the dissemination of useful information found by one goal to all of the other goal modules that might be interested" [8].

Good Programming Practice

While we may agree that short direct procedures which carefully check the state of the world before changing it are a contribution to good style it is not so obvious that distributed control is of similar benefit. The very fact that there is no global control structure means that large scale systems with many thousands of procedures will be extremely difficult to understand and evaluate. This may not be of any immediate consequence for experimental AI but it certainly goes against the principles of good software engineering. The main thrust of software engineering is to impose structure on large projects so that performance goals can be achieved and systems can be evaluated and controlled.

Nevertheless, all this is a matter of style; a more serious objection concerns the flow of information between procedures. If Winston's maxim that "Procedures should presume as little as possible about the situation in effect <u>when they are called</u>" [3] is too optimistic, as we have argued, then the emphasised words should be replaced by <u>"while they are active"</u>. This ultimately leads to a knowledge crisis because any new procedure must frequently check that its conditions have not been altered by <u>any</u> of the other procedures. If all n(n-1)/2 interactions in a system of n procedures are possible then a highly defensive style of programming is needed. We can't check every procedure to examine its possible side effects because these depend upon innumerable data driven relationships. Also we can't assume any global data to be static during our procedure execution; so we must check out <u>all relevant conditions immediately before any action takes place</u>. We may also need to check that conditions are being maintained <u>during</u> actions. This will give longer, highly complex procedures and undermines the original maxims.

SEARCH BASED PLANNING

Another approach to planning in the blocks world is to view a given task as a particular path through the space of all possible actions. This assumes that actions are treated as operators which are specified in terms of state transformations and the task is well defined in terms of start states and goal states. The planning problem then becomes one of searching for an acceptable path (not necessarily optimum) from a start state to a goal state. The path represents a sequence of operator applications that produce the desired state transformation and thus is equivalent to a plan of actions.

For our simple blocks world we can describe a given configuration such as:-

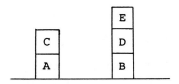

by a data base of facts or assertions:-

(on C A)
(on A TABLE)
(on E D)
(on D B)
(on B TABLE)
(clear C)
(clear E)

The operator PUT-ON can then be defined in terms of preconditions and transformations on existing facts:-

<u>Operator</u> (PUT-ON x y)   where $x \neq$ TABLE
<u>Preconditions</u> (clear x) AND (clear y)
<u>Actions</u> (on x z) is retracted
(on x y) is asserted
(clear z) is asserted IF $z \neq$ TABLE
(clear y) is retracted IF $y \neq$ TABLE

Although the facts are held in a different data structure than the previous system (which used property lists) it is clear that exactly the same information is available but in a different form. The one exception is that positional data is no longer used. Indeed, no geometric data is handled or processed. Consequently, there is no need for functions like FIND-PLACE: the positions of blocks are either undefined if on the table or are defined relative to other blocks by "on". Notice that only one operator has been specified and so a solution will be a sequence of several PUT-ON applications, each of which can be directly expanded into the sequence of sub-operators - MOVE-HAND, GRASP, MOVE-OBJECT, UNGRASP. Thus, the plan can be generated in the same form as for GOALPLAN. There is no need for operators like MAKE-

SPACE or CLEAR-TOP because the preconditions for PUT-ON ensure that it is only applied in the case of free block moves.

It is important to realise that this type of representation is very closely related to both logical methods and rule based techniques. If the facts are seen as axioms and the operators as rules of inference, then automatic theorem proving techniques could be used to search for a solution. The series of proof steps would give the path to the desired theorem (the goal). Alternatively, the data base could be viewed as the working memory of a rule based or production system and the operator(s) could be coded in terms of IF-THEN rules. The solution would then be generated by a particular pattern of rule firings and associated assertions.

In order to explore these near equivalent formulations an automatic search procedure SEARCHPLAN was written in Franz Lisp, (as before, a Prolog version was also written). The search generation method is depth-first search with exact goal state testing and user controls for maximum depth limits. An exhaustive blind search (i.e. without heuristics) was used in order to investigate the general shape and other characteristics of the state space for the blocks world domain.

The search begins at the start state and one of the possible instantiations for variables x and y is generated for PUT-ON. This leads to a new state and further operator applications are pursued. Eventually, either the goal state is reached or the maximum depth limit initiates backtracking through other branches of the state space. The search process ensures that all valid combinations of PUT-ON are pursued down to a given depth limit. In a sense, SEARCHPLAN can be seen as a sort of inverse of GOALPLAN; the former generates operator sequences to reach a given state while the latter generates state change sequences to satisfy a desired operator application.

To use SEARCHPLAN it is necessary to set up the assertions to define the start state and goal state. The configurations below show two sample start states, EASY and HARD, and also a GOAL state.

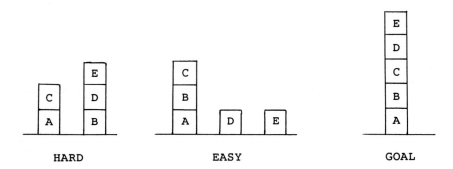

HARD   EASY   GOAL

It can be seen that EASY only requires 2 PUT-ON actions to achieve the GOAL, whereas HARD requires 7 applications. To initiate a search the function SEARCH is called with arguments STATE, GOAL, EFFORT and LAST. The first two arguments are states, EFFORT is the depth limit on search levels and LAST is the last most recently moved block. LAST (set to nil initially) is included to prevent the same block being moved in consecutive operations. The essential structure of the search program is shown below.

Basic Search Strategy

```
Function Search (State, Goal, Effort, Last)
    - if State = Goal then exit (success)
    - if Effort = 0 then exit (failure)
    - generate Successor-list from State (excluding Last)
    - for each in Successor-list - assert new state
```

- decrement Effort
- record Last
- recursively Search

The program was tested on a series of trials using both start states and with effort values from 1 to 8 inclusive. A summary of the results is given below.

TEST RESULTS

Max. search depth EFFORT	Number of states examined		cpu time s = seconds m = minutes		No. of steps in solution		Actual Solution	
	EASY	HARD	EASY	HARD	EASY	HARD	EASY	HARD
1	7	4	0.55s	0.27s	–	–	–	–
2	28	18	1.9s	1.0s	2	–	D->C E->D	–
3	153	85	9.4s	4.9s	2	–	as above	–
4	795	433	44s	24.8s	2	–	as above	–
5	2971	2204	2.7m	2m	5	–	C->TABLE D->E C->B D->C E->D	–
6	13164	10297	12m	9.4m	5	–	as above	–

7	2089	8080	1.9m	7.3m	7	7	C->E	C->TABLE
							B->D	E->TABLE
							C->TABLE	D->E
							B->A	B->A
							C->B	C->B
							D->C	D->C
							E->D	E->D
8	9566	10737	9.9m	11.2m	8	8	C->E	C->TABLE
							B->D	E->C
							C->TABLE	D->TABLE
							B->A	B->A
							D->E	E->TABLE
							C->B	C->B
							D->C	D->C
							E->D	E->D

Critique of SEARCHPLAN

The results generated for EASY show a basic property of depth-first searching; that the first solution found will rarely coincide with the shortest path solution. Having found a solution in 2 steps for EFFORT=2, the system will return the much worse 5 step solution for EFFORT=5 because this is discovered <u>before</u> the 2 step solution is reached. These "extra" steps in solutions appear in most of the results. Thus depth-first searching not only finds non-optimal solutions with "wasted" moves, but also takes longer to find them, if allowed to do so. The close correlation between EFFORT and the number of steps in the solution indicates that there are many repetitions of the goal state throughout the state space. This effect is heightened because we are searching through a cyclic graph as though it were a tree structure. Breadth-first search would be superior in this respect and it would also guarantee to find the shortest path solution. However, breadth-first search can require enormous storage overheads in any non-trivial problem. For a counter-intuitive example of a case where directed depth-first search is superior to breadth-first see Elliott & Leask[9]. In this map navigation example, a small amount of application knowledge reverses the performance rating of theoretical recommendations.

The other shortcoming of our search is that, because repeated states are not detected, our search tree is potentially infinite. The use of a best-first technique would solve most of these problems by maintaining a set of unique "best yet" states in a check list. However this introduces heuristics, as "best" must be defined in terms of a computed nearness-to-goal function. Besides being very application dependent (and quite difficult to define in any general way for a blocks world), the introduction of such knowledge will dramatically transform the nature of our problem, even to the extent that searching may become unnecessary. For example, we could try the algorithm - make any random move that increases the "nearness-to-goal" value.

It is interesting that the solution to HARD at level 7 takes less work than the full examination of the state space for the failure at level 6. Also HARD generates fewer states than EASY in the early trials because there are less free tops in HARD than EASY and hence less scope for operator applications. This illustrates the artificial nature of searching for this problem. Consider the case of all blocks lying on the table. This can be one of the worst start states for a search procedure as there are so many options open. However, for a human this might well be considered to be the best starting position for a block stacking task.

These experiments tell us about the shape of the state space but indicate that blind searching is really only useful as a last resort when no other data are available. Without any guiding heuristics, mechanical search is a rather incompatible technique for an everyday problem. In any realistic application, there would be more operators and each would have long and elaborate specifications. This is because each operator must handle all the assertions and deletions that are required to avoid the frame problem. With an operator like "move the box" there would be extensive processes to manipulate the facts about the contents of the box at the time it is moved.

The possibility of using heuristic knowledge to produce an "informed search" hinges on the quality of heuristics that can be found. In our case, a measure of "nearness-to-goal" is not easy to compute. (The number-of-correct-assertions can't be used as a heuristic measure because this takes no account of their important inter-relationships.) Experience shows that the amount of searching required is inversely related to the quality of the heuristic. However all heuristics are application dependent and so we would have to re-write them for each different task. This is generally unacceptable in industrial situations.

Other techniques for improving search based planning include context or filter mechanisms that select and restrict sets of applicable operators at branch points. There are also planners that search through the space of partial plans rather than through state spaces. These higher level planners, such as NOAH [10], are able to resolve conflicts between plan segments and provide protection mechanisms. Other enhancements include the careful incorporation of sensory data at selected plan points so as to efficiently match the search to the task [11]. Search seems most effective when it is used to support other techniques rather than as a primary mechanism.

BLOCKS WORLDS VERSUS REAL WORLDS

The original blocks world arose from early vision research where simplified objects and environments were used to reduce the complexity in visual scenes. A block is an archetype of an object that can be represented by a collection of restricted edges and vertices. The attractions of this formalism include a certain abstract elegance that reduces robot planning problems to a form of symbolic puzzle solving. While planning strategies can be usefully investigated in this framework, we must not be seduced into thinking that we have, even approximately, a simulation of a real life robot situation. Consider the knowledge that is contained in the blocks world model. All that is known of the problem domain is a few facts about block supports, free tops and locations and a set of conditions describing the application of certain operators. This is very similar to the minimal knowledge required to play games like chess. Systems with such limited information are able to play mechanically but are quite unable to "understand" the game or even play strategically. In real life robotics applications there are many more relevant pieces of data that should be incorporated into our systems. For example, the generalised concept "move" subsumes several different types of detailed movement; sliding, pushing, lifting, falling. If we expand our model to include these significant distinctions we will be forced to consider concepts like friction and lubrication, properties of materials and the effects of tolerances on dimensions. These are only a few of the possibilities, but the important point is that introducing even one such factor not only makes the model more realistic but offers the potential to solve planning problems by completely different and possibly more powerful methods. We have what Hayes calls an incomplete "conceptual closure" [12]: every time a new feature is added, a dramatic change is forced on the model. This indicates the inadequacy of the blocks world. When we have better models that allow closures over a given application domain we will be able to represent and process all sorts of different events without the constant introduction of new concepts. Of course, this approach will not be without its problems. Hayes points out many likely obstacles; but being able to model the richness of the real life application domain must be vital for future progress.

As an example of this alternative approach to our assembly task problem we briefly describe our recent work on causal and functional modelling [13]. Figure 3 shows part of a network expressing the enabling and causal relations between stages of the assembly. To achieve any given state, (shown as boxes) all of the relevant enabling conditions, (the entering links) must be available. Thus, if washer g cannot be located properly, either it won't clear the peg or component f is not fully home. In the latter case, the three linked conditions describe possible causes for f failing to reach its correct location. Many other causal reasoning chains can be followed in this type of model. Such models have great expressive power for describing authentic events in the world and provide a guiding framework for reasoning about actions, plans and planning failures.

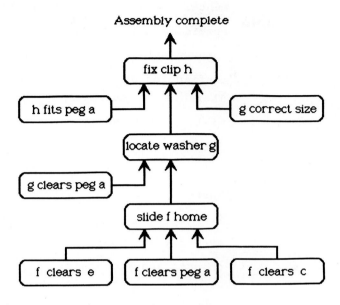

Figure 3.

Finally, we notice another dangerous planning assumption. This is the assumption that plans are fixed instructions to be followed during an execution phase. The world is not a passive puzzle that requires a solution. There are reactive agents and unstable conditions that make real life problems look like dynamic adversaries. We need to treat planning as a real-time activity by integrating plan generation with execution monitoring, and allowing revision and modification of current plans. As McDermott has pointed out [14] current plans have far more importance than future plans, which are really only estimates of what _might_ happen.

We can summarise the limitations of the blocks world under three main headings:

Complete Information — In real life, objects do not have precisely defined shapes and the environmental relationships are not fully specified. There will be considerable unknown data, inaccurate data and even conflicting data.

Full Control — There is an assumption that actions will always result in particular well defined operations on the environment. This is not true in real worlds. There will be external agents, irreversible events, non-observable events, as well as irrecoverable objects and processes.

No Feedback Monitoring — Most systems operate in a dead reckoning mode. Just a small amount of feedback data can reduce computational loads enormously. Monitoring would also reduce the uncertainties in the current plan.

CONCLUSIONS

This paper has described a series of experiments designed to investigate the use of simple planning techniques in an industrial robotics setting. We have argued that these methods address the wrong issues for practical tasks in an industrial assembly world. Goal directed planning will fail because many real life tasks do not decompose into completely independent subgoals and there are often strong interactions between most component parts of the task. Search based methods also suffer from the same difficulties. In order to avoid the frame problem both techniques must burden their operators with enormous responsibilities to check out the side effects of their actions. Both planning methods are designed on the assumption that the changes caused by an operator will involve only a small proportion of the total facts describing the world. When there are strong sub-task interactions these methods become unworkable. Rather than try and modify these planners we argue that their knowledge of the problem domain is so sparse and remote from real applications that the introduction of small amounts of domain specific knowledge will so change the task that entirely different techniques will emerge. These are likely to be based on naive physics models.

Finally, we have argued that the blocks world model is a misleading and inappropriate framework for representing robot actions. We wish a planner to be seen as a dynamic, interactive agent that is willing constantly to revise it's estimates and incorporate new information. In order to support this, the world model should be much more authentic with many real life features being represented. This will allow a better form of "common sense" reasoning akin to that proposed by naive physics workers [15]. We are encouraged that some of the current research in AI is pursuing these issues and look forward to new styles of planner for realistic robot tasks.

ACKNOWLEDGEMENTS

Thanks to Dr Fred Long for writing the Prolog versions of the programs and for many useful discussions. Thanks also to several generations of students who helped with the experiments and tolerated being used as guinea pigs for these ideas.

REFERENCES

[1] Winston, P.H.,"The MIT Robot", in Machine Intelligence 7, (Eds) B. Meltzer and D. Michie, Edinburgh University Press, 1972.
[2] Winston, P.H., "Artificial Intelligence", Addison Wesley, 1984.
[3] Winston, P.H. and Horn, B.K.P., "Lisp", Addison Wesley, 1984.
[4] Winston, P.H., "Heterarchy in the MIT Robot", Vision Flash 8, A.I. Lab, M.I.T,, 1971.
[5] Waldinger, R., "Achieving Several Goals Simultaneously" in Machine Intelligence 8, (Eds) Elcock, E.W. & Michie, D., Ellis Horwood, 1977.
[6] Raphael, B., "The Frame Problem in Problem Solving Systems", in Artificial Intelligence and Heuristic Programming, (Eds) Findler, N.V. and Meltzer, B., Edinburgh University Press, 1977.
[7] Fikes, R.E. & Nilsson, N.J., "STRIPS: A New Approach to the Application of Theorem Proving to Problem Solving", Artificial Intelligence, Vol. 2, No. 1, pp. 189-208, 1971.
[8] Fahlman, S.E., "A Planning System for Robot Construction Tasks", Artificial Intelligence, Vol. 5, No. 1, pp. 1-49, 1974.
[9] Elliott, R.J. & Leask, M.E., "Routing Finding in Street Maps by Computers and People", Proc. AAAI-82, Pittsburgh, 1982.
[10] Sacerdoti, E.D., "Planning in a Hierarchy of Abstraction Spaces, Artificial Intelligence, Vol. 5, No. 2, pp. 15-135, 1974.
[11] Brooks, R.A., "Symbolic Error Analysis and Robot Planning", Int. J. Robotics Research, Vol. 1, No. 4, pp. 29-68, 1982.
[12] Hayes, P.J., "The Second Naive Physics Manifesto", in Hobbs & Moore, 1985.
[13] Cheung, L.Y. and Lee, M.H., "The Role of Casual Modelling in Automatic Assembly", Proc. 7th International Conf. on Assembly Automation, Zurich, pp. 335-340, 1986.
[14] McDermott, D.V., "Reasoning About Plans", in Hobbs & Moore, 1985.
[15] Hobbs, J.R. & Moore, R.C., (Eds) "Formal Theories of the Commonsense World", Ablex, New Jersey, 1985.

APPENDIX

-> (pic)

a	(1 0)	table	(c)	
b	(3 0)	table	(d)	
c	(1 1)	a	nil	
d	(3 1)	b	(e)	
e	(3 2)	d	nil	
table	nil	nil	(a b)	
hand	(5 10)	nil	nil	

hand is grasping nil

```
2        |     | E |
1        |  C  | D |
0        |  A  | B |
   0  1  2  3
   initial state
```

```
plan is nil
-> (put-on 'a 'd)
   ===============
    put-on a d
        get-space d
            find-space d
            make-space d
                clear-top d
                    clear-top e
                    get-rid-of e
                        find-space table
                            find-table-place
                                clearp (0 0)
                        put-at e (0 0)
                            grasp e
                                move-hand (3 2)
                            move-object e (0 0)
                                remove-support e
                                move-hand (0 0)
                                add-support e (0 0)
                                    get-object-under (0 0)
                            ungrasp e
        put-at a (3 2)
            grasp a
                clear-top a
                    clear-top c
                    get-rid-of c
                        find-space table
```

```
                              find-table-place
                                  clearp (0 0)
                                  clearp (1 0)
                                  clearp (2 0)
                              put-at c (2 0)
                                grasp c
                                  move-hand (1 1)
                                  move-object c (2 0)
                                    remove-support c
                                    move-hand (2 0)
                                    add-support c (2 0)
                                      get-object-under (2 0)
                                ungrasp c
                          move-hand (1 0)
                        move-object a (3 2)
                            remove-support a
                            move-hand (3 2)
                            add-support a (3 2)
                              get-object-under (3 2)
                          ungrasp a
-> (pic)
a       (3 2)       d           nil
b       (3 0)       table       (d)
c       (2 0)       table       nil
d       (3 1)       b           (a)
e       (0 0)       table       nil
table   nil         nil         (c e b)
```

2			A
1			D
0	E	C	B
	0	1 2	3

final state

hand   (3 2)   nil   nil

hand is grasping nil

<u>plan is</u>

((move-hand (3 2)) (grasp e) (move-hand (0 0))
 (move-object e (0 0)) (ungrasp e) (move-hand (1 1))
 (grasp c) (move-hand (2 0)) (move-object c (2 0))
 (ungrasp c) (move-hand (1 0)) (grasp a)
 (move-hand (3 2)) (move-object a (3 2)) (ungrasp a)
)

ANALOGICAL REASONING BY INTELLIGENT ROBOTS

Nicholas V. Findler and Laurie H. Ihrig
Artificial Intelligence Laboratory
Computer Science Department
Arizona State University
Tempe, AZ 85287; USA

ABSTRACT

Robotics research has produced economically and organizationally satisfactory tools for industry, and exploration of and manipulation in outer space, under the ocean and other dangerous or difficult-to-access places. Intelligent robots, however, are still largely a promising possiblity around the horizon. The adaptation of Artificial Intelligence methodology for robots seems to be a difficult and lengthy process. Both general-purpose and domain-specific techniques are needed. In this paper, we investigate some fairly universal concepts within the block world context.

Analogical reasoning (AR) has long been recognized as an important component of problem solving. In general, AR involves applying the (possibly modified) solution of one problem to a second problem which is in some sense analogous to the first. The prerequisite the two problems have to satisfy is that they have the necessary number and type of important features in common. The task is to discover automatically what the important features are. We discuss at length some general ideas, two basic models and a few advanced processes relating to AR.

Our program generates specific solutions to a number of similar problems that share several properties. The problems are to build certain three-dimensional bodies which satisfy a number of geometrical requirements and constraints. Problem situations are then generalized in the manner of concept formation. Those problems that have similar solutions are replaced with a single concept -- the type definition of a class of problems. Our program, itself, identifies new (hidden or "chunked") properties it has determined to be essential.

Frames are used to describe problem situations. Four conceptual levels of frames are distinguished: (i) The situation level contains slots for situation properties, the types of available objects, the goal and the eventual solution. (ii) The object level has slots for specific object properties and for lists of possible components that can make up the object. (iii) An unlimited number of component levels look like the object level and represent the components of components...of the objects. (iv) Finally, the property level can contain properties of situations, objects or components.

The underlying learning is a three-stage process. In the first, shaping stage, heuristic search techniques are used to find a solution to a particular problem. The resulting plan is an action sequence which is then associated with the problem situation. In the second, AR stage, problems with similar action sequences are grouped under a single situation class. A class definition is established which is sufficient to distinguish its members from all other situations. Rules are generated which connect the situation classes and action sequences to be performed in them. The final, <u>consolidation stage</u> compiles the rules into a <u>decision graph</u>. The variables determining the situation class are re-ordered on the decision graph so that the action plans can be retrieved the most efficiently.

> "...no such thing as a false analogy exits: An analogy can be more or less detailed and hence more or less informative." (Nobel Prize lecture by K.Z. Lorenz, 1973)

1. INTRODUCTION

The human ability of analogical reasoning (AR) is usually considered to contribute to the success of relating new problems to old ones and solving them efficiently, of assimilating knowledge about new items and new relations with reference to similar ones already known, of making a decision in a situation that shares some properties with another in which a previous decision produced the desired result. It is natural to attempt to provide this powerful tool for AI systems. However, the relatively few projects in this area [1-3,5-16,19] have not yet generated the power and generality needed for a unifying computational model, as shown in two excellent review reports [17,18].

One of us (NVF) has been involved in several studies relating to AR. The Associative Memory, Parallel Processing Language, AMPPL-II [4], could return solutions to simple AR word problems, based on A:B::C:X, such as 'Jeanne is to French (her mother tongue) as Jose to what?' (X=Spanish) or 'What property is shared by G.B. Shaw's wit and a razor's edge?' (X=sharpness) or 'Who are related to Joe the same way as Tim and Neal are to Bob?' (X=Ron, Justin and Nathan -- they are Joe's uncles).

In another project [5,7], we investigated the effect of experience gradually accumulated on the peformance of an AR system doing construction and theorem proving in plane geometry. Further, [6] describes a set of simulated cooperating robots which can make inductive discoveries using AR.

2. SOME GENERAL IDEAS ABOUT ANALOGICAL REASONING

AR can contribute to problem solving in two distinct ways. In both cases, we are given an old problem or task, <u>P1</u>, whose solution <u>S1</u> is known, and an analogous new one, <u>P2</u>, whose solution, <u>S2</u>, is sought. First, we assume that we are able to compute a mapping function <u>M1</u> that would transform <u>P1</u> to <u>P2</u>. It is hoped then that if we apply <u>M1</u> to <u>S1</u>, the resulting <u>S2*</u> is not "too far" from the desired solution <u>S2</u>. Similarly, it <u>is</u> assumed in the second paradigm that we can find a mapping function M2 that would transform <u>P1</u> to S1. Applying <u>M2</u> to <u>P2</u> results in <u>S2**</u> that again is not too "far" from <u>S2</u>. (Figures 1/a and 1/b show these ideas.) Another technique, such as the classical means-end analysis, would then modify <u>S2*</u> or <u>S2**</u> to obtain a satisfactory solution, <u>S2</u>.

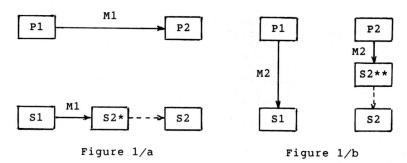

Figure 1/a   Figure 1/b

Two paradigms of problem solving with analogical reasoning.

The numerous mechanisms of the learning process enable us not only to rely on a large stock of situations, but also to extract their essential characteristics. By means of these characteristics, we can then select a small number of situations from our "total experience," which are the most similar to a new situation we are faced with. This fact represents the equivalence between intelligence and problem solving ability.

Similarity is a very general and loosely defined concept. Different objectives may invoke different types of similarity. The relevance of resemblance is determined by the context of the problem given. Is, for example, a high chair similar to a playpen? A possible answer is no since they are made of different materials and have different shapes. On the other hand, the answer may be yes because they both are baby furniture. It is obvious that experience provides the foundation for the discovery and the recognition of meaningful similarity.

Each situation can be characterized by a combination of features. We shall use the term feature to represent one or several chunked properties. Whereas properties are atomic and directly measurable, features can in general be observed as present or absent only. Features are important because they reduce the information processing load [20]. At the simplest level, resemblance can be measured by the weighted sum of common features two situations share. Namely, not all features of a situation contribute equally to the selection of some appropriate action for coping with the situation. Further, there are numerous higher order similarities people discover and make use of. For example, in plane geometry, point and line (theorems of duality), circle and line (line being a circle of infinite radius), etc., are components of meaningful resemblance between certain problems. They serve in calling for similar theorems in proof procedures or construction tasks. Again, experience is used in discovering such higher order similarities.

We can be concerned only with task domains that have or can be expected to have a well defined process for determining solutions to problems. In other words, task domains in which the solutions and problems are statistically or randomly related do not belong in this study.

In any nontrivial task, a search-and-test cycle can be identified. A measure of intelligence would be expressed by the number of search-and-test cycles required to obtain a satisfactory solution, averaged over a variety of tasks of different domains. This train of thought leads to the conclusion that AR is an indispensable component of an intelligent system and no problem solving activity can be efficient if it does not make use of AR in some fashion.

Our usual desire for parsimony demands that the mechanism of AR, whether to be recognized in the human brain or to be implemented in a computer problem solving system, should be "fairly" task-independent. Further, we would also prefer for the same reason that it should not require any a priori information concerning the probable characteristics of a task domain in order to achieve gradually better performance.

One would like to establish a programming system the AR component of which could discover analogies at different levels between problems and would yield a very short list of potential solutions to another component that would then try these out, one after the other, until a sufficiently powerful and economic method is found.

A few plausible working hypotheses constitute the rationale of AR. These include the following.

1. Each problem is describable as an (ordered) collection of certain, possibly overlapping, fundamental features.

We have put the word _ordered_ in parentheses to indicate that it is a desirable goal because it reduces the search space in general but may not always be possible to accomplish. The decision as to what constitutes a feature is left to the user although we can envisage a system that starts working in different problem domains using a trial-and-error method and as experience is accumulated, it selects more and more appropriate features.

2. Solutions are associated with respective problems in a well-defined, deterministic manner.

This assumption is stronger than it sounds. Causality, inherently underlying all scientific investigations, does not imply that one can measure all relevant variables, and with sufficient precision, to establish reliable explicatory and predictive relationships. Here we have assumed that the features of the problems are identifiable and are strongly enough correlated with the solutions so that the latter can be directly derived from the former.

3. In the task domains of interest to us, similar problems have similar solutions.

Similarity, of course, must not be defined in a circular manner (i.e., "problems are considered similar if their solutions are similar") but must be measured along certain dimensions that depend on a priori features of problems on one hand and of solutions on the other.

4. When two problems have similar solutions, the features present in one problem but not in the other are likely to be of lesser importance.

5. In turn, features shared by problems which have similar solutions are likely to be important.

The last two hypotheses can be strengthened quantitatively in the sense that the more problems, which have similar solutions, share a feature the more important it is likely to be. Also, the more features are shared by two problems the more similar their solutions should be.

Admittedly, not all problem domains of interest satisfy completely the stipulations implied by the above hypotheses. Further, the total mechanization of every phase of AR for many problem domains that do satisfy the stipulations would present technical difficulties that may be beyond our present capabilities. It is also true that the procedure one can arrive at will have a flavor of approximation for two reasons. First, the knowledge base on which the search for similarity operates is necessarily limited. Second, for nontrivial problems, the solution steps will have gaps in between themselves, which have to be filled in by some heuristically guided trial-and-error method, or control must be transferred to another problem solving component of the program at that stage.

3. ON THE CONTRIBUTIVE AND HIERARCHICAL MODELS OF ANALOGICAL REASONING

We wish to study the problem of what information should be extracted from "raw" experience consisting of descriptions of problems and solutions, and how this information can then be used in determining solutions to new problems. Let us go back to the second paradigm illustrated on Figure 1/b.

The mapping function $M2$ uses the features of a problem as independent variables and its value is the solution to the problem. The gradual construction of this function consists of recording the number of associations over many problems between particular features and solutions. In general, the solution will be represented by an ordered sequence of steps (for example, applications of transformation operators). This way, it is not necessary to accept the whole solution as a starting point in the attempt to solve a new problem but tentative solutions will be composed of only solution segments of varying length. The frequency of prior usage (score) of solution segments associated with a given problem feature provides an important heuristic guide. Namely, the solution segments are ordered and offered for testing according to the score values. The testing for adequacy in a solution may be done by another system component using, for example, logical or analytical techniques.

We wish to point out how similar this approach is to the development of empirical sciences. In medicine, for example, the long-term process of forming causal relations between symptoms and diagnoses, and the ordering of components of cure to diagnoses seem to have taken place along such lines. The learning process in our model is also analogous to the real life event. Every time a new problem is solved, its features are separated and identified, and the knowledge base is updated accordingly.

We call the above paradigm the <u>contributive model</u> because we consider the relevant features of the problem to have contributed to the selection of solution (components). The contributions have a cumulative effect. The more problems with a given feature have used a certain solutions component, the more likely new problems with the same feature will use that solution component. The feature in question quasi demands the use of that solution component and competes with other features offering <u>their</u> contributions.

It is also to be noted that it is usually necessary to decompose a complex problem into conjunctive and disjunctive subproblems in order to make the search process more efficient and, also, the consequent learning process more meaningful. The decomposition is, of course, task-dependent and is the responsibility of another component of the problem solving system.

In many task domains, certain features of problems have a dominant role, their absence or presence may determine a large part of the solution regardless of the contributions of other features. Problems in econometrics or meteorology come into mind, in which there is a well-defined hierarchical relationship among the features to be considered in order to identify the solution correctly. The contributive model would prove to be extremely inefficient in such cases; we have to find an additional source of power for such complicated search processes.

Let us turn now our attention to the first paradigm shown on Figure 1/a. Remember, the function $\underline{M1}$ represents mappings between problems whose solutions are "similar." On an intuitive basis, one could say that two similar problems will be close to each other in the problem space defined by a judiciously selected set of features. One can also expect the respective solutions to be close in a corresponding solution space. If we can find a mapping function between a problem with a known solution and one without a solution, it is plausible to try to apply the same mapping function to the known solution to obtain a tentative solution to the new problem.

In order to measure the "distance" in the problem space, we have to define a metric. Let the distance be the weighted sum of the number of features shared by the two problems. The weighting should express the relevance of the features with regard to the solution. The whole system of weights of relevance can be well expressed by a hierarchical structure -- hence the name <u>hierarchical model</u>. Without any prior knowledge or help from the user, the program has to assume all possible permutations among the features of a new problem, only one of which will be correct. The ordering of features in a linear string for a given problem represents a part of the hierarchy of features characteristic of the whole problem domain. (Note that the constributive model becomes a special case of the hierarchical one if in the experience gathering phase, combinations rather than permutations of features are recorded.)

The system will return that sequence of solution steps which has been associated in the knowledge base the highest number of times with the longest string of matching features, starting with the assumedly most important feature in the leftmost position. If the knowledge base is "sufficiently" large, individual features and ordered feature sequences are uniquely associated with solution steps and solutions step sequences; so the direct application of the mapping function $\underline{M1}$ is possible. Otherwise, the same problem-solving philosophy prevails concerning the list of potential solutions supplied by the hierarchical model as with that supplied by the contributive model. Namely, these lists are ordered according to the plausibility level defined by the score of prior usage (the number of times the same solution step is recommended by different feature entries). Tentative solutions are completed and tested by other analytical or logical program components.

The learning process works in two directions. First, irrelevant permutations of problem features become gradually deleted. Second, new entries consisting of ordered features and solution steps are added to the knowledge base.

4. SOME ADVANCED PROCESSES RELEVANT FOR ANALOGICAL REASONING

We have outlined above the importance of similarity between new and old problems in regard to problem-solving. There are, however, many types and levels of similarity and also many ways in which they are associated with the selection of proper actions. The two models of AR described above provide the fundamental ideas about discovering and utilizing the characteristics of situations but by themselves, they do not suffice in trying to establish high level relationships between problems and solutions, and between problems, and between solutions. The following ideas point to more sophisticated but still relatively easily implementable processes.

For most, if not all, nontrivial problem domains, it is possible to chunk properties according to several different principles. Although the asymptotic behavior of the problem solving system should not greatly depend on how the features are composed of individual problem properties (but, of course, memory and time requirements can vary widely), some chunking techniques could result in much faster knowledge acquisition than others. Therefore, the program should construct <u>several knowledge bases</u> for a given problem domain and see which one proves the most efficient and most effective.

Another idea of pragmatic value is to <u>start out with the contributive model</u> because it presents lesser processing demands. If some features seem to emerge as decisive factors in the selection of solution steps, these should be separated and placed in the hierarchical model.

A very important high level learning process is the <u>pruning of the knowledge base</u>. Various inductive and deductive inferences can be generated, which can reduce the domain of search for solutions. For example, the separation of "influential" and "uninfluential" features would be guided by the numbers of prior usage. There are, however, two types of irrelevant features which also need to be identified. One type has a fairly uniform contributive effect which can be easily detached. The other type causes random and fluctuating contributions to the scores, particularly at the early stage of knowledge acquisition. We have found in an earlier study [5] that such an irrelevant feature, in contrast with "influential" ones, has the following characterisitics. If one forms the ratio of the scores of two solution steps associated with this feature, the ratio assumes widely different values as experience accumulates. This ratio is, however, fairly constant for "influential" features. Furthermore, one can discard those entries in the knowledge base of the hierarchical model in which some permutation of features has suggested wrong solutions more than a certain number of times. This latter number, of course, decreases as the knowledge base increases. Combining features into higher order ones (further chunking), when justified, should also streamline the knowledge base.

It is known that humans discover and utilize, mostly on the basis of previous experience, higher order similarities also, such as theorems of duality, structural, semantic, functional, and thematic similarities. The incorporation of these concepts in problem solving programs is a major challenge. Our contention is that the two fundamental models of AR can deal with such high order similarities as long as sophisticated feature extraction programs can be set to work in cooperation with the AR component.

Finally, we note that the paradigm of AR should be capable of discovering efficient problem solving strategies. The flexibility and modularity of system components responsible for knowledge acquisition, feature extraction, feature chunking, and composition of tentative solutions would make it possible to experiment with different strategies for given problem domains and to come up with one that appears to be optimal according to some set of criteria.

5. THE 'WORKER' AND ITS TOOL KIT

Next, we describe the first few steps already made in a planned sequence of experiments of gradually increasing complexity to illustrate AR as part of a problem-solving strategy.

A simulated robot, WORKER, has the task to build reasonably complex structures out of simple building blocks. Superficially, it is similar to Fahlman's BUILD [21]. Unlike BUILD, WORKER does not start with a complete specification of the goal object to be built -- the goal description contains only its "critical" properties. WORKER must plan which blocks to use and a best-first search determines the least-cost sequence of actions to attain the goal. On the other hand, whereas BUILD is given stacked blocks which need to be freed before use, in our case the blocks are placed separately on a plane. Further, we are not concerned about testing for stability and finding movable sub-assemblies -- some of the problems BUILD faces.

WORKER plans a solution and, when verified, attaches it to the relevant problem description. When the same goal has been attained several times with a variety of building blocks, WORKER transforms the solution sequence of actions into an ordered set of if-then rules. The AR module then decreases the size of the rule base by replacing the individual situations by situation classes. This revised rule base is transfomed into a decision graph, which is finally reduced to its simplest form by eliminating all redundancies. The organization of the system is shown on Figure 2.

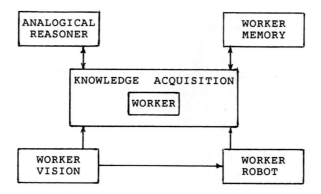

Figure 2
The organization of the system

We shall now discuss these procedures in some detail.

5.1. THE WORKER

The representation of knowledge about the blocks and the goal uses the frame format. We identify four conceptual levels of frames as follows.

(i) The total problem situation is at the top level. Slots pertain to the properties of the situation as well as the available object types. The values of the latter, namely the number of objects of different types available, can be provided either directly by the user or by a program, WORKER VISION, which uses the digitized image of a TV camera. A slot is associated with the goal and another with the solution once it is obtained.

(ii) The representation of the properties and, when applicable, components of objects is at the next lower level.

(iii) Since components, themselves, may have components through an unlimited number of levels, these frames are like the above object frame.

(iv) Finally, properties of situations, objects or components are at the bottom conceptual level.

The construction plan is a sequence of steps applying either put operators (put-on-top, put-on-side, put-on-end) for similar objects (sharing some dimensions) or join operators (join-on-top, join-on-side, join-on-end) for non-similar objects. The best-first search strategy has a gradually diminishing number of operators and objects as the goal is getting closer. WORKER continually checks whether the current situation has been encountered before, and, if so, what plan was generated for it.

A number of heuristics narrow the list of applicable operators and objects. Such include the following.

H1: Build complex objects by first building columns and rows of objects.

H2: If the (sub)goal specifies height but not width or length, then use the operator put-on-top.

H3: If the length and/or width of the goal object is greater than those of the available blocks, generate appropriate subgoals.

The plan generated is displayed to the user for confirmation before placing it into the action slot of the problem situation frame. Alternative plans may be generated at other times or provided directly by the user. These are ordered according to cost, i.e., the number of steps involved. Further, the slot prec (precondition) holds functions which test the applicability of the plan (properties of the goal and object components matched with the current situation). The slot perf (performance) contains a function which executes the highest priority plan in the action slot.

5.2. The AR Module

AR takes place at two levels: (i) at the object level, the equivalence of different objects is discovered, and (ii) at the situation level, different problem situations are judged equivalent.

(i) Equivalent objects can replace each other in the construction of (at least one) goal object. The program assigns equivalent objects to a class whose definition is specific to the goal object. A particular class of blocks contains also such blocks that have different measurements along a "don't care" dimension, implied by the goal specification. Other goals may be specified by the user as having a measurement "less than X" or "greater than Y". These also contribute to the formation of classes.

The numerical value of object equivalence is found by exchanging the objects in the rule base. It is the percentage of the number of rules with unchanged precondition that have unchanged actions. This measure, therefore, estimates the probability of a correct response if two equivalent objects replace each other. Whenever the equivalence value is above a threshold, a new frame is created to define the new object class.

AR checks the completeness and consistency of alternative definitions of object classes. If a single property suffices, others will be ignored as redundant. Otherwise, combinations of properties are tried in a systematic manner until, eventually, an exhaustive Boolean combination of all properties may be arrived at as a last resort. Also, all totally correlated measurements (e.g., always identical width and length) are reduced to one variable -- another method of eliminating redundancy.

(ii) Equivalent situations have the same number of objects in each category of relevant objects. (The number of non-relevant objects may differ in equivalent situations; these do not contribute to construction plans, anyway.) The relevance of an object G to a particular goal can be determined by dividing the rule base into subsets of rules in each of which the number of objects other than G are held constant. In each subset i, we find the number of non-equal actions A_i. If the number of rules in the subset is M_i and T is the total number of subsets, then the relevance of object G is

$$R_G = 1/T \cdot \sum_{i=1}^{T} (A_i - 1)/(M_i - 1)$$

Assuming a uniform probability distribution of situations, R_G is a measure of the correlation between the actions and the number of objects of the category in question. The program compacts the rule base by eliminating objects of low relevance from the condition parts.

5.3. WORKER Memory Processes

Beyond compacting the rule base, further memory savings can be achieved by compiling the rules into a decision tree. In fact, to save computing time, this is done each time a given number of new rules have been established.

There are other methods that optimize decision trees. First, individual values of decision variables are combined into subranges as large as possible without any information loss. More importantly, the order of the decision variables from the root matters a great deal. With N decision variables, there are $N!$ possible orderings, requiring $(N! - 1)$ comparisons to find the tree with the least number of nodes and links, in an exhaustive search. Instead, we have developed an algorithm [22] that requires the construction of only $N(2^{N-1} - 1)$ decision trees. By taking into account the so-called adjacency restriction [23], this number can be further reduced to $(N - 1)(2^{N-1} - 1)$.

We note that we have developed also some heuristic techniques that provide satisfactory but not necessarily optimum memory savings in rearranging the level order of decision variables on a decision tree.

5.4. The User-WORKER Interaction

The KNOWLEDGE ACQUISITION (KA) module transforms WORKER into an expert system by allowing the user to input his own rules for constructing objects. This greatly enhances the peformance of the system as complicated substructures can be included and later used as components for difficult goal objects. A menu enables the user to do the following.

(i) **Define a new object**
The user is prompted for the name, properties and their values of a new building block or goal object.

(ii) **List of available blocks**
The user provides object names of blocks to be used.

(iii) **Provide a new solution**
The user provides the name of the goal object and the set of building blocks available. After that, he is asked for the sequence of operators, one at a time, for the whole plan.

(iv) **Delete a plan**
The user is prompted for the name of the goal object and of the plan to be deleted.

(v) Define a new heuristic

The user provides the name, the condition and the action of the heuristic. Unless the latter involves only predefined functions, this task is to be left to the programmer.

(vi) Delete a heuristic

Again, this is to be left to the programmer.

(vii) Display an object or solution

The frame description of an object or of a problem situation with its attached solution will be displayed. The user is prompted for the name of the item to be displayed.

(viii) Start

This initiates program performance.

(ix) Stop

This gets the user out of the KA module.

The KA module also includes an explanation facility. The user, upon seeing the partial or full sequence of operators leading eventually to the construction of the goal object, can ask WHAT? ("What did you just build?") and WHY? ("Why did you build it this way?"). The system answers by displaying the frame description of the object built up to that point or, in the second case, of the problem situation that contains the action sequence employed.

The second type of question is very useful in both debugging the system, and in enlarging and refining its knowledge base. Namely, the user can delete a particular plan and replace it with one or several new ones. For this very reason, the system must be able to detect and correct human error, as well as determine the relative adequacy of alternative plans.

5.5. The WORKER-Vision and the WORKER-Robot

The task of the WORKER-Vision is to identify the blocks on the table, and their position and orientation. Its input from a matrix camera is the grey levels in 128x128 pixels, ranging from 0 to 255.

The input to the WORKER-Robot is the action sequence planned by the WORKER in terms of operator steps. WORKER-Robot transforms this into a sequence of approach, pick-up and put-down commands, which it then performs.

6. SOME RESULTS AND CONCLUSIONS

Success is measured by the degree of improvement in the performance of the system over trials in which the same goal object is attempted with different sets of randomly chosen building blocks. A faster response time indicates savings gained by retrieving solution segments to recognized subgoals, rather than generating a plan from scratch.

We have considered learning here to be a three-stage process. In the first stage, termed the <u>shaping</u> stage, search techniques are utilized to find a solution to a particular problem situation. The resulting plan is an action sequence which is then linked to the situation. In the second, <u>analogical reasoning</u> stage, problems associated with similar action sequences are grouped under a single situation class and a class definition is established which is sufficient to distinguish its members from other situations encountered. At the end of stage two, there exists a set of rules each of the form: <u>if</u> we have a member of a certain situation class <u>then</u> perform this action sequence. In the third, <u>consolidation</u> stage, the rules are compiled into a decision <u>graph</u>, which is then simplified by eliminating redundancies on it and ordering the decision variables on it in an optimum manner so that the desired plans can be retrieved most efficiently.

One must realize that there exists a trade-off between efficiency and accuracy. If AR is involved "too early", i.e. when the knowledge base contains only a small sample of problems and solutions, the retrieved solution segments will be worse than least-cost. The error rate of this type is decreased when the invocation of AR is delayed. One could, in principle, optimize the overall cost of obtaining solutions if an appropriate cost function on search and plan execution could be established. We intend to study this problem further.

In spite of the relatively simple task environment, we were able to pose and answer some very difficult questions. We feel, however, that the experiments performed served a much more important purpose; namely, we have become aware of the depth and breadth of the problems involved in AR. We now realize that we must yet go very far and we can even see a little ahead of us along that road.

7. REFERENCES (In chronological order)

[1] Evans, T. G.: A program for the solution of a class of geometric analogy intelligence test questions (In M. Minsky (Ed.) <u>Semantic Information Processing</u>. MIT Press: Cambridge, MA, 1968).

[2] Becker, J. D.: The modelling of simple analogic and inductive processes in a semantic memory system (<u>Proc. IJCAI-1</u>, pp. 655-668; Washington, DC, 1969).

[3] Kling, R. E.: A paradigm for reasoning by analogy (<u>Artificial Intelligence</u>, 2, pp. 147-178, 1971).

[4] Findler, N. V.: The language AMPPL-II (In N. V. Findler, J. L. Pfaltz and H. J. Bernstein: <u>Four High-Level Extensions of FORTRAN IV: SLIP, AMPPL-II, TREETRAN and SYBOLANG</u>. Spartan Books: New York, 1972).

[5] Findler, N. V.: Analogical reasoning in problem solving (<u>Proc. IJCAI-77</u>, vol. I, pp. 345-346, 1977).

[6] Findler, N. V. and J. N. Shaw: MULTI-PIERRE -- A learning robot system (<u>Computers and Graphics</u>, 3 pp. 107-111, 1978).

[7] Findler, N. V. and D.-T. Chen: Toward analogical reasoning in problem solving by computers (J. of Cybernetics, 9, pp. 369-397, 1979).

[8] Winston, P. H.: Learning and reasoning by analogy (Comm. ACM, 23, pp. 689-702, 1980).

[9] Gick, M. L. and K. L. Holyoak: Analogical problem solving (Cogn. Psych., 12, pp. 306-355, 1980).

[10] Carbonell, J. G.: Experimental learning in analogical problem solving (Proc. AAAI-82, pp. 168-171; Pittsburgh, PA, 1982).

[11] Winston, P. H.: Learning new principles from precedents and exercises (Artificial Intelligence, 19, pp. 321-350, 1982).

[12] Gentner, D.: Structure mapping: a theoretical framework for analogy (Cogn. Science, 7, pp. 155-170, 1983).

[13] Burstein, M. H.: A model of learning by incremental analogical reasoning and debugging (Proc. AAAI-83, pp. 45-48; Washington, DC, 1983).

[14] Winston, P. H., T. O. Binford, B. Katz and M. Lowry: Learning physical descriptions from functional definitions, examples, and precedents (ibid, pp. 443-439).

[15] Carbonell, J. G.: Learning by analogy: formulating and generalizing plans from past experience (In R. S. Michalski, J. G. Carbonell and T. M. Mitchell (Eds.): Machine Learning: An Artificial Intelligence Approach, Tioga: Palo Alto, CA, 1983).

[16] Gick, M. L. and K. J. Holyoak: Schema induction and analogical transfer (Cogn. Psych., 15, pp. 1-38, 1983).

[17] Hall, R. P.: Analogical reasoning in artificial intelligence and related disciplines (Tech. Report, Dept. of Comp. and Info. Sciences, Univ. of California at Irvine, 1985).

[18] Kedar-Cabelli S. T.: Analogy -- from a unified perspective (Tech. Report, Dept. of Comp. Science, Rutgers Univ., 1985).

[19] Carbonell, J. G.: Derivational analogy: a theory of reconstructive problem solving and expertise acquisition (In R. S. Michalski, J. G. Carbonell and T. M. Mitchell (Eds.): Machine Learning: An Artificial Intelligence Approach, Vol. II, Morgan Kaufmann: Los Altos, CA, 1986).

[20] Miller G. A.: The magical number seven, plus or minus two: Some limits on our capacity for processing information (Psych. Rev., 63, pp. 81-97, 1956).

[21] Fahlman, S. E.: A planning system for robot construction tasks (Artificial Intelligence, 5, pp. 1-49, 1974).

[22] Bickmore, T. W., N. V. Findler, L. H. Ihrig and W.-W. Tsang: On the heuristic optimization of a certain class of decision trees. (Submitted for publication).

[23] Smith, D. E. and M. R. Genesereth: Ordering conjunctive queries (Artificial Intelligence, 26, pp. 171-215, 1985).

KNOWLEDGE-BASED REAL-TIME CHANGE DETECTION, TARGET IMAGE TRACKING AND THREAT ASSESSMENT

L. F. Pau
Technical University of Denmark
DK-2800 Lyngby
Denmark

ABSTRACT

This paper describes the overall hardware and software architecture of a knowledge-based change detection, target tracking and threat assessment system. It ultimately assigns threat level and threat scenario labels to the observed scene, based on time dependent target features or changes in the scene, as well as target maneuvers and target number. The aims are: reduced scene interpretation workload, shorter reaction time, reduced false-alarm rates, and simple adaptation to changing fields of view/sensors.

1. INTRODUCTION

   1.1 Target Understanding

   Significant work has been carried out in the past, aiming at using scene background, map and scenario related knowledge, to carry out image understanding tasks, e.g. evaluate the meaning of the configuration of objects in the scene in relation to image feature extraction. Therefore, besides image segmentation and feature extraction, we must address the problems of knowledge representation, knowledge-based search, and the resolution of conflicting information. Among the better approaches are a distributed architecture and control structure, with parallel non-directional search algorithms (1, 2).

   1.2 Temporal Knowledge in Target Understanding

   In the field of change detection, target detection, tracking and classification, the classical approach is to segment out the areas of change (targets), and classify them one at a time or jointly in each image frame (3).

   This approach does, however, not allow for making use of object-to-object relations, frame-to-frame relations (also called disparity) (4), object-to-background relations, nor target motion estimation. More generally, temporal context knowledge can be introduced into target

tracking and classification, noticeably for threat assessment and false
alarm reduction. Threat assessment is the extension of target image
understanding to image sequences, by allowing for increments of symbolic
information over time. The previously mentioned distributed architecture
and parallel search then apply to subsequent time instants or time
intervals.

Because procedural or symbolic information is used, target and change
understanding go beyond the field of motion analysis (14), which
emphasizes primarily the computation of observables (points, lines,
contours, range, optical flow) in a sequence of images, and time-relations
between these same observables, e.g. one may want to get the object
skeleton in one frame and relate it to the next skeleton.

Tasks similar to target and change understanding have been studied in
relation to automobile traffic flow, blood cell motion (15), computer
based lip reading, motion compensated image coding, manufacturing control (16), heart wall motion.

1.3 Control Structure

The research will serve the purpose of carrying-out, in both top-down
and bottom-up modes, the matching of the detected motion/target features
with the knowledge bases in order to infer threat levels. For example,
a large change/target area may trigger the application of certain texture
measures to evaluate smoke, screens, chaff or other related phenomena,
Or, as another example, the number of detected change areas/targets or
their density, may be such that frame-to-frame procedural relations
will label most as noise or clutter. In the simplest approaches, pairwise symbolic matching computes disparity between target pairs in two
different image frames; for each candidate target, a local search is
made. The transformation that gives the best matching target in the
search area is mapped and clustered in the Hough domain; the location
information about the matching target pairs enables motion detection.
Yet pairwise symbolic matching does not embody knowledge about a
feasible, suspected or impossible movement in time and space.

1.4 Scope

Therefore, the present paper reports about research and implementation
of a hierarchical knowledge based real-time target tracking and threat
assessment system, which is characterized by knowledge based processing
of the temporal information.

The lower level performs moving change area/target preprocessing, background subtraction, object labelling, clustering and segmentation, target feature extraction, and position related risk assignment.

The medium level performs target tracking using background chaining from current positions, with forward chaining of motion estimation constraints. This procedure includes target/track split or merge handling.

The highest level, implemented in compiled PROLOG, contains three sets of threat assessment procedures, all subject to global unification of the variables in each set:

i) position independent threat class classifications of all change areas/targets, using current observations only,
ii) penetration scenario predicates, covering different types of target maneuvers in variable time-intervals, and operating on i),
iii) threat assessment metarules operating on the type ii) predicates to merge/delete/fill/substitute (penetration scenario, time interval) objects, and generate threat situation output (not directly target related).

KNOWLEDGE REPRESENTATION (KR)

The target related KR is critical for the system correct classification and threat assessment probabilities, including false alarm rates. However, it is also critical in terms of processing speed from frame acquisition through threat assessment completion; but the two requirements are conflicting. In this system, the compromise KR, in view of the available knowledge base, is as follows:

2.1 Target Features (F(a))

These features, resulting from the low-level image processing, are:

```
F(a)=(f1(a)----fn(a))     feature vector
a=target number
f1(a)=target area (estimated size)
f2(a)=target risk label, which is one of the threat levels r1,--,
      rn defined below
f3(a)=x coordinate of target centroid
f4(a)=y coordinate of -      -
f5(a)=variance in x direction of target
f6(a)=variance in y direction of target
f7(a)=cross variance of x-y target
f8(a)=first invariant moment of target
f9(a)=second  -              -     -    -
f10(a)=emissivity.
```

The threat levels (ri) are a position-only characterization of the threat posed by a target located in a region labelled by this level r_i. By convention, the threat level of a target which poses the highest possible danger by virtue of its position only, will be r_n. The lowest threat level will be r_1.

2.2 Threat Scripts

The KR selected is based on scripts intrinsic to variable time-intervals, in order to characterize the threat as a result of the target movements, and of the context.

```
Script = "Threat Type" (           )
Theater=(         )
Target role=(           )
Point-of-view=(Detection, Evade, Deception, Counter-measures)
Time-interval=(Longest interval TM)
Sensor=(           )
Event-sequence=first:   Time-intervals=(Subinterval)
                        Position=(f3(a),f4(a))
                        Size=f1(a)
                        Shape=(f8(a),f9(a))
                        Risk=f2(a)
                        Emissivitiy=f10(a)

           thereafter:  Time-interval=(Subinterval)
                        Maneuver=
                        Position=
                        Size=
                        Shape=
                        Risk=
                        Emissivity=
                        Direction of movement=(m)
                        Speed=(v)

           until:       Time-intervals=(Subinterval)
                        Maneuver=
                        Position=
                        Size=
                        Risk=
                        Emissivity=
                        Direction of movement=
                        Speed=
```

In this script:

- The time sub-intervals are subintervals of the largest interval specified in the script head.
- The direction of movement (m) is estimated as the slope of the oriented position vector from the previous time-interval in the event sequence, to the current time-interval.
- The speed (v) is estimated likewise to (m).

2.3 Script Representation

All scripts are represented as first order logic predicates instantiated by a time trigger. The scripts in the knowledge base are thus predicate facts, coded in PROLOG.

2.4 Threat Type Representation

The threat type is the head of each script. The threat types are selected in a finite, application specific, list (tl, l=1,p).

3. THREAT ASSESSMENT KNOWLEDGE BASE

The threat assessment knowledge base is using first order predicate logic for the threat scripts. It is decomposed into three specialized knowledge bases.

3.1 Current Observations

The corresponding predicates are simple classification rules applying to each target, which are only target size, shape, and position related. These classification rules classify each target into threat classes using the current target features f1, f2, f3, f4, f8, f9, f10 provided by all instantiated scripts. Examples for threat classes are:

* large threat   - target in high threat level area (e.g. (f2(a) >rL)
            OR - large target (e.g. f1(a)>sL)
            OR - specific shape (e.g. <uL, (f8, f9))> >0, where uL is a discriminant mapping)
* medium threat - target in medium threat level area (rm1<f2(a)<rm2)
            OR - unknown shape
            OR - medium size (e.g. sm1<f1(a)<sm2)
            OR - specific emissivity f10(a)
* little threat - defined specifically in a similar way

3.2 Maneuvers

Maneuvers which cover one time interval at the time, up to the current instant, but apply to all targets observed in that time interval. A penetration predicate is only true if a maneuver has indeed happened within that time interval, involving targets characterized only by the threat class allocation and by the observation time. The intervals are thus predicates applying only to the relative time ordering and number of threat class observations. Such general maneuvers include:

* approach
* move-away
* hide
* surge
* crowd-increase
* crowd-decrease
* evade

Each such predicate is defined by alternative logical conditions. The maneuvers are slots in the script.

3.3 Scenarios

Each scenario is defined as a sequence of maneuvers related to adjacent time intervals. The scenarios are specified in the knowledge base through metarules in predicate form, applying to the merger/deletion/ substitution of the scripts describing maneuvers in the scenario. These metarules are supplemented by other rules applying to:

- the time-consistency of target tracking information
- the stability over time of certain target behaviours to make them consistent with a given scenario
- the cumulated target occurrences and threat levels over the scenario time interval
- the reassessment of the inferred target classifications and maneuvers in view of persistence/disappearance of specific target features.

3.4 Remark

As a result of the selected knowledge representation and knowledge base structure:

* the inferred scenarios do not depend explicitly on specific targets, images or background
* the maneuvers do not depend explicitly on specific targets, images or backgrounds, but can only be characterized in a given background
* the target threat classes are target dependent, but not frame or background dependent
* the target features are independent of the background reference image.

4. TARGET FEATURE EXTRACTION

The target features of Section 2.1 are obtained by applying a feature extraction preprocessing to the scene images. It consists of a training stage, and in a subsequent feature extraction stage, The result is a binary image with the targets, and several target feature lists. The targets may be colour labelled and counted, each target having a separate colour/number.

4.1 Training of the Feature Extraction

The training of the feature extraction consists of:

a) A reference background image IO, updated every time the overall integral illumination level changes by more than p% (typically 3-5%) and at the latest every time period T (sensor related).

b) A signal-to-noise ratio a (typically a>4 dB) in the difference image dI=I(t)-I(t-1), used for fixing the thresholding level of dI on the basis of the dI histogram of the grey levels.

c) A risk-zone characterization of the scene IO, whereby (possibly overlapping) closed regions in IO get assigned a priori threat levels (ri), rl>---->rn, through a graphic tablet or another man-machine interface. As a result of this definition, a given pixel in IO may have no, or several threat level assignments. If colour values are assigned to the threat levels (ri), and each closed region is filled according to these colours, IO leads to a coloured threat level image R characterizing the overall threat from position information only.

4.2 Target Feature Extraction

Assuming the:

* reference background image IO to be stored in the image plane 2
* the moving target image MTI (t) to be stored in image plane 3

the basic target feature extraction algorithm is:

F0 = Zero image planes 1, 3, 4; store IO in 2,
F1 = Acquire at time (t) and digitize the sensor image in image plane 1, and display it,
F2 = Subtract (scaled subtraction) the reference image in image plane 2 from 1, and display the result from image plane 1,
F3 = Apply low-pass filtering to 1, and assign result to image plane 4,
F4 = Threshold the image in 4 according to the SNR ratio a (see 3.1),
F5 = Apply a high-pass filter to 4, and assign the result to image plane 1,

F6 = Subtract the moving target image MTI (t-1) in 3 from 1, and assign result to 1,

F7 = Threshold 1 with a fixed threshold to generate a binary target image, and assign the result to 3 to obtain the updated MTI (t),

F8 = Apply a line-by-line target labelling procedure to 3; locate and count all subtargets at a time (t),

F9 = Apply a clustering procedure to the results of F8, to group together into targets connected subobjects; locate and count the number of targets,

F10= Calculate by line-by-line recursions the target features:

f1(a) (t) = target (a) area

f3(a), f4(a) (t) = target (a) coordinates

f5(a), f6(a), f7(a) (t) = target (a) covariance matrix elements

f8(a), f9(a) (t) = target (a) invariant moments

F11= Assign the target risk label list f2(a) (t), the threat levels existing in the risk zone characterization R, at the target centroid locations (f3(a), f4(a)),

Repeat F1

5. INFERENCE OF THE THREAT LEVEL

5.1 Script Instantiation

The inference of the threat level is carried out each time t, on the basis of the longest past time-interval (t-TM, t). This inference is carried out using the knowledge representation of Section 2, where the script slots are filled for each target (a) by the target features (fj(a), j=1, 10) at all instants in the subintervals of (t-TM, t) in which (a) was observed.

5.2 Inference Engine:

The selected inference engine applying to the threat assessment is the predicate unification and backtracking of PROLOG; it does however include elimination predicates applying to repeat answers to queries, as well as postponed conditional predicate evaluations ("dif" predicate).

5.3 Tracking Algorithm:

Target or change area tracking is a by-product only of the target understanding system. Tracks are determined separately, and displayed in overlay mode to the displayed sensor data. The approach is to apply a multiple criteria graph routing algorithm, implemented in PROLOG, which will ensure that frame-to-frame relations between target positions satisfy a dominance relation. This dominance relation must hold jointly to the three criteria; average speed, object size, risk. Merge rules apply to these criteria when valuating alternative routes between locations at either instant.

6. REAL-TIME IMPLEMENTATION

The operational environment is composed of:

1) A 32-bit high speed bi-processor computer system; the front end processor carries out target feature extraction concurrently with the main processor which runs the inference procedure. The shared memory contains the knowledge base, as well as target feature files and tracking data. The front-end code was implemented in Assembler after translation from FORTRAN-77, and also with a selection of commands to be executed on microcoded firmware. The main processor runs compiled C-PROLOG. The labelled targets are also colour-indexed.

2) A DMA between the host and the image processor.

3) A general purpose image processor, with four 512x512x8 bit image planes, a sensor data digitizing board, and a MC 68000/VME board with additional RAM for object code developed on EXORMACS (from original application specific macrocommands and Assembler code). The image processor implements the feature extraction algorithm (F), and triggers the sensor data acquisition/build-up.

The processing speeds achieved are, as an example:

    Number of targets: Maximum 8
    Number of types of maneuvers: 14
    Number of scenarios: 5 (with each up to 20 maneuvers)
    Target feature extraction
       time (incl. sensor data acquisition): 0,2 s.
    Threat inference time: 200-900 ms (least for the highest
                                                      threat levels)

This environment especially relies on PROLOG-FORTRAN interfaces, as well as a set of drivers and synchronization routines imbedded in the real-time operating system.

7. DEVELOPMENT ENVIRONMENTS

Three development environments have been developed to test separately parts of the system, or to allow for simulation and knowledge acquisition:

* Simulation of target movements and corresponding target feature extraction, and script instantiation, with or without tracking.
* Simulation of scenarios, from simulated target features changing over time, and corresponding knowledge base management as well as inference
* Integrated development environment using a VAX 780/VMS instead of the bi-processor computer system
* Graphics package to acquire the threat level zones through a graphic tablet or cursor

8. CONCLUSIONS

This system achieves high speed through a knowledge based control of the target and change area evaluation algorithms, which get selected dynamically in response to specific queries or inferences. This knowledge based change analysis, target tracking and threat assessment implements a knowledge based controller tightly related to operational expertise. Especially, the use of temporal context increases the capability of such a system, by better threat assessment over time. Finalized implementations require special purpose circuits, faster memory access, micro programming, pipelining and distributed knowledge base organization.

9. REFERENCES

1. R.A. Brooks, Model based 3-D interpretation of 2-d images, IEEE Trans. on PAMI, Vol. 5, 140-150, March 1983.
2. V. Kumar, L. Kanal, Parsalled branch & bound formulations for AND/OR tree search, IEEE Trans. on PAMI, February 1984.
3. SPIE, Infrared technology for target detection and classification, Proc. of SPIE, Vol. 302, August 1981.
4. S.T. Barnard, W.B. Thompson, Disparity analysis of images, IEEE Trans. on PAMI, Vol. 2, Nov. 1980.
5. D.B. Reid, An algorithm for tracking multiple targets, IEEE Trans. on AC, Vol. 24, p. 843, 1979.
6. Y. Bar-Sholom, Tracking methods in a multitarget environment, IEEE Trans. on AC, Vol. 23, p. 618, 1978.
7. C.B. Chang, K.P. Dunn, L.C. Youens, A tracking algorithm for multiple targets in clusters, MIT Lincoln Lab, TR-687, 1984.
8. A.V. Forman, P.J. Rowland, W.G. Pemberton, Contextual analysis of tactical scenes, Proc. of SPIE Application of Artificial Intelligence, Vol. 485, p. 189, 1984.

9. E.M. Drogin, AI issues in real time systems, Defense Electronics, p. 71, August 1983.
10. A. Barr, E.A. Feigenbaum, P.R. Cohen (Eds), The handbook of artificial intelligence, William Kaufmann, Los Altos, CA, 1981.
11. Proc. Workshop on applications of image understanding and spatial processing to radar signals for automatic ship classification, Naval Electronic Command, Washington, D.C., February 1979.
12. R.P. Bonasso, Analyst: an expert system for processing sensor returns, MITRE Corp., MTP-83W00002, 1983.
13. L.F. Pau, M.El Nahas, Infrared image acquisition and classification, RSP, Wiley, London/N.Y. 1984.
14. J. Aggarwal (Ed), Proc. of IEEE Workshop on motion representation and analysis, Kiawah Island, Charleston, S.C. May 7-9, 1986.
15. M. Levine, A rule based system for the characterization of blood cell motion, to appear, CRC Press, Boca Raton, FL. 1986.
16. G. Borchardt, A complete model for the representation and identification of physical events, M.S. thesis, University of Illinois, Champaign, IL., 1985.

FIGURE 1 :   MOTION TRACKING

APPROACH: CONSISTENCY VERIFICATION IN FRAME-TO-FRAME CORRESPONDENCE THROUGH MULTIVALUED GRAPH ROUTING

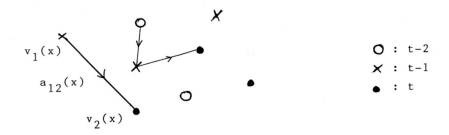

○ : t-2
✕ : t-1
● : t

ARC ATTRIBUTES:

ATTRIBUTE OF ARC a_{ij}:   SPEED   MERGE RULE : AVERAGE
                              OBJECT SIZE   MINIMUM
                              RISK   MAXIMUM

ROUTING ALGORITHM: DOMINANCE RELATION BETWEEN PATHS WITH ARCS HAVING INCOMPLETE FEATURE/ATTRIBUTE LISTS

IMPLEMENTATION: . FINITE SLIDING MEMORY
                  . PROLOG UNIFICATION
                  . EXTENSION TO MULTIPLE SENSORS, AND CLUSTER TO CLUSTER CORRESPONDENCE

FIGURE 2: THREAT UNDERSTANDING ARCHITECTURE

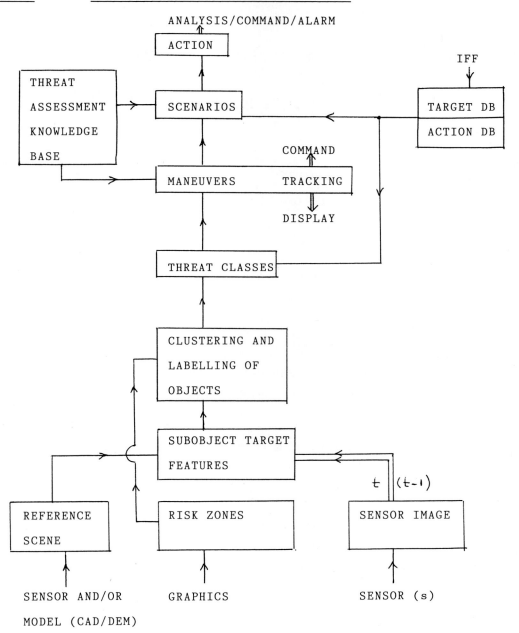

FIGURE 3:   INFERENCE OF THREAT LEVEL

1. FILL SCRIPT SLOTS
2. CONTROL STRUCTURE :
   Prolog top-bottom and left-right unification of facts, scripts (in the world) with knowledge base
3. INFERENCE:
   Expresses threat build-up by relating to scenarios detected; this search requires:
   i. ELIMINATION PREDICATES FOR REPEAT SCENARIOS, OR ESCAPE STATES
   ii. POSTPONED PREDICATE EVALUATIONS ("dif")
   iii. INFINITE TREE UNIFICATION FOR UNKNOWN STATES

4. THREAT LEVELS : ●   ●   ●   ●   ●

THREAT ASSESSMENT TEST SEQUENCE

t = 0
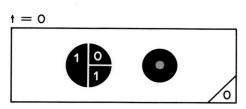

t = 5
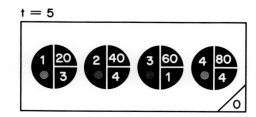

t = 1
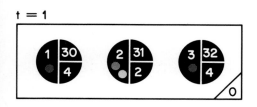

t = 6
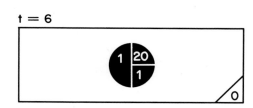

t = 2
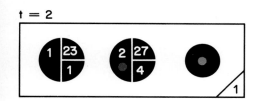

t = 7
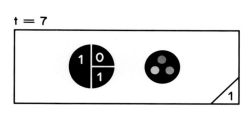

t = 3
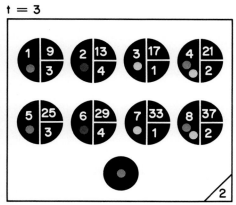

t = 8
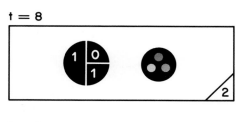

t = 9
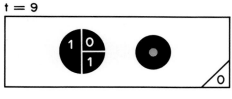

t = 4
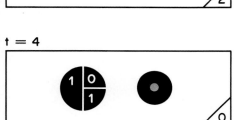

LEGEND:
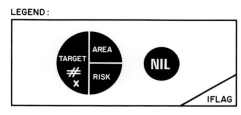

TASK AND PATH PLANNING FOR MOBILE ROBOTS

Raja Chatila
Georges Giralt
Laboratoire d'Automatique et d'Analyse des Systemes du CNRS
7, Ave. du Colonel Roche
F-31077
Toulouse Cedex, France

ABSTRACT

Autonomous mobile robots represent today a perfect paradigm for third generation robots and possibly the clearest case study for machine intelligence. They imply multi-level environment perception and modelling, decisional autonomy ranging from general planning to specific task operating, autonomous mobility capacity, sophisticated high level man-machine interface, and efficient execution control systems.

This provides for a very large set of research issues as well as for the start up of a broad domain of real-world applications including projects which are outdoors oriented such as ALV and indoors or around-buildings oriented such as HILARE.

Most of the concepts and results presented in this paper are general and can be extended such as to cover man-made and natural-like environments. We have chosen to use a factory-like environment in which we hypothetically place HILARE, an autonomous mobile robot, as a case study. We also describe very briefly some of the main features of HILARE Mark II, the new experimental robot we have designed.

In this framework we present a world model which includes:

(i) a set of operators, which correspond to elementary tasks, e.g. primitive actions that will be used in the man-machine communication interface.

(ii) a three-level environment model: geometrical, topological and semantic

(iii) the functional capabilities of the machine, represented by a set of Specialized Processing Modules (SPM) and Specialized Decision Modules (SDM).

The following section presents decision processes organization and inference mechanisms to achieve task and path planning. We stress the importance to produce at this stage an execution model that allows to implement an efficient execution control.

To conclude, in the last section we consider the role of automatic acquisition and learning processes to build and update environment models, and to instanciate and to refine operator parameters.

1. INTRODUCTION; 3rd Generation Robots

Research in the field of robotics has centered since the 1980's on the concepts, methods and systems which make up third generation robots, i.e. machines that exhibit "intelligent connection of perception and action" [Brady 85].

In this research trend, we witness today a broad, internationally spread-out effort to design and build robots which are based on highly complex signal and information processing systems, aiming to achieve the highest possible level of operating autonomy [Giralt 84a].

Two objectives are thus actively pursued: basic research in machine intelligence, and real-world applications in which important gains are expected, such as:

- better man-machine ergonomics;
- improved operating efficiency;
- increased operating flexibility.

Autonomous mobile robots represent today a perfect paradigm for third generation robots and possibly the clearest case study for machine intelligence. They imply multi-level environment perception and modelling, decisional autonomy ranging from general planning to specific task operating, autonomous mobility capacity, sophisticated high level man-machine interface, and efficient execution control systems.

This provides for a very large set of research issues as well as for the start up of a broad domain of real-world applications including projects which are outdoors oriented such as ALV [Ieeer&a 86] and indoors or around-buildings oriented such as HILARE [Giralt 79], [Chatila 82].

In these applications, mobility and decisional autonomy range from partial (man-supervised robots) to complete. Examples of special interest are:

- wireless vehicles in flexible manufacturing systems
- industrial cleaning
- construction sites
- farming (fruit picking, etc.)
- mining
- nuclear plants (inspection, repair, etc.)
- underseas
- public safety (fire fighting, rescue, etc.)

At the core of the research issues to be addressed, lies the development of a path and task planning system.

Our approach and the implementation we have made in the scope of the HILARE project are presented in the following five sections using a case study (section 2). Sections 3, 4 and 5 cover respectively the interrelated aspects of world modelling, decisional processes for task and path planning, and execution control. Section 6 is devoted to learning functionalities, a very important extension of the decisional structure.

2. CASE STUDY

Most of the concepts and results presented in this paper are general and can be extended such as to cover man-made and natural-like environments. We have chosen to use a factory-like environment as a case study (Figure 1) in which we hypothetically place the autonomous mobile robot HILARE. Let us consider the main features of this case study:

(a) The world layout is composed of three rooms (Ri), a corridor (C), a pillar (P), and six connecting doors (Di). Walls

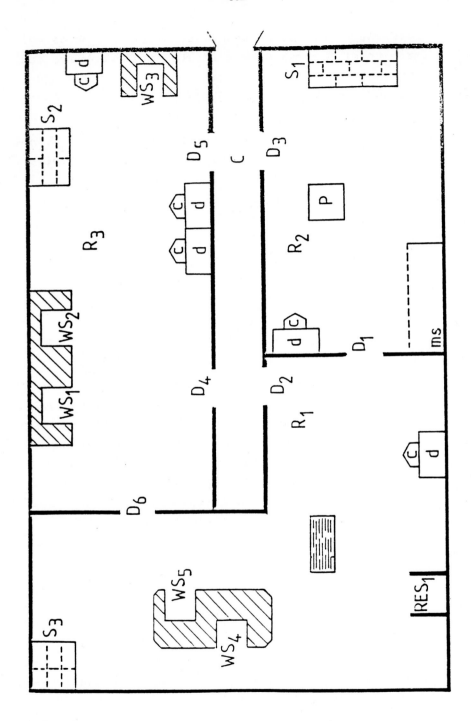

FIGURE 1 Case Study: Workshop-like Environment

are physical barriers which segment potentially traversable space, thus generating a natural topological space structure (section 3.2).

(b) A set of built-in working sites is distributed within the rooms. It includes five workstations (WSi), three storages (Si), one refueling station (RES), and one robot maintenance site MS.

(c) Non transportable objects of various kinds, such as boxes, tables, desks (d), chairs (c), may lay on the floor in any room. Some will only be present occasionally, and some others will have a more permanent location, but still changing from time to time. They will be considered as obstacles to be avoided by the robot while carrying out a task. Finally, there are transportable objects (Oi), that can be moved by the robot from or to some WSi or Si.

(d) The set of tasks to be executed includes object transportation, refueling and robot repair. While operating, the robot receives three different sorts of information inputs:

- orders, i.e. a partially ordered set of goals
- signals, i.e. updatings of the perceived state of the world
- data directly provided to the machine by the human operator or other systems, modifying or updating the world model (e.g. inventory status for storage Si, door Di open or closed)

In order to be able to properly operate, we will endow the robot, i.e. a specific version of HILARE, with the following physical features (Figure 2):

(a) A distributed processing architecture of 4 on-board microprocessors (Intel 8086's and 80286) radio-linked to a ground based computer.

(b) A belt of sixteen fixed ultrasonic emitter/receivers distributed on the robot and providing range data up to 2 meters.

(c) A panoramic scanning laser range-finder which provides range data up to 10 meters.

(d) Two optical path encoders for odometry and trajectory control.

FIGURE 2: HILARE

We will also briefly discuss the use of stereovision.

3. THE WORLD MODEL

The world model consists of the robot's operating knowledge, including its functional procedures, and of the environment representations it possesses.

3.1 Operating Knowledge

We will present in this section the necessary basic functions that the mobile robot should possess.

3.1.1 Functional Description

We shall stress the obvious fact that autonomous mobility is the central problem to be addressed. The robot will have to move in a variable and dynamic environment, even if the basic structure of space is rather stable; it will have to understand and execute orders, in the real world. It follows that this machine should be able to perceive its ever changing environment, to build suitable models of it, to manipulate these models for path finding and localization, and to update the models when necessary.

It is conceivable to load the known space models to the machine from blueprint architect maps. But this is, of course, impossible for the "furniture" (tables, chairs, etc.). On the other hand, the actual building and the blueprints may differ. This points out the necessity for the robot to have the capacity to build the environment model by itself. On the other hand, especially in this kind of environment, it is desirable that the model as built by the robot possesses some features, such as workstation labelling, specified navigational or forbidden areas, and that the robot have the capacity to simplify this model in some cases (as it will be shown in the examples of section 3.2). Therefore, model construction should be an interactive process. HILARE design includes these three modalities.

Furthermore, the robot system is functionally decomposed into a set of so-called specialized processing modules (SPMs) that are used by a set of specialized decisional modules. In our case study, we shall consider only two SDMs: one for navigation and motion control, and the other is a planner for the tasks the robot is assigned to. The SPMs that will be described are those used for navigation, the other ones being application specific, such as manipulation or loading tool control.

3.1.2 Operators, Elementary Tasks, Tasks

- We define a task, or a plan, as a sequence (possibly with parallelism) of operators and elementary tasks.
- An elementary task is a precompiled procedure and is itself a sequence of operators and/or robot functions.
- An operator is a basic action performed by one or several SDMs, making use of SPMs. The operators are the primitives of the command language.
- A robot function is performed by a SPM.

3.1.2.1 Elementary Tasks

The distinction between these four levels comes from the fact that the robot possesses some basic aptitudes that are expressed by the SPMs and the operators, on top of which we can define application-oriented capacities expressed by the elementary tasks. A specific task is then carried out by a plan generated by the robot.

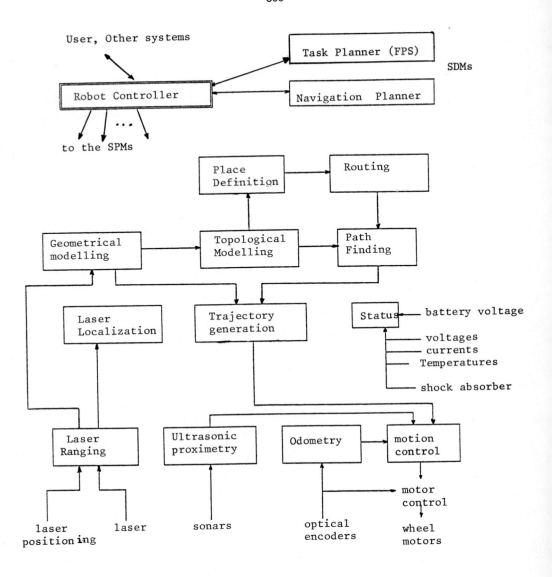

Fig 3. Functional description

In our case study, the elementary tasks are:

1. STORE_PART
2. RETRIEVE_PART
3. DOCK_WORKSTATION
4. BUFFER_OBJECT
5. PARK
6. REFUEL

3.1.2.2   Operators

The operators fall into two categories: elementary operators and macro-operators. The elementary operators are the basic motion and robot operating primitives, and are directly executable. A macro-operator is a specified sequence of elementary operators whose execution is context dependent (its parameters are dynamically instanciated by the planner).

An operator is written as a LISP function and is defined as follows:

(OP_NAME (OP_PARAMETERS)
          (OP_BODY)
          (OP_MONITORS))

where OP_NAME is the name of the operator;
OP_PARAMETERS are set by the planner or the user;
OP_BODY contains the SPMs calls and other operations;
OP_MONITORS are the conditions to be checked while executing this operator and the actions to be performed when they are true. OP_MONITORS are instanciated by the execution controller and will be detailed in section 5.

Example operators are:

GOTO (FRAME LOCATION) where FRAME is a place name to which is attached a coordinate frame, and LOCATION is specified in the coordinate frame. If FRAME is NIL, then the GOTO is within the place where the robot is.

MOVE (DX DY DANGLE)

STATUS: The system gives the current physical status such as battery voltage level.

DOCK (STATION): Dock to a station specified by a configuration of sensory readings. Robot motion relatively to the station, during docking, is monitored by the sensors to detect the desired configuration.

ROUTE (FRAME): Path finding to a given frame symbolically expressed.

MAP_PLACE: Acquisition of space structure in an exploration/modelling mode within a given place.

LOCALIZE: Robot localization.

LOAD/UNLOAD(OBJECT)

Most operators are only activated as part of the plan produced in order to carry out a task. Some of them may be called asynchronously by the robot execution controller. This is the case of LOCALIZE for example.

Before describing how these operators can be used to carry out tasks, let us examine the environment models necessary for executing them.

3.2 Environment Models

3.2.1 Environment Perception

Since autonomous mobility must be achieved, an efficient perception system has to be implemented. A multisensory structure is a central concept of the HILARE Project which has become widely favoured [Giralt 84b], [Paul 85], [Harmon 86]. However, environment modelling could make use of a preferential medium such as stereovision or laser range-finding.

Stereo and laser radar are complementary in the sense that ranging is useful for solving stereo ambiguities, and that stereo, and more precisely, edge stereo is useful for providing a rather fast and low-cost 3D representation of a scene [Saint Vincent 86]. Indeed, a ranging device with the same kind of output is out of price for a factory application. On the other hand, the robot needs some perceptual means that quickly provides data on its surroundings for motion control. Ultrasonic ranging is to our opinion very much suitable for this purpose.

We define 3 kinds of environment models: a geometrical model, a topological model, and a semantic model.

3.2.2 The Geometrical Model

It is built from sensory data, and contains object shape, dimensions and location expressed in local coordinate frames. It is used to build the other two models, and to execute the physical actions.

3.2.2.1 Building the geometrical model from range data

The range data are provided by a scanning laser range finder. The scanning is a horizontal scanning only. Indeed (in the framework of our application), most objects are vertical, and a scanning at a given height is sufficient to build ground projection maps. It could be argued that some features could not be perceived, but we emphasize the two following facts:

- Most of these features wouldn't be perceived either if a 3D scanning was performed. This is the case of small proturbances.
- The modelling process is interactive and done under human supervision. Therefore, the user has the possibility to modify and correct the model.

The modelling is done by performing a segmentation on the range points, then a least-squares approximation of each subset of points thus obtained. This produces a polygonal approximation of each subset of points thus obtained. This produces a polygonal approximation of the parts of objects perceived from the current robot location (Figure 4).

We will see how the user may modify this model after describing the model process from stereo data. More details on range data processing can be found in [Boissier 85].

3.2.2.2 Building the geometrical model from stereo data

Edge stereo provides a 3D representation of objects. As most of the scene elements possess vertical boundaries, we approximate the objects by vertical planes. This gives a first representation, that could be refined interactively, and that may be a basis for object identification and robot localization.

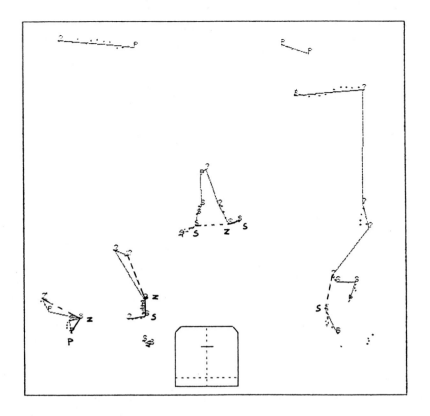

Figure 4 Laser range data and polygonal approximation

On the other hand, the ground projection of edges is used to build a model for navigation by introducing hulls bounding these edges (Figure 5) [Saint Vincent 86]. Stereo, motion vision, and other augmented facilities for sensing as well as for locomotion and on-board processing are included in HILARE Mark II; old HILARE will retire early in 1987.

3.2.2.3 Consistency of the geometrical models

The geometrical model is built gradually by the robot in an initial phase, before beginning to operate in its environment. During this phase, some objects or portions of objects will be seen more than once, from several locations of the robot. The problem is how to combine these different views together to produce a consistent and accurate model of the environment, taking into account the sensor inaccuracies and the uncertainties on the robot's position [Chatila 85a].

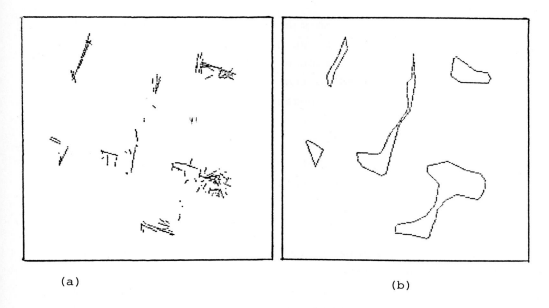

(a)   (b)

Figure 5   Edge stereo data ground projection and hull approximation

Multiple views are combined together at the geometrical primitive representation level, i.e. the polygonal line from laser sensing and the planar surface from stereovision. This choice is made because these primitives are stable with respect to the perception process, to the approximation algorithm, and to the robot's position because they intrinsically represent the objects, whereas range points or stereo edges are very unstable. For each object, a local frame is introduced wherein all its elements are referenced. This enables to have a more accurate representation because the uncertainties on robot location are considered only differentially between consecutive perceptions instead of accumulating position errors. The different local frames are related together and to a global frame. When the modelling process is interactive, the operator may improve the accuracy of the transformation matrix between frames.

3.2.2.4   Interactive modelling

By building its own environment models, the robot will introduce some complex structures that might be simplified. A typical example is given in Figure 4 where the polygonal approximation is in some cases too much saw-like. The actual world features that give birth to such representations are called textured regions. They frequently are irrelevant to navigation, especially when compared to the robot's dimensions, and add complexity to the model. The human user supervising the robot can point out such zones in order to simplify their representation (dashed line in Figure 4).

The user also helps the robot to label and identify objects and places, and to build the semantic model (section 3.2.4).

3.2.3 The Topological Models

Several topological models are useful for path planning and space structure understanding. We introduce the hierarchical concept of place defined as a portion of space that has some geometrical, topological or semantic attributes. At the lowest level, a place is a quasi-convex polygonal cell (qcpc). Free space is decomposed into qcpc [Chatila 82] [Chatila 85b], and the cell connective graph is the first topological model. Space decomposition is based on the obstacle edges in the 2D ground projection obtained from the geometrical model. The strictly convex cells decomposition would produce in some cases very narrow polygons that do not have any spatial property. Figure 6 gives an example of cell decomposition.

A first topological model can be autonomously built by the robot and used for path finding together with a cost function that is usually based on distance (section 4.2), but in which other criteria may be included, such as the difficulty of crossing certain areas due to ground nature or excessive maneuvers. Higher level topological models are based on concepts of places such as room, corridor, hall, workshop, storey. They are built by the robot itself by applying graph-decomposition techniques on the cell-graph after introducing topological and geometrical definitions of these concepts (section 6). Topological models for robot navigation represented by hypergraphs are also used by Wong in this book.

The nodes in the graph representing the topological model at a given level have attributes and properties, and are represented

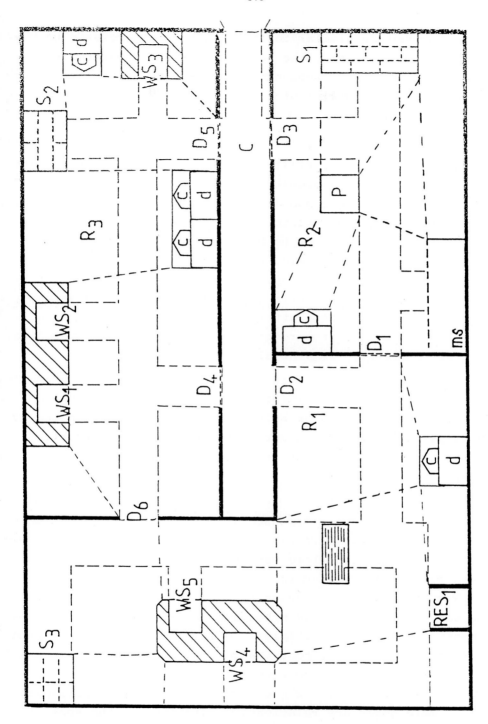

Fig 6. Cell decomposition of the case study environment

by a data structure similar to frames with slots related to the properties. For example, the "contains" slot associated to a given place lists the object names that are in this place. For example, the contains slot of room R1 is the following list: (WS4, WS5, S3, RES1).

3.2.4 Semantic and Task-Oriented Models

The semantic model is partially expressed in the labels and attributes included in the geometrical and topological models which represent the corresponding information on navigable space. However, attributes on navigability degree as well as on other criteria are very important to be equally expressed in a semantic model (e.g. road vs field, track vs marsh). Therefore, it is necessary to provide the operator with the ability of defining navigable space attributes such as ways, forbidden areas, and approach zones. These attributes may be related to some physical features that the robot's sensors are able to perceive, or may be completely "immaterial" and unobservable.

Again, these structures may or may not coincide with the robot's own space decomposition. Therefore, a mapping between these two representations is done by introducing attributes in the cell graph, and marking the ways, if any, within each cell.

4. TASK AND PATH PLANNING

4.1 Task Planning

Task planning consists, in our case study, in producing the sequence of operators and elementary tasks in order to accomplish a given goal. The planner is FPS [Sobek 85]. FPS is a rule-based system. It reasons about its own operators and contains rules for its internal task scheduling, operator selection, goal decomposition and plan criticism. Operator description contains pre and post conditions, continuation conditions for execution control, and the actual robot operators and elementary tasks to be used to execute them. The environment model is a predicate list. Simple goals are relational predicates, or the application of a given operator to a goal.

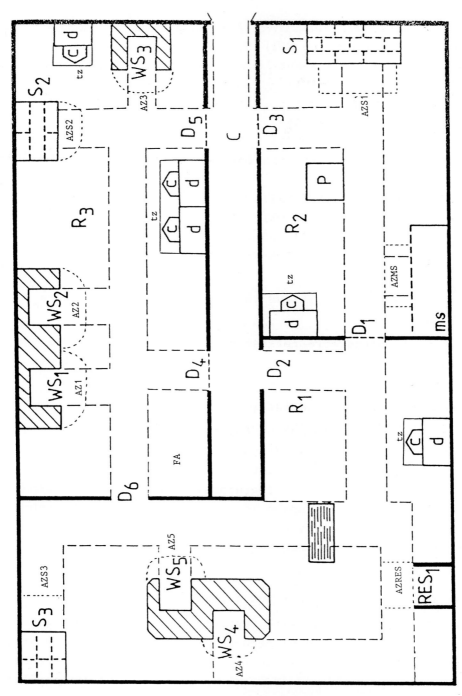

Fig 7. Case study environment with Ways, Approach zones, Forbidden areas, and textured zones.

Example:

The task   (CONVEY Ok to WS1)

with the following relevant knowledge:

        (INSTORE Ok S1)

        (INROOM S1 R2)

        (INROOM WS1 R3)

        (INROOM ROBOT R1)

is decomposed by FPS into the following operators and elementary task sequence:

        (GOTO R2 (XS1 YS1))

        (RETRIEVE_PART Ok)

        (GOTO R3 (XWS1 YWS1))

        (DOCK_WORKSTATION WS1)

        (UNLOAD Ok)

The elementary task RETRIEVE is composed of the operators DOCK and LOAD.

The plan is furthermore refined by the decisional module "navigation and motion control" as we will see in the following section.

4.2 Path Planning and Navigation

Path planning is performed by the Navigation and Motion Control System. It is a rule-based system for planning and execution control in the domain of robot navigation, and makes use of a subset of the SPMs and the operators. This system performs all that is related to space modelling, path planning, robot localization, trajectory control and obstacle avoidance. We already saw how space modelling is done.

4.2.1 Robot Localization

Three modes are used for robot localization:

- Absolute localization with respect to external known beacons. The first mode is useful because it could be very accurate, but it necessitates a structuring of the robot's environment. This may be considered in a workshop, but it is difficult to achieve in an outdoors environment, except for satellite-based navigation aids for example. We will not dwell further on this rather simple mode.

- Dead-reckoning by trajectory integration. This mode is very important since it provides permanent knowledge on the robot's position. Odometry appears easy and quite convenient to use here; inertial platforms also give the same information also but will not be considered here. The major drawback of this mode is the cumulative error which arises from trajectory integration, and that necessitates a resetting by means of an external reference.

- Localization with respect to the environment's natural features [Boissier 85]. This mode is by far the more complex and the one that gives the largest autonomy to the robot. Robot localization is achieved by matching a local scene with the global world model. In our case, both local and global models are ground maps obtained by laser panoramic scanning. The matching process is based on hypothesizing a first correspondence between the model and the perception by considering some particular features (longer obstacle edges, walls, the pillar in room R2), and then propagating this first hypothesis on other elements thus validating or invalidating it. This approach enables to gain in efficiency compared to the approach in [Drumheller 85]. Figure 8 shows an example of laser-based localization.

To gain efficiency, position information as given by the odometry, although inaccurate, is taken into account whenever available.

After docking to a workstation, position updating is also done. In this case, the robot's position inaccuracy is that of the workstation's, augmented by the docking error.

Furthermore, robot localization returns a numerical value (x,y, angle) and a symbolic value (robot in room R3).

4.2.2 Path Planning

Robot position is permanently updated, and is part of the NMC system knowledge. The operator (GOTO R2 (XS1 YS1)) in our example is decomposed by the NMC planner making use of the topological and geometrical representations of the environment as follows:

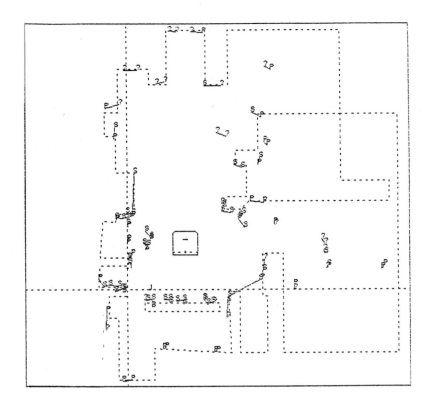

Figure 8 Robot Localization by Local and Global Scene Model Matching

```
(GOTO (XD1 YD1))
(GOTHRU D1)
(GOTO (XS1 YS1))
(DOCK S1)
```

where the goals of the GOTO operator are the coordinates of door D1 and storage S1. There shouldn't be any confusion between the GOTO produced by FPS and the GOTO produced by NMC. The output of FPS is a goal for NMC.

After producing this plan, path planning itself is basically a graph search procedure applied on the hierarchy of topological space models. Within a given level of representation, a cost function guides the search. The graph attributes introduced to point out preferential navigation zones, ways or forbidden areas are taken into account by the cost function. When navigation is within completely known space, as is generally the case in our application case study (except for the modelling phase), the search produces the best path with respect to the distance and the navigability criteria. Otherwise, the cost function contains an estimate of the traversal cost of unknown areas.

4.2.3 Trajectory Execution

A geometrical trajectory corresponds to the topological path. The optimality of the topological path does not generally correspond to the shortest trajectory, because the primitives selected to represent it do not strictly correspond with the topological shortest distance. These primitives are straight lines, circle arcs and clothoids. The clothoids enable to smoothly change the curvature radius of the trajectory [Chochon 83], [Kanayama 85], [Khoumsi 86].

Figure 9 shows an example of trajectory execution.

4.2.4 Local Obstacle Avoidance

During robot motion, unexpected obstacles may occur. The ultrasonic system is used to detect and avoid such obstacles, We will see in the section on execution control how this is achieved.

5. EXECUTION CONTROL

Mobile robot control is a very important issue. However, it is still not very much addressed. Interesting approaches are presented in [Brooks 85] and by Harmon in this book.

The planned actions and motions are to be carried out in the real world where the robot is not the unique actor, under time constraints. The inadequacy between the models and reality should also be taken into account.

The control system actually operates the robot. It sequences the elementary tasks, the operators and the processing modules, and monitors the execution of the robot's actions. It has to achieve two important objectives:

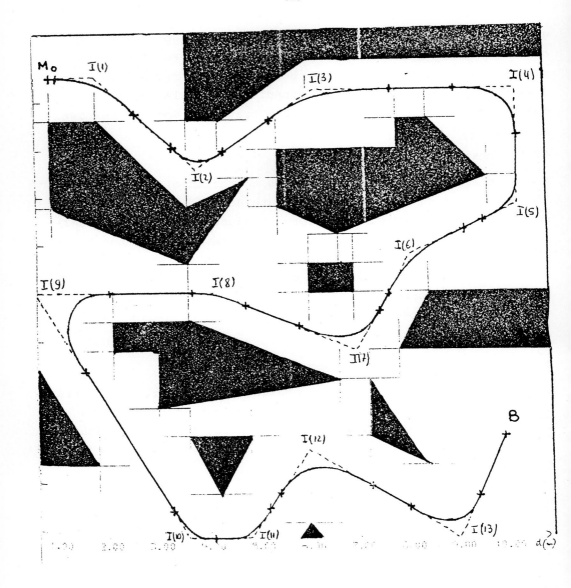

FIGURE 9 Trajectory Composed of Straight Lines and Clothoids

- It has to operate the robot in real time. This means that external or internal asynchronous events should be taken into account at their own time scale.
- It has to decide for the course of action when such events occur.

The controller should be able to check if the planned actions can be executed, given the actual environment and robot state. If any action fails, it should provide enough information to the planner for replanning (diagnosis) or at least attempt for a local plan mending when possible.

Together with the plan to be executed, an execution model is transmitted to the controller. This execution model contains the world state as it is supposed to be during the execution and after the completion of each plan step. It includes the world model elements, relationships and predicates that are directly concerned with the current plan (e.g. topological relationships, storage status). The execution model also contains a plan mending flag for each plan step that authorizes or not local corrective actions by the controller in case of failure of the plan step, for instance, local avoidance of unexpected objects without updating of the area's space model.

Finally, the controller provides for the interface between the user and the robot system (section 5.3).

5.1 Controller Sequence

To meet all these requirements, we propose the following structure for the robot controller:

- a decisional kernel, written as a rule based system,
- a set of distributed monitors.

5.1.1 The Decisional Kernel

The decisional kernel is a rule based system containing knowledge on robot operation and failure diagnosis [Ghallab 84]. Rule systems enable granularity of knowledge and are very suited for expressing operating knowledge.

The controller's rules are simple and can be written in propositional logic. Indeed, in our context, the state space of the robot is finite, and the possible failures are also known before hand. Therefore, this rule system can be precompiled into a decision tree. This enables to achieve efficiency of operation while flexibility of expression is kept.

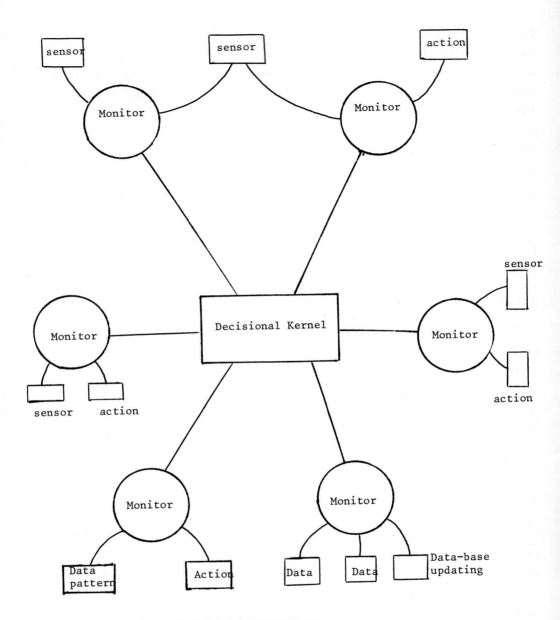

FIGURE 10   ROBOT CONTROLLER STRUCTURE

5.1.2 The Monitors

The monitors enable real time reaction to asynchronous events. They are semantically equivalent to some of the kernel's rules, and are instanciated by it. For instance, they enable to monitor robot actions through the sensors. Their condition side is either a sensory information or data patterns in the execution model, and their right hand side is an action to be executed immediately. The decisional kernel is then informed of the monitor's execution. Several levels of monitoring can be defined, ranging from guarded commands to dynamic modification of the robot's behaviour (navigation mode selection, sensor use, etc.).

5.2 Examples

A. Unexpected obstacle detection and avoidance [Khoumsi 86].

Before robot motion, monitors are programmed by the kernel in order to detect any object lying on the robot's trajectory by the ultrasonic system. When the distance is under a given threshold, the robot stops. The monitor can be expressed very simply by the rule:

   $<$(LESSP (RANGE sonar$_i$) threshold$_i$)$>$
   -- $<$(STOP_MOTION)$>$

The decisional kernel then proceeds to analyzing the situation by scanning with the ultrasonic sensors, building a local model of the environment, and if local plan mending is enabled, it selects an avoidance trajectory, taking into account the known environment model. It programs new monitors that will be operational during the local obstacle avoidance to detect termination conditions on success or failure. When these monitors are triggered, and in case of success, the previous action is resumed, unless other events occur.

Figure 11 shows an example of local obstacle avoidance. The avoidance trajectory is a circle that is updated by sonar readings.

B. Robot localization and robot refueling

During normal operation, odometry readings provide the robot's position. But as the robot moves, the uncertainty on these readings grow higher. Therefore, a monitor compares the estimated position error (that is a function of the travelled distance) with a threshold, and sends a message to the kernel when it is exceeded. The kernel decides for a laser localization, depending on the context: if the robot is during obstacle avoidance or docking, laser localization is inhibited by some rules. Similarly, battery voltage level is also monitored and when it is below a given threshold, battery power supply becomes an imperative task. Again, the execution of this task is dependent on the context as expressed by some kernel rules.

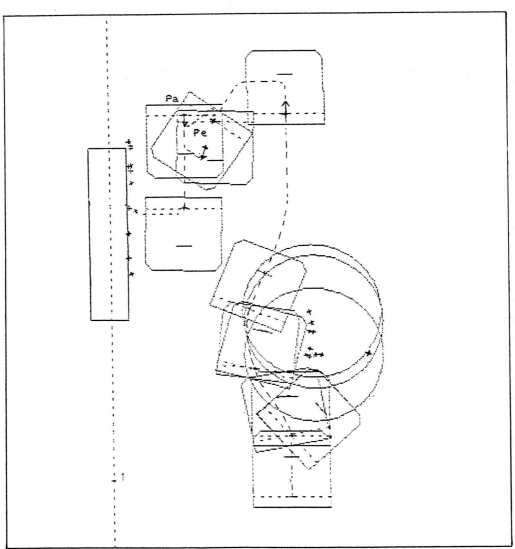

FIGURE 11 Robot Control: Unexpected Obstacle Detection and Avoidance Using Sonars, and Workstation Docking

C. Early detection of workstation

The docking procedure consists in performing some predefined maneuvers. In the usual mode of operation, the robot navigates to a given point, defined as the workstation entry point, and then begins to dock. However, for several reasons such as odometry drift or change of the initial trajectory during local obstacle avoidance, the robot could be in the vicinity of the workstation before reaching the entry point. This is also detected by a monitor that compares the sensory readings (sonars, ...) with a specified model pattern. The decisional kernel then may update the sequence of actions in order to generate a new trajectory leading directly to the entry point.

D. Monitoring of knowledge updating

The robot should be able to take into account knowledge updating by the user or other systems, both during planning and during execution. We will focus on the second case, the first one being by itself a very important issue, beyond the scope of this paper.

While the robot is operating, the execution model is used by the controller to dedicate monitors to the surveillance of the status of items that are relevant to the plan under execution. For example, if the plan specifies that the robot has to cross door D1, then the status (open or closed) of door D1 is monitored in the world model, even when D1 is not in sight, because this model might be updated by an external system.

When the updated information is "negative", e.g. door D1 is closed instead of being open, the controller is in some cases able to do plan mending, for example to select an alternate route without the need of replanning the whole task. The case of "positive" updating, i.e. information that modifies the world model by making it closer to the goal, requires generally replanning, except in very simple cases.

5.3 User/Robot Interface

The controller provides for the user/robot interface. The user can give orders to the robot or may want to access some inform-

ation. We define three categories of orders:
- Indicative orders: these are normal orders (tasks, elementary tasks or operators) that the robot will execute.
- Sequenced orders: the user may define a partially ordered sequence of tasks to be carried out. If the robot is executing a sequence, then a new indicative order is executed at the end of the sequence.
- Imperative orders: orders to be executed at once with the highest priority, unless robot security is jeopardized. Furthermore, the user may inhibit the security explicitly.

6. LEARNING PROCESSES

Learning is considered within the scope of this case study in two different aspects:

1. Space structure understanding and model refinement

Space structure understanding is based on analyzing the geometrical and topological models in order to build higher level representations [Laumond 86a]. The concepts of door, corridor, room are defined by a set of rules using metrical and topological notions such as:

"a door is an articulation node (in the cell graph) representing a small-area cell and separating two biconnected components representing large-area places".

These rules are used to label the cell graph after decomposing it into substructures such as articulation nodes and pairs, biconnected components. The robot is thus able to produce by itself a higher level representation of its environment. The stability of this structure with respect to changes in the environment, such as modification of object location, introduction or removal of objects, opening or closing doors, enables to deduce the environment invariant structure. Thus, as the robot moves and executes tasks, it modifies its geometrical model and cell graph. Each new graph is decomposed and labelled to produce a representation with the higher level concepts. A comparison of the models at the room/door/corridor level enables to deduce the stable features.

Figure 12 shows two decomposition of the same environment with different objects, as produced by the system, and the associated graphs.

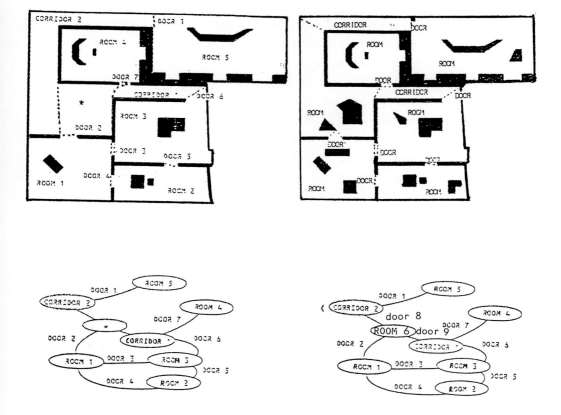

Figure 12   Environment Invariant Structure Acquisition

The environment's structure understanding and the introduction of hierarchical topological models improves the behavior of the robot. Path finding for example becomes hierarchical, and therefore, more efficient. Spatial relationships between objects become also easier to express: object1 is_near object2 means that object1 and object2 are in the same room, and that euclidean distance between them does not exceed a given threshold.

Finally, man-machine communication is also improved because the robot's spatial representations are close, and can be made to coincide with the user's.

Model refinement is experience-based labelling of the robot's models. Navigability attributes, and the introduction of secondary routes when the usual paths are closed are typical examples. The first case is a mere correspondence between the robot's graph representations and the attributes such as preferential navigation zone, way etc. that are introduced or updated by the human operator as the robot executes its tasks. The second case is based on monitoring the frequency of unexpected obstacles on a given usual trajectory or way, and deciding for an alternative way when a given threshold is exceeded. The new way is selected using the usual cost function, including an expression of the obstacle avoidance trajectory for the previous path.

2. Elementary task generation

This aspect includes parameter instanciation for some elementary tasks, and the deduction of environment and task-dependent macro-operators. A usual example is learning the docking procedure, i.e. a specific set of operators, for a given workstation layout. In this case, not only parameter instanciation will be better achieved, but also the definition of the specific set of operators will benefit from the interaction between in-built geometrical reasoning operators [Laumond 86b], and experimentation.

7. CONCLUSION

Typically, autonomous mobile robots have to be able to perform unique tasks in largely unknown work conditions. They constitute a paradigm for research on decision and perception for robot operation in unstructured environments, i.e. third generation robots and all real-time aspects of machine intelligence and the key to novel application fields. Although many of these new applications are of the out-of-factory type, we have chosen to make our presentation on a factory-like case study. We have already argued about the generality of the concepts and methods which can be developed upon that sort of environment.

We believe there is a second and very important reason to put forward: indoors or around building environments stress the time factors, on board technology and efficiency aspects; they look forward to today or near future applications ranging from FMS to surveillance.

We have focused our paper on research issues that we believe are at the core of this domain:

- Space models enabling the robot to understand its environment enough to achieve efficient autonomous mobility.
- Operators and elementary task primitives to provide for a convenient planning capacity and user/machine interface.
- Production-rule based Pattern Directed Inference Systems with structures and algorithms adapted to robot operation.
- An execution control system that carries on plans and handles discrepancies between real-world status and knowledge bases (space models, data bases, ...). This system integrates a world execution model and multisensory perception.

Important resources and efforts must be brought in to further investigate all these research issues. Among the most challenging aspects, we believe are reliable 3D perception, efficient learning functionalities and the capacity to reason about time for robot planning and execution control.

REFERENCES

[Boissier 85] L. Boissier, "Modelisation de l'environnement et localisation du robot mobile Hilare par telemetrie laser", Theses 3eme cycle, LAAS-UPS, Toulouse, Dec. 85.

[Brady 85] M. Brady, "Artificial Intelligence and Robotics", AI Journal, Vol. 26, No. 1, April 1985.

[Brooks 85] R. Brooks, "A layered intelligent control system for a mobile robot", 3rd ISRR, Gouvieux, France, Oct. 1985.

[Chatila 82] R. Chatila, "Path planning and environment learning in a mobile robot system", European Conf. on Artificial Intelligence, Orsay, France, July 1982.

[Chatila 85a] R. Chatila et J.P. Laumond, "Positioning referencing and consistent word modelling for mobile robots", Proc. IEEE Int. Conf. on Robotics and Automation, St. Louis, March 1985.

[Chatila 85b] R. Chatila, Mobile robot navigation: space modelling and decisional processes, 3rd Int. Symposium of Robotic Research, Gouvieux, France, October 1985.

[Chochon 83] H. Chochon & B. Leconte, "Etude d'un module de locomotion pour le robot mobile Hilare", Rapport de fin d'etudes ENSAE, LAAS, June 1983.

[Drumheller 85] M. Drumheller, "Mobile robot localization using sonar", AI memo 826, MIT, January 1985.

[Ghallab 85] M. Ghallab, Task execution monitoring by compiled production rules in an advanced multi-sensor robot. In Robotics Research, the Second ISRR, MIT Press, 1985.

[Giralt 79] G. Giralt et al. "A multi-level planning and navigation system for a mobile robot; a first approach to HILARE", 6th IJCAI, Tokyo, August 1979.

[Giralt 84a] G. Giralt, "Mobile Robots", NATO ASI Series, Vol. 11, Edited by M. Brady, L. Gerhardt and H. Davidson, 1984.

[Giralt 84b] G. Giralt et al, "An integrated navigation and motion control system for autonomous multisensory mobile robots", in Robotics Research, the 1st ISRR, MIT Press, 1984.

[Harmon 86] S. Harmon et al. "Sensor data fusion through a distributed blackboard", IEEE Conf. on Robotics & Automation, San Francisco, April 1986.

[Ieeer&a 86] IEEE Conf. on Robotics and Automation, Special sessions on the ALV Project, San Francisco, April 1986.

[Kanayama 85] Y. Kanayama & N. Miyake, "Trajectory generation for mobile robots", 3rd ISRR, Gouvieux, France, October 1985.

[Khoumsi 86] A. Khoumsi, P. Migaud, Amelioration des capacites comportementales d'Hilare - Pilotage et controle d'execution de mouvements. Rapport LAAS 86062, Toulouse, 1986.

[Laumond 86a] J-P. Laumond, A learning system for the understanding of a mobile robot environment. European Working Session on Learning, LRI, Orsay, France, February 1986.

[Laumond 86b] J-P. Laumond, "Feasible trajectories for mobile robots with kinematic and environment constraints", to appear in the Proc. of the Int. Conf. on Autonomous Systems, Amsterdam, Dec. 1986.

[Paul 85] R. Paul et al, "A robust distributed sensor and actuation robot control system", 34d ISRR, Gouvieux, France, October 1985.

[Saint Vincent 86] A. Robert de St. Vincent, "A 3D perception system for the mobile robot Hilare", IEEE Conf. on Robotics and Automation, San Francisco, April 1986.

[Sobek 85] R.P. Sobek, "A robot planning system using production rules", 9th IJCAI, Los Angeles, August 1985.

Task Planning and Control Synthesis for Flexible Assembly Systems

A.C. Sanderson and L.S. Homem-de-Mello

Department of Electrical and Computer Engineering
and The Robotics Institute
Carnegie-Mellon University
Pittsburgh, PA 15213

ABSTRACT

The development of planning tools which facilitate computer-aided design of assembly systems is important to making the systems implementation process more efficient and cost-effective. Design of flexible assembly systems requires a representation of the assembly task, a planning strategy to select feasible assembly sequences, and a control synthesis procedure to schedule operations and resources. This paper describes a nonlinear planning approach which proposes feasible assembly sequences based on a relational model of part contacts and attachments. The resulting plans are consolidated into an AND/OR graph representation which provides a basis for efficient scheduling of operations. Scheduling efficiency using this representation is compared to fixed sequence and precedence graph representations for a simple example.

1 Introduction

The design and implementation cost of an automated assembly system often overshadows the capital cost of equipment, and the development of planning tools which facilitate computer-aided systems design and evaluation is key to making this implementation process more efficient and cost-effective. The planning approach described in this paper emphasizes assembly task decomposition and task sequencing and demonstrates the key role of representation in the planning process. Assembly sequence planning provides a basis for scheduling of operations and systems resources, and permits evaluation of device implementation alternatives.

Planning of assembly tasks may be decomposed into a series of steps as shown in figure 1. The product design describes the geometric and physical assembly. For a given product design, assembly sequence planning proposes feasible assembly sequences and may search over the space of feasible sequences using a cost function to weight alternatives. In this paper, the assembly sequence planning process uses a compact relational graph representation of geometric and physical relations of products, and yields an AND/OR graph representation of feasible plans.

An assembly sequence is a system independent description of the assembly procedure. The steps in the assembly sequence are mapped onto a set of operations (for example, GRASP, MOVE, INSERT, ATTACH). Such task-oriented operations may be treated as discrete operations with continuous attributes. Control or scheduling of discrete operations must be consistent with a

---

This research is supported in part by Conselho Nacional de Desenvolvimento Científico e Tecnológico (Brazil) and by the Robotics Institute of Carnégie-Mellon University.

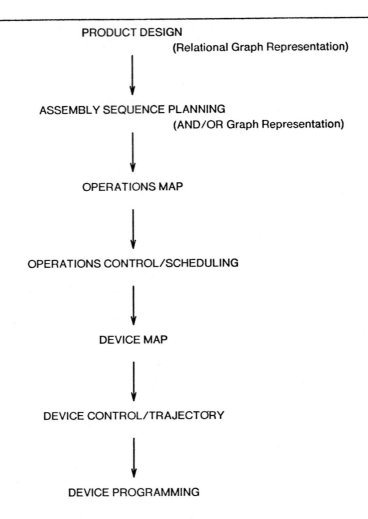

Figure 1: Steps in assembly system planning

feasible assembly schedule and attempts to optimize system performance in terms of throughput, cost, reliability and flexibility. Approaches to the discrete scheduling problem have been described in operations research [1] and constraint-directed reasoning [2]. The control synthesis problem will not be discussed here, but has been addressed in [17].

Planning strategies have been important in artificial intelligence and a number of programs have been written which attempt to derive feasible operations sequences for 'blocks world' problems. Examples such as BUILD [3] and STRIPS [4,5] generate linear plans that sequence operations in order to match current states to goal states. The application of means-ends analysis to isolated subgoals in these systems often produces inefficient plans which may do and then undo intermediate states. The representation of plans in these systems is based on ordered lists of preprogrammed operations. STRIPS [4] uses a triangle table to store a plan which improves the capability to recover from errors within a fixed sequence. More recent planning systems such as NOAH [6] add steps which resolve conflicting operations and eliminate redundant operations. NOAH represents plans as a partially ordered sequence of operations with respect to time. As discussed in section 3, these programs do not easily adapt to the sequence planning problem for real assemblies since they do not accommodate the more complex task representation.

Implementation requires mapping of operations to specific devices, formulation of control, and programming of devices. Specific trajectory planning and motion control is required to implement operations. Approaches to automatic derivation of these motion control programs from geometric description of parts and specification of operations have been described in [7,8,10,11,12,13] Strategies for fine motion planning [14], and physical models of part interactions [15,16] are also important ingredients of the problem at this level.

Assembly planning in the broad sense suggested by figure 1 provides tools for integrated design of products, systems, and control structures. While it is currently not feasible to automate all of these processes this framework may provide a set of interactive tools which will assist designers in development and implementation. The choice and configuration of systems is a major element of such a design process, and the capability to evaluate alternatives such as modular assembly lines versus robotic assembly cells will be a major feature of such a system.

2 Relational Model of an Assembly

A computer-aided design database provides a complete geometric description of an assembly in terms of volume or surface primitives of individual parts and their spatial relations. Such a description may be thought of as a representation of the goal state of an assembly planning system, but in practice such a purely geometric representation is very difficult to use for planning sequences or trajectories. Work on trajectory planning and task-oriented programming [7,8,10,11] has most often used a relational representation of assemblies as a basis for deriving relative position and orientation information about parts and subassemblies. Our approach outlined below defines relations and attributes needed to plan feasible assembly sequences. More detailed geometric information is imbedded in layered attribute lists for access in different phases of the planning problem.

An assembly consists of a set A of rigid objects, $a_j \ \epsilon \ A$, connected in a stable, rigid (our current assumption) geometric configuration. Each object has a coordinate frame attached to it, and P_j is defined as a homogeneous transformation between the world coordinate frame W and the coordinate frame of a_j in its current state. In the final assembly configuration each object must reach some goal state $P_j = P_j^*$ as specified by the geometric design.

In practice, it is impossible to manipulate a solid object into an arbitrary final position without considering constraints on the allowed motion imposed by other objects. We will consider two categories of such constraints:

- *Local or blocking constraints* - adjacent objects or surfaces prohibit *incremental* motion near final position.

- *Global or trajectory constraints* - obstacle objects prohibit global trajectory access to final position.

Local or blocking constraints are determined by surface *contacts* between objects. Such contacts are represented by relations between objects. Contact relations have TYPES which describe basic geometric categories. Contact relation TYPES include: planar, cylindrical, spherical, groove, slot, and screw. Contact relations have two sets of attributes. Contact attributes specify the direction and range of the contact blocking relations such as normal vector or axis vector, or range. The contact attributes are sufficient to determine local blocking constraints without completely describing the contact geometry. The surface attributes attach a coordinate frame to each contact surface and specify complete geometric relations between contact surface and object coordinate frames. The data structure associated with a contact is specified by:

CONTACT: ⟨object 1⟩, ⟨object 2⟩, ⟨type⟩,
⟨surface 1⟩, ⟨surface 2⟩,
⟨contact attribute list⟩;

OBJECT: ⟨object name⟩,
⟨goal state⟩,
⟨surface list⟩;

SURFACE: ⟨surface name⟩, ⟨object name⟩,
⟨surface attribute list⟩;

CONTACT ATTRIBUTE:
⟨TYPE⟩,
⟨normal⟩, ⟨axis⟩, ⟨range⟩, ⟨radius⟩.

SURFACE ATTRIBUTE:
⟨surface name⟩, ⟨object name⟩
⟨position⟩, ⟨surface model⟩;

As example of a relational graph showing the contact data structure for a simple example is shown in figures 2 and 3. Such a structure provides explicit representation of local constraints and implicit representation of global constraints.

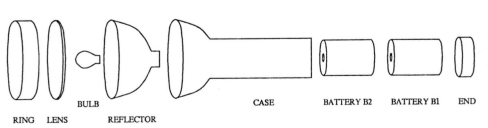

Figure 2: A simple flashlight

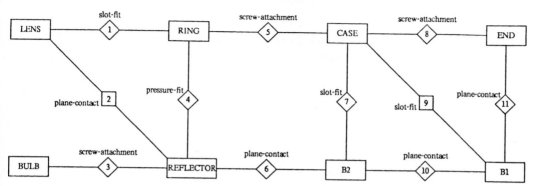

Figure 3: Contact data structure for the flashlight shown in figure 2

While contact relations are sufficient to capture the geometric properties of an assembly, they are not sufficient to describe some physical properties of assemblies. In particular, assemblies are held rigid by *attachments* which impose forces on contacts, and limit the relative motion between objects. We model an attachment as an *operator* on contacts which fixes relative motion. Attachments are mapped onto the contact date structure as an *overlay operator* with the following structure:

ATTACH (⟨type⟩, ⟨contact list⟩, ⟨attribute list⟩)

where ⟨type⟩ refers to a set of physical attachment types such as screw, snap, lock, clamp, glue, etc., each with its own attribute list structure. The same attachment may often operate on multiple contacts. The attachment operator provides a very powerful tool in the planning process since it defines the relation between contacts and models, the forces which maintain the assembly in a rigid state. Figure 3 includes the screw attachment operators for the simple example of figure 2.

While the contact/attachment representation is sufficient to plan assembly sequences, one additional feature has proven useful to simplify the examination of global constraints. A virtual contact relation VCONTACT provides a means of expressing a blocking relation by an obstacle which is not in contact. VCONTACT may be used in an interactive mode to simplify trajectory planning. VCONTACT defines a virtual object which extends to contact an existing object.

3 Planning for Assembly

Planning the assembly of one product made up of several component parts can be seen as path search in the state space of all possible configurations of that set of parts. The initial state is that configuration in which all parts are disconnected from each other, and the goal state is that in which the parts are properly joined to form the desired product. The moves that change one state into another correspond to the assembly operations since they change the relative position of at least one part. There may be many different paths from the initial state to the goal state. Krogh and Sanderson [17] present an overview of task decomposition and operations.

In this paper, any set of parts that are joined to form a stable unit is called an *assembly*. A component part is also an assembly, with a special property. The word *subassembly* refers to an assembly that is part of another, more complex assembly, and it always carries the subset/set connotation.

Because there are many configurations that can be made from the same parts, the branching factor from the initial state to the goal state is greater than the branching factor from the goal state to the initial state. A backward search, therefore, will be more efficient than a forward search for the assembly planning problem. The problem of finding how to assemble a given product can be converted to an equivalent problem of finding how the same product can be *disassembled*. Since assembly operations are not necessarily reversible, the equivalence of the two problems will hold only if the operations used in *disassembly* are the reverse of a feasible assembly operation regardless of whether these reverse operation themselves are feasible or not. The expression *disassembly operation*, therefore, refers to the reverse of a feasible assembly operation.

The backward search suggests a decomposable production system in which the problem of *disassembling* one product is decomposed into distinct subproblems, each one being to *disassemble* one subassembly. Each decomposition must correspond to a *disassembly operation*. If solutions for both subproblems that result from the decomposition are found, then a solution for the original problem can be obtained by combining the solutions to the subproblems and the operation used in the decomposition. For subassemblies that contain one part only, a trivial solution containing no operation always exists. Usually there will not be a unique way to decompose the problem, or to *cut* the assembly, because there may be several different ways to assemble the same product.

We consider the problem of assembly sequence planning as a *goal-directed* planning problem which looks for ways to *cut* the relational graph representation in a sequence of feasible cuts to disassemble a product. By definition, inverting this operations sequence defines a plan to assemble the product.

3.1 Linear Planning

A linear planning approach to the disassembly problem would compare current state to goal state and produce an ordered list of subgoals. In this case an initial subgoal might be

> REMOVE (object A)

Linear analysis would expand each goal by ordered preconditions such as:

> DETACH (object A, state 1)
> TEST LOCAL CONSTRAINTS (Object A, state 1)
> TEST GLOBAL CONSTRAINTS (Object A, state 1)
> TEST STABILITY (subassembly, state 2)
> REMOVE (object A, state 2)

We will not describe detailed procedures for testing preconditions in these planning approaches. Briefly, the approach includes:

- Attachment Preconditions - each attachment operation has specific precondition state requirements which are tested directly without geometric reasoning.

- Local Blocking Constraints - Blocking constraints may be tested by *constraint configuration space diagrams* which provide a representation of local degrees-of-freedom of movement of air object. The union of constraint conditions in the configuration space of the object summarizes all blocking conditions. Search of configuration space for unblocked incremental motion determines the feasibility of the disassembly operation.

- Stability - Stability of the resulting subassembly is tested by examining incremental blocking relations for all parts in the subassembly. If incremental motion subject to gravitational forces is permitted, then the resulting subassembly is unstable.

- Global Blocking Constraints - An obstacle avoidance path among all objects in the current state must be found. This problem must be solved using complete geometric models as opposed to local contact constraints.

The planning approaches described in this section result in ordered sets of operations which constitute DETACH and CUT operations on the relational graph model. Representation of the complete search space of feasible assembly plans is described in the next section.

This sequence may not be feasible and a next step will either (a) abandon this search path or (b) expand the constraints further. For example, if object A is blocked by object B:

>   DETACH (object A, state 1)
>   REMOVE (object B, state 1)
>   TEST LOCAL CONSTRAINTS (object A, state 3)
>   TEST GLOBAL CONSTRAINTS (object A, state 3)
>   TEST STABILITY (Subassembly, state 4)
>   REMOVE (object A, state 4)

This approach would further expand REMOVE (object B) and sequentially set subgoals

>   REMOVE (object E)
>   REMOVE (object D)
>   REMOVE (object C)
>   REMOVE (object B)
>   REMOVE (object A)

For typical assembly processes this linear planning procedure becomes extremely cumbersome due to the complex interrelations among parts. It becomes an inefficient method of searching the goal-directed graph.

3.2 Nonlinear Planning

The nonlinear planning strategy looks in parallel for feasible REMOVE operations without stacking subgoals, as illustrated in figure 4. Preconditions are examined in parallel in order to test feasibility as shown in figure 5. In this approach the tree of feasible operations may be searched using any number of standard algorithms and may be used to find preferred plans by weighing the cost of operations.

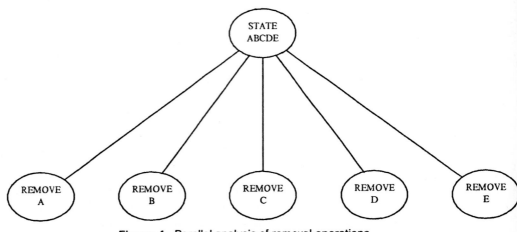

Figure 4: Parallel analysis of removal operations

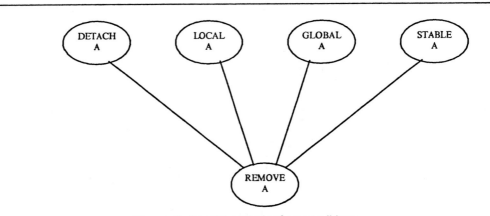

Figure 5: Parallel analysis of preconditions

3.3 Hierarchical Planning

A principal difficulty in either the linear or nonlinear planning operations discussed above is the computation of global blocking constraints, since that step requires testing of general geometric relations. We are examining a hierarchical planning strategy which computes full plans based on partial constraints at one level, then refines the plans based on full constraints at a second level. This hierarchical strategy is schematically:

```
LEVEL 1:    PLAN (FIRST LEVEL CONSTRAINT,
                  BROAD COST FUNCTION)

LEVEL 2:    PLAN (SECOND LEVEL CONSTRAINTS,
                  NARROW COST FUNCTION)
```

where in our current work we consider First Level Constraints:

- Attachment
- Local Blocking
- Stability

Second Level Constraints:

- Global Blocking

Clearly the effectiveness of this hierarchical strategy will depend on the particular assembly problem. However, increasing emphasis in design for assembly yields designs which minimize global blocking or obstacle avoidance issues, and we feel that these designs will be particularly effective since local blocking constraints will eliminate most of the unfeasible assembly sequences.

4 Representation of Assembly Plans

Planning, in the sense described in the previous section should yield all possible sequences of operations that can be used to assemble a product. That information is the input to the scheduling process, which in real time selects one of those sequences and assigns the machines that will do each operation.

Fox and Kempf [18] represented assembly plans by precedence diagrams which would actually encompass several possible sequences of operations that would perform the task of assembling a given product. In real time, depending on the conditions at the shop, the intelligent robot would pick the most appropriate sequence. Using Fox and Kempf notation, the selection of one sequence, and the assignment of operations to specific machines is what is commonly referred to as the scheduling process. Since that selection process involves much less computing time than the planning process, no degradation in the efficiency of the robot operation should occur.

The problem with the precedence diagram formalism, as Fox and Kempf themselves point out, is that for most products no single partial order can encompass every possible assembly sequence. The assembly of the simple product shown in exploded view in figure 6, for example, may be completed by following one of the ten different sequences of operations that are represented graphically in figure 7. It is possible to combine some sequences into one partial order using precedence diagrams. Figure 8 shows three possible ways to combine two of the first four sequences in figure 7; the only restriction is that the insertion of the stick cannot be the last operation. It is possible to combine three of those four sequences into one partial order by using a dummy operation, but it is not possible to combine the four sequences into one partial order, nor it is possible to combine any of those sequences with the other six sequences in figure 7.

A closer look at the partial ordering representation of plans, in the light of the above assembly example, shows another deficiency of that solution. Two distinct feasible sequences, A-B-C and B-A-C, for example, do not differ simply by the sequence of the operations. Inserting the stick first is not the same operation as inserting it after the receptacle and the cap have been screwed together. The latter operation is probably easier to execute. Similarly, screwing the receptacle and the handle with the stick inside is probably easier to do if the receptacle and the cap are

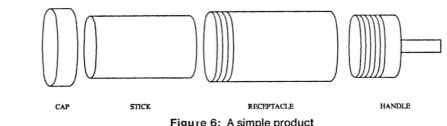

CAP   STICK   RECEPTACLE   HANDLE

Figure 6: A simple product

screwed, than otherwise. The partial ordering approach, however, does not capture this subtle difference. This section will describe another approach to the representation of plans that captures this difference, and that can combine all possible assembly sequences.

Structures called AND/OR graphs [19], or *hypergraphs*, are useful in representing decomposable problems and they have been used to represent the *disassembly* problem. The nodes in such a hypergraph correspond to assemblies; nodes corresponding to assemblies that contain only one part are the terminal nodes. The hyperarcs (or k-connectors, k being any integer greater than zero) correspond to the *disassembly operations*. Each hyperarc that leaves one node corresponds to a *disassembly operation* applicable to the assembly of that node, and the successor nodes to which the hyperarc points correspond to the resulting subassemblies produced by the *disassembly* operation. Because for most products the assembly operations usually mate two subassemblies, the hyperarcs in the corresponding AND/OR graph are usually 2-connectors. There are cases, however, of operations that mate more than two subassemblies (e.g., assembling a hinge with two wings and one pin), as well as operations that involve only one subassembly (e.g., drilling a hole in a part). Hyperarcs in AND/OR graphs can represent all those possibilities.

A *solution tree* from a node N in an AND/OR graph is a subgraph that may be defined recursively as either N itself if N is a terminal node, or N plus one of its outgoing hyperarcs plus the set of solution trees from each of N's successors through that hyperarc. This definition assumes that the graph contains no cycle as is true in the *disassembly* problem. There may be none, one, or several solution trees from a node in an AND/OR graph.

The useful feature of the AND/OR graph representation for the assembly problem is that it encompasses all possible partial orderings of assembly operations. Moreover, each partial order corresponds to a solution tree from the node corresponding to the final (assembled) product. This feature is demonstrated through following example in the next section.

4.1 A Simple Example

Figure 9 shows the AND/OR graph for the product in figure 6. Each node in that graph is labeled by a database that corresponds to an assembly. In figure 9, the databases are represented by exploded view drawings, whereas in a computational implementation, the databases are relational data structures. To facilitate the exposition, both the nodes and the hyperarcs in figure 9 have identification numbers.

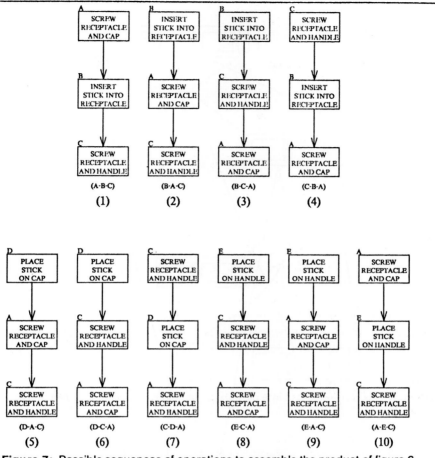

Figure 7: Possible sequences of operations to assemble the product of figure 6

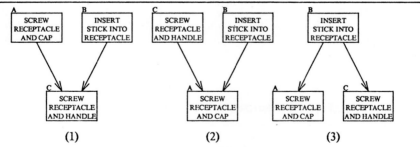

Figure 8: Precedence diagrams: *(1)* combines A-B-C and B-A-C; *(2)* combines C-B-A and B-A-C; *(3)* combines B-A-C and B-C-A

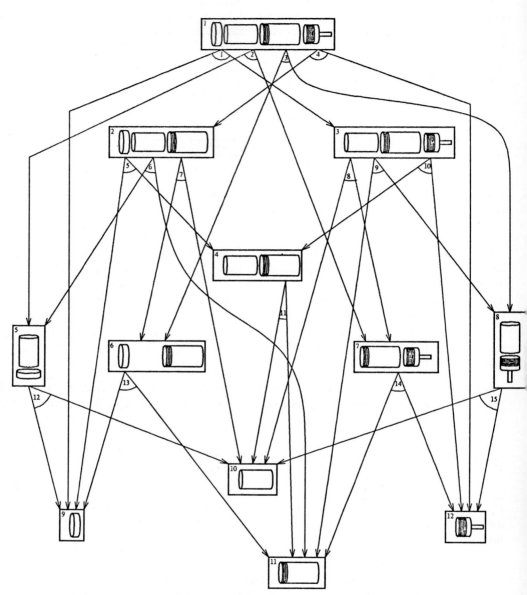

Figure 9: AND/OR graph for the product of figure 6

The root node in figure 9 (node 1) is labeled by a database that describes the assembled product. There are four hyperarcs leaving that node. Each of those four hyperarcs corresponds to one way the whole assembly can be *disassembled* and each one points to two nodes that are labeled by databases that describe the resulting subassemblies. Similarly, the other nodes in the graph have a leaving hyperarc for each possible way in which their corresponding subassembly can be *disassembled*.

Any subassembly that can be made up of the component parts may appear only once in the graph, even when it may be the result of different *disassembly* operations. The subassembly of node 4, in figure 9, for example, may result from two different operations, which correspond to hyperarcs 5 and 10. Moreover, those two hyperarcs come from two distinct nodes.

Nodes corresponding to component parts (nodes 9, 10, 11 and 12) are the terminal or goal nodes since they correspond to *disassembling* problems for which a (trivial) solution is known. There are eight solution trees from the root node (node 1) and they are shown in figure 10.

One important feature of the solution tree representation shown in figure 10 is that the distinction between operations becomes apparent because distinct operations correspond to distinct hyperarcs. In other words, two distinct assembly sequences include the same operation only if the two corresponding solution trees include the hyperarc corresponding to that operation. Hyperarc 1, for example, is present in the solution trees (a), (b), and (c); therefore, the same assembly operation is part of three distinct sequences. Conversely, the operations SCREW THE RECEPTACLE AND THE CAP in sequences A-B-C, B-A-C, and B-C-A of figure 7 correspond to hyperarcs 1, 5, and 13 in figure 9; therefore, they are three different operations. The sequence diagrams in figure 7 and the precedence diagrams in figure 8 fail to make this distinction.

Solution trees (d) and (e) in figure 10 correspond, each one to two sequences, but unlike the precedence diagrams of figure 8, the operations are exactly the same, regardless of the order in which they are executed.

5 Finding the Best Plan as an AND/OR graph search

To solve problems that require optimization, such as the selection of the best assembly plan, one must be able to traverse the space of all candidate solutions, regardless of the method used to solve the problem. The choice of the representation is critical since it is often difficult to delimit the set of potential solutions in a form which enumerates all the elements.

The AND/OR graph representation encompasses all possible ways to assemble one product, and therefore allows one to explore the space of all possible plans. Since plans correspond to solution trees in the AND/OR graph, the selection of the best plan can be seen as a search problem. Any such search problem requires a criterion to compare plans. One possibility is to assign to the hyperarcs weights proportional to the difficulty of their corresponding operations, and then compute the cost of a solution tree from a node, recursively, as:

- zero, if the node has no leaving hyperarc; or

- the sum of the weight of the hyperarc leaving the node and the costs of the solution trees from the successor nodes.

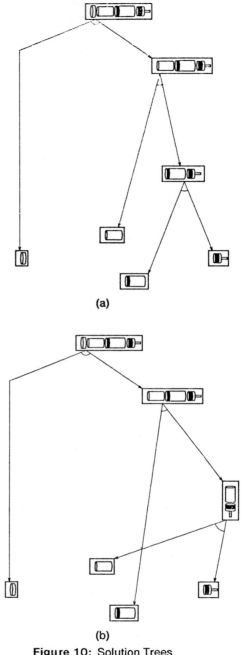

Figure 10: Solution Trees

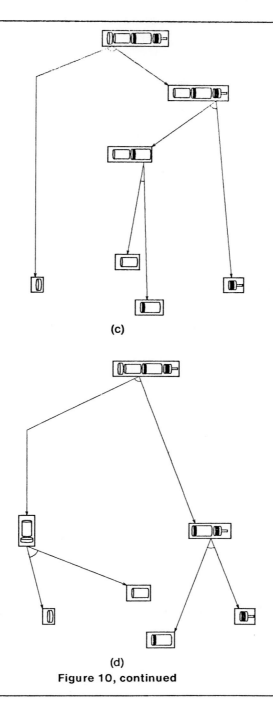

(c)

(d)
Figure 10, continued

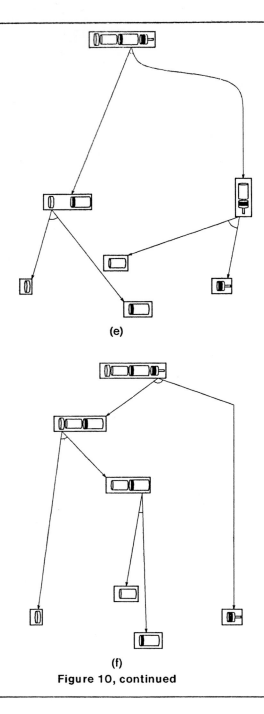

(e)

(f)
Figure 10, continued

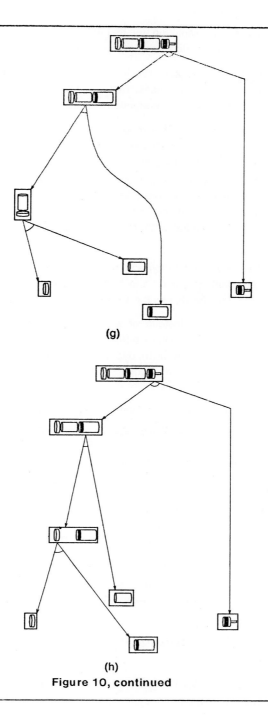

(g)

(h)
Figure 10, continued

The best plan corresponds to the solution tree that has the minimum cost. The search for the best plan can be conducted using generic algorithms such as the AO*[18].

A variety of factors might be considered in assigning weights to hyperarcs, including time duration of their corresponding operations, requirements for reorientation of fixturing, cost of resources needed, reliability, as well as production priorities and constraints.

For the product in figure 6, the AND/OR graph (figure 9) has 15 hyperarcs, which correspond to 15 different assembly operations. Table 1 shows one possible assignment of weights to hyperarcs. Those weights have been computed by adding two factors. The first factor is the type of assembly operation, with screw operation weighing 4, insertion 2 and placement 1, in accord with typical time, fixturing and manipulation requirements. The second factor taken into account is the difficulty of handling the participating subassemblies, and is proportional to their number of degrees of freedom; subassemblies with more degrees of freedom are more unstable, and therefore more difficult to handle.

Using that assignment of weights to hyperarcs, the total cost for the solution trees of figure 10 can be computed. The solution trees (a) and (h) have the minimum cost of 11; the solution trees (c), (d), (e), and (f) have total cost 13; and the solution trees (b) and (g) have the highest cost of 14.

For more complex assemblies, instead of a complete enumeration as done above, search algorithms can be used to reduce computation. For the product in figure 6, a search using AO* will yield one of the solution trees (a) or (h), depending on how the partial solutions and tip nodes are ordered for expansion.

Table 1: Assignment of weights to hyperarcs

factor	hyperarcs														
	1	2	3	4	5	6	7	8	9	10	11	12	13	14	15
operation	4	4	4	4	4	4	2	2	4	4	2	1	4	4	1
subassembly degrees of freedom	1	4	4	1	2	4	0	0	4	2	0	0	0	0	0
total	5	8	8	5	6	8	2	2	8	6	2	1	4	4	1

6 Opportunistic Scheduling Using the AND/OR Graph Representation

To evaluate how the use of AND/OR graph representation for assembly plans affects assembly efficiency, a comparative analysis among the three representation schemes discussed in this paper has been conducted.

The product in figure 6, and the robot workstation of figure 11 have been used as examples. The workstation is equipped with two manipulators and the parts are presented in random order. It is assumed that a cap, a stick, a receptacle, and a handle always come together, varying only in their order. It is also assumed that both manipulators are controlled by the same central unit and they both are able to execute the following actions:

- <u>acquire</u>: fetching, by one of the manipulators, of one part from the part feeder
- <u>buffer</u>: temporarily storing one part into a fixed location within the workstation
- <u>mate</u>: joining two subassemblies which are currently held by the manipulators
- <u>retrieve</u>: fetching, by one of the manipulators, one part known to be in the parts buffer

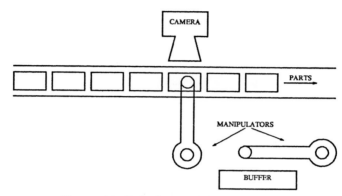

Figure 11: Robotic workstation

The efficiency of this assembly station depends on the capacity to handle parts in random order. This requires on-line scheduling of system resources depending on the order of parts arrival. The relative impact of plan representation schemes on assembly efficiency can be compared by the average number of operations needed; a smaller average number of operations corresponds to more efficiency.

The first sequence of figure 7 (A-B-C) has been used as an example of fixed sequence representation and the first precedence diagram of figure 8 (which combines A-B-C and B-A-C) as an example of precedence graph representation. Similar results will be produced using the other fixed sequences or precedence graphs. The number of operations that would be performed for each one of the 24 possible orderings in which the four parts of the simple product can be acquired is shown in Table 2. At least 7 operations are necessary: four acquisitions and three matings; depending on the order in which the parts are presented, buffering, and therefore retrieving may also be necessary.

When using the fixed sequence representation of plans, extensive buffering is necessary. For example, if the order the parts come is S H R C (stick, handle, receptacle, and cap) both the stick and the handle must be buffered since they are not used in the first operation; adding two bufferings and two retrievings to the four acquisitions and three matings that are always necessary yields 11 operations. The average number of operations for all 24 possible orders is 9.8.

Table 2: Number of operations needed to assemble the product of fig. 6 for all the sequences in which the parts may be acquired, and for the three schemes of plan representation
C = cap S = stick R = receptacle H = handle

sequence	first sequence of figure 7	first precedence diagrams of figure 8	AND/OR graph of figure 9
C S R H	9	9	7
C S H R	11	11	9
C R S H	7	7	7
C R H S	9	9	9
C H S R	11	11	9
C H R S	9	9	9
S C R H	9	9	7
S C H R	11	11	9
S R C H	9	7	7
S R H C	11	9	7
S H C R	11	11	9
S H R C	11	9	7
R C S H	7	7	7
R C H S	9	9	9
R S C H	9	7	7
R S H C	11	9	7
R H C S	9	9	9
R H S C	11	9	7
H C S R	11	11	9
H C R S	9	9	9
H S C R	11	11	9
H S R C	11	9	7
H R C S	9	9	9
H R S C	11	9	7
average	9.8	9.2	8

Using precedence diagrams for the representation of plans avoids some of the buffering and reduces the average number of operations to 9.2. For the sequence S H R C, for example, only the handle must be buffered since the insertion of the stick into the receptacle may be the first operation.

Using the AND-OR graph representation of plans, however, avoids most of the buffering, and yields the average of 8 operations. For the same S H R C sequence, for example, no buffering is needed because the robot can follow the sequence of operations corresponding to the solution tree (b) in figure 10.

7 Conclusion

This paper describes an approach to assembly systems planning which emphasizes selection of assembly sequences, representation of sequences, and use of this representation for operations scheduling. This approach is part of a framework which would provide interactive tools for design, configuration, and implementation of assembly systems. The AND/OR graph representation allows one to efficiently search the space of feasible plans and therefore is well-suited to problems where dynamic scheduling is important. The relational graph representation of assemblies has been studied for simple cases and found to be useful when local constraints dominate the choice of feasible sequences. This representation is being extended and evaluated for more complex assemblies. The use of the AND/OR graph representation for scheduling problems is demonstrated here, but no general approach to development of the control algorithm itself was described. The development of algorithms for opportunistic scheduling suitable for real-time application is being pursued. For complex cases, the choice of operations may depend on embedded cost functions. Strategies that depend on such evaluation functions and perhaps do not possess the recursive properties used here will be of interest.

ACKNOWLEDGEMENTS

The authors would like to express their thanks to Karen Lewis for her assistance in preparation of the manuscript.

References

1. Bellman, R. et al. *Mathematical Aspects of Scheduling and Applications*, Pergamon Press, 1982.

2. Fox, Mark S. *Constraint-Directed Search: A Case Study of Job-Shop Scheduling*, Ph.D thesis, Carnegie-Mellon University, December, 1983. Also published as technical reports CMU-CS-83-161 and CMU-RI-TR-83-22.

3. Fahlman, Scott Elliott "A Planning System for Robot Construction Tasks", *Artificial Intelligence* 5, No. 1 (1974): 1-49.

4. Fikes, Richard E. and Nilsson, Nils J. "STRIPS: A New Approach to the Application of Theorem Proving to Problem Solving" *Artificial Intelligence* 2, (1971): 189-208.

5. Fikes, Richard E. et al. "Learning and Executing Generalized Robot Plans" *Artificial Intelligence* 3 (1972): 251-288.

6. Sacerdoti, Earl D. *A Structure for Plans and Behavior*, Elseiver North-Holland, 1977.

7. Ambler, A.P. and Popplestone, R. J. "Inferring the Positions of Bodies from Specified Spatial Relationships" *Artificial Intelligence* 6 (1975), 157-174.

8. Popplestone, R.J., Ambler, A.P., and Bellos, I.M. "An Interpreter for a Language for Describing Assemblies" *Artificial Intelligence* 14 (1980), 79-107.

9. Popplestone, R.J., Ambler, A.P., and Bellos, I.M. "RAPT: A Language for Describing Assemblies) *The Industrial Robot*, Vol. 5, No.3, (1978):131-137.

10. Lieberman, L.I., and Wesley, M.A. "AUTOPASS: An Automatic Programming System for Computer Controlled Mechanical Assembly" IBM J. Res. Development, 21 July, (1977), 321-333.

11. Wesley, M.A., Lozano-Perez, T., Lieberman, L.I., Lavin, M.A., Grossman, D.D. "A Geometric Modeling System for Automated Mechanical Assembly, IBM J. Res. Development vol 24, No. 1, (Jan, 1980) 64-74.

12. Lee, K., and Gossard, D.C. "A Hierarchical Data Structure for Representing Assemblies, Part 1, *Computer-Aided Design*, 17, 1, Jan/Feb, 1985, 15-19.

13. Lee, K. and Gossard, D.C. "Inference of the Positions of Components in an Assembly, Part 2, *Computer-Aided Design*, 17, 1, Jan/Feb, 1985, 20-24.

14. Mason, M.T. "Manipulator Grasping and Pushing Operations, Ph.D. Thesis, MIT, (1982).

15. Peshkin, M.A., and Sanderson A.C. "Robot Manipulation of a Pushed Sliding Object: Part I No Contact Friction", CMU Robotics Institute Technical Report, CMU-RI-TR-85-18, (1985).

16. Peshkin, M.A. and Sanderson, A.C. "Robot Manipulation of a Pushed Sliding Object: Part II Contact Friction, CMU Robotics Institute Technical Report, CMU-RI-TR-86-7, (1986).

17. Krogh, Bruce H. and Sanderson, Arthur C. "Modeling and Control of Assembly Tasks and Systems" CMU-RI-TR-86-1, Robotics Institute Technical Report, CMU-RI-TR-86-1, (1985).

18. Fox, B.R. and Kempf, K.G. "Opportunistic Scheduling for Robotics Assembly", *In 1985 IEEE International Conference on Robotics and Automation*, pp. 880-889, IEEE Computer Society, 1985.

19. Nilsson, Nils J. *Principles of Artificial Intelligence*, Springer-Verlag, 1980.

MARS: AN EXPERT ROBOT WELDING SYSTEM

Pierre Sicard and Martin D. Levine
McGill Research Center for Intelligent Machines
Computer Vision and Robotics Laboratory
McGill University
3480 University Street
Montréal, Québec, Canada H3A 2A7

1. Introduction

Arc welding is considered to be one of the most promising applications of intelligent robots. This situation first stems from a low manual productivity due to the severe environmental conditions resulting from the intense heat and fumes generated by the welding process. Second, arc welding is the third largest job category behind assembly and machining in the metal fabrication industry.

Automation of the welding process opens a very challenging area of research in such fields as robotics, sensor technology, control systems and artificial intelligence. The research presented here, which is part of a proposed project for building an expert welding robot, is concerned with the adaptive control of the welding robot equipped with a vision system.

Existing welding robots require that the trajectory be programmed prior to the process. Also the process parameters are preset according to the existing conditions and cannot be changed during welding. These robots suffer from an evident lack of flexibility. If the parts to be welded are slightly misplaced or if the joint dimensions are not constant along the weld line, the resulting weldment is of very poor quality.

The capabilities of an adaptive welding robot address two main aspects. The first aspect concerns the control of the end effector's path and orientation so that the robot is able to track the joint to be welded with high precision. This seam tracking problem has generated a large amount of research in the past five years. The second aspect deals with the control of the welding process variables in real time. For example, it is important to control the amount of metal deposited into the joint in accordance with the dimensions of the gap separating the parts to be welded.

The on-line control of the welding process would conventionally involve the determination of a mathematical model of the process, the latter constituting the heart of the feedback loop. Adaptive control, as understood in this work, is not exactly what is commonly referred to as truly "adaptive", where the process parameters are estimated on-line. On the other hand, the great complexity of the welding operation implies that a fixed mathematical model may not be powerful enough to be used as the basis for a control operator. This has lead to an approach that is called "learning control" [Nevins et al. 84] and which is depicted in Figure 1.

The feedback operator is selected from a database to suit the actual conditions of the process. This operator remains the same for one welding procedure, but may be changed between successive procedures to cope with the initial conditions of each. Thus, the memory block in Figure 1 is not active in the real-time feedback loop. Nevertheless, it can still be called "adaptive" since the robot adapts to the varying environment of the welding process, in opposition to preprogrammed welding robots that use fixed parameters for the whole process.

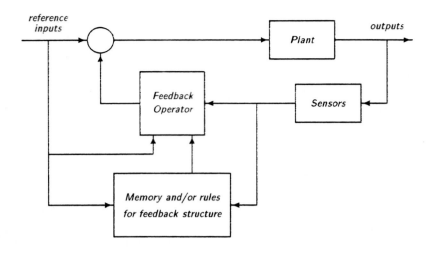

Figure 1   Learning control (from [Nevins et al. 84])

The purpose of this research is to identify the various relevant variables of the welding process and to investigate how these are interrelated. This may lead to the determination of a mathematical model to be used in the feedback control of process variables for a welding robot. Furthermore, it is desirable to construct a computerized so-called expert welder which is capable of approaching the capabilities of a human welder. The proposed system is hierarchical in nature.

The next section discusses the specification of the process variables and parameters. This is followed in Section 3 by an outline of the hierarchical controller and the requirements of modularity. Section 4 deals with the model to be used by the expert welder. Readers are referred to [Sicard and Levine 86] for a detailed literature survey and justification of the model.

2. Process Variables and Parameters

2.1 Process Variables

The design of a welding process control system involves, in the first place, the specification of the process variables. The welding process can be viewed as a system whose inputs are the process variables and whose outputs are the final characteristics that define the weld quality.

The process variables or system inputs can be broken down into three categories: those that can be varied on-line during the process, those that are set prior to the process and finally those that can not be modified. These categories are addressed respectively by [Dornfeld et al. 82] as the primary manipulatable inputs, the secondary manipulatable inputs and the unmanipulatable inputs.

For the GMAW process, the variables that can be altered independently on-line include voltage, wire feed rate and torch speed. Any changes in these variables produce an immediate effect on the weld quality. Because of the nature of the process, the current cannot be controlled independently from the wire feed rate. For example, an increase in the current causes the electrode tip to melt more rapidly, thereby requiring a higher wire feed rate to maintain a constant arc length [Lindberg and Braton 76]. Thus it is possible to control the current or the wire feed rate, but not both at the same time.

The variables that are set prior to the process include shielding gas flow and composition, torch angle, and electrode material and size. These variables remain constant during the process but may vary from one process to the other. They also differ from the first category in the sense that they need to be varied widely before they have a significant effect on weld quality. Their selection is done using charts giving the appropriate values depending on the type of application.

Variables in the last category define the process environment. They include plate thickness, joint geometry and physical properties of the base metal. All of these variables must be considered carefully for the appropriate selection of the secondary variables and for the adequate control of the primary variables.

2.2 Weld Quality Criteria

The parameters relevant to weld quality result from geometrical and metallurgical considerations. The former addresses the mass balance of the process which requires that the exact amount of metal be deposited to fill the joint. The latter deals with the heat balance which governs the metal fusion. One must maintain an adequate fusion in spite of variations in energy input and heat dissipation.

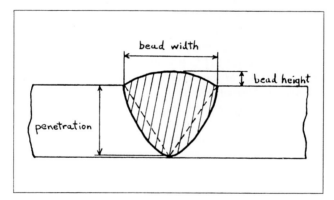

Figure 2 Bead cross-sectional area.

The cross-sectional area of a single weld bead, which is the solidified metal deposited in the joint, is depicted in Figure 2. The basic geometrical parameters that determine an acceptable weld shape include penetration, bead width, bead height and cross-sectional area. Depending on the type of joint these parameters may be specified differently. In the case of a fillet weld (Figure 3) for example, the geometrical parameters consist of the leg lengths and penetration [Linberg and Braton 76]. For the purpose of this discussion, basic geometrical parameters such as those in Figure 2 are considered.

In the case of penetration, it is generally defined as the depth to which the weld metal combines with the base metal as measured from the surface of the base metal [Althouse et al.

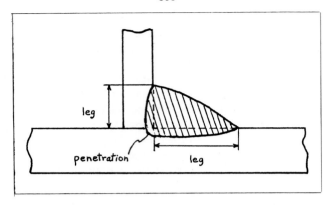

Figure 3 Fillet weld.

80]. The final bead configuration is crucial in defining the weld quality. Constraints on weld dimensions are specified prior to the process and depend on the type of application. These constraints must be respected in order to produce a good quality weld. Penetration remains the most difficult parameter to control since there are no means to measure it directly during the process.

The weld quality also addresses metallurgical and mechanical considerations, including such parameters as hardness, strength, soundness and residual stress. These characteristics are hardly measurable on-line and they result from a number of complex factors. However, they all depend on the energy or heat involved in the welding process. It is thus possible to obtain satisfactory metallurgical properties by controlling the thermal events that take place during the process. These thermal events may be quantified using the following criteria : the peak temperatures in the metal, the distribution of temperatures and the metal cooling rates. They are classified as secondary weld parameters by [Dornfeld et al. 82] or as intermediate indirect weld parameters by [Cook 80, Cook 81] [1]. It is interesting to measure these variables on-line because they reflect the behavior of the process. The peak temperatures indicate the extent to which the metal fuses and certain temperatures must be reached to assure good fusion. The temperature distribution indicates how the heat is being dissipated in the workpiece. For instance, this distribution may be perturbed by the presence of clamps that act as heat sinks. Thus it is important to maintain a uniform temperature distribution centered over the joint. Finally, the cooling rates are largely responsible for the final metallurgical properties of the weldment.

A schematic diagram of the variables and parameters is presented in Figure 4. The adequate measurement of thermal events during the process, along with the detection of the varying environment, are the keys to successful robotic welding.

2.3 Detection of Process Features

In the past ten years, the research in sensory methods for robotic arc welding has been aimed at detecting the following features:

---

[1] It appears that the thermal events could also be considered as outputs since they result from a combination of process inputs. However, they are dynamic rather than static outputs because they cannot be measured once the process is terminated. This is in distinction to the geometrical characteristics of the final weld which are shown as outputs in Figure 4.

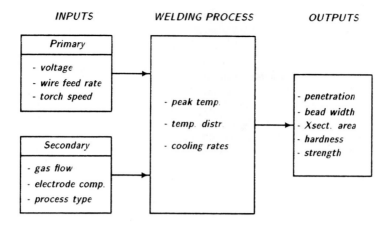

Figure 4 Weld variables and parameters.

1. joint geometry

2. weld pool geometry and location

The former is needed to perform seam tracking and to cope with variations in joint dimensions. The latter is useful to control thermal events during the process and to achieve fine tuning of the torch position when necessary.

2.3.1 Joint Detection and Seam Tracking

The most important function of an arc welding robot is its capability of following precisely the joint to be welded. Especially in the welding of large structures such as in the naval or car industry, it is practically impossible to always position the parts with the desired accuracy. Non-adaptive welding robots follow a preprogrammed path and may deviate slightly from the joint, leading to poor weld quality. On top of this, if the dimensions of the joint gap vary at one point, the process variables should be altered to deposit the exact amount of metal to adequately fill the joint. One is thus interested in implementing a joint recognition capability for the robot to cope with small variations in parts positioning and joint preparation.

One approach suggested by [Cook 83a] uses arc characteristics to achieve seam tracking. The method, based on "through-the-arc" sensing, exploits the fact that the welding current is proportional to the electrode-to-workpiece spacing. By weaving the arc back and forth across the joint, its geometrical profile may be obtained. However, joint detection can be more efficiently achieved using a vision sensor. In the field of passive vision, a television-based system for the tracking of butt joints has been developed [Richardson and Anderson 83]. The vision device is embedded in the welding torch barrel such that an image of the weld pool and the joint is obtained. Others [Arata and Inoue 73] use the light emitted by the arc to obtain information about the joint. The sensor is located in front of the torch and the shape of the emitted light is correlated with the dimensions of a square groove joint.

In the field of active vision, structured light techniques provide an opportunity to detect the location of the joint as well as its geometry. By projecting a stripe of light onto a butt joint for example, it is possible to locate it and determine the gap width [Inoue et al. 80]. Stripes of light are also useful to determine the angle of weld [Morgan et al. 83]. By projecting bright and dark stripes on the workpiece [Arata et al. 77], one can also achieve shape recognition in the case of fillet joint welding.

The use of laser light coupled with image sensors has given the best results for joint detection. In such systems, a light sectioning technique by laser is used for imaging the sectional pattern of the joint groove [Dufour and Bégin 83, Kawahara 83, Dornfeld and Tomizuka 84, Dufour and Cielo 84, Agapakis et al. 85, Clocksin et al. 85, Corby 85, Smati et al. 85]. The laser diode scans the workpiece transverse to the joint and the information is detected by a CCD array. The range data can be obtained using triangulation techniques. This is the most widely used method at present. An alternate technique to obtain range data is based on "time-of-flight" measurements. It determines the range from the time needed for the laser light to travel to the target and back to the receiver [Page et al. 81].

An algorithm that recognizes lap, T and butt joints from the sectional pattern obtained by the laser is described in [Clocksin et al. 85]. This operation takes place during a survey pass prior to welding and serves to build a database of models. Joint tracking is performed in real-time by comparing the current image to the appropriate model. Discrepancies are treated as error signals and are fed back to the control system.

The three-dimensional geometry of the joint and workpiece may be obtained by inferring the surfaces between successive cross-sectional patterns. This knowledge can be very useful in determining the volume of molten metal to be deposited, as well as to track the joint. In the latter case, control of the torch vertical displacement can be achieved by knowing the inclination of the part in the weld direction.

2.3.2 Weld Pool Detection

The features sought through weld pool observation are its geometry and its location. The three-dimensional shape of the molten pool determines the final bead dimensions. The depth cannot be obtained directly from front-face measurements, but surface geometrical features can be detected. These features consist of width, length and area (Figure 5). Thus, the task is to use surface geometry characteristics to estimate the resulting bead dimensions.

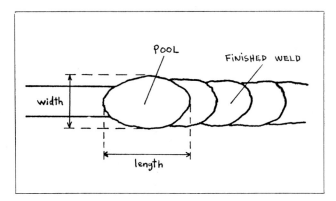

Figure 5 Surface of weld pool.

The weld pool must at all times be located on the centerline of the joint. Indeed, the fact that the torch precisely follows the joint does not necessarily mean that the weld pool is well centered. Under certain conditions, the arc is not stationary but shows a motion that will cause the weld pool to wander about the joint [2]. By measuring the weld pool position, appropriate control of the torch lateral position can be achieved by making sure that the weld pool remains centered.

Weld pool detection can be achieved by optical sensing. The observation of the molten metal area by optical sensors such as photodiodes or CCD arrays exploits the fact that these sensors have a certain response in the near infrared. Figure 6 compares the response of a CCD array with the radiations emitted by the molten metal [Rider 83, Bonvalet et al. 85]. The three curves on the right show the radiated energy around the melting point of the metal (T_{melt}). It can be seen that the maximum emission wavelength of the radiated energy at the weld pool edge is 1.6 μm. This is just at the limit of the CCD array response and it is estimated that only 4 % of the energy can be detected. However, the array does produce a steep response at the boundary between the solid metal and the molten pool because of the sudden change of emissivity at that location. Thus, the weld pool edges can be fairly easily detected.

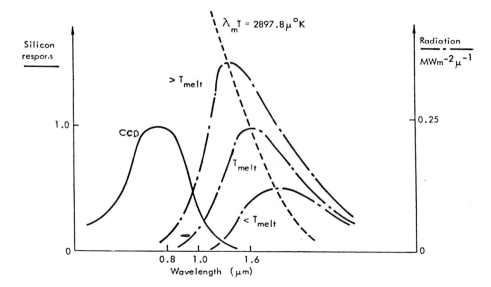

Figure 6 CCD array response vs. metal radiation (from [Rider 83])

The intense radiation emitted by the arc causes great interference for weld pool observation. There exist two major ways to overcome this problem. The first method consists of momentarily extinguishing the arc to record the CCD array response [Rider 83]. This requires perfect synchronization between the sensor and the control system. The main drawback is that it can only be applied with GMA in the short-circuiting transfer mode. The second approach uses a narrow band filter to block the arc light [Sweet et al. 83, Corby 85]. Another method [Richardson et al. 82, Richardson and Anderson 83] consists of mounting the vision system such that the optical axis is aligned with the welding electrode. The bright light of

---

[2] This is particularly true in the case of butt joints where the pool lies on a flat surface. If the weld pool moves away from the joint center at one point, the molten metal may not fuse with the base metal on one side. This phenomenon of incomplete fusion may lead to a severe weakness of the weldment.

the arc is then masked by the electrode shadow. However, this technique is only applicable to the GTA welding of butt joints.

Two-dimensional weld pool geometry analysis usually requires sophisticated image processing techniques that must be performed in real-time [Arata and Inoue 72, Arata et al. 76, Inoue 80, Richardson and Richardson 83].

An alternate approach for measuring the weld pool consists of using IR vision. The purpose of the infrared detector is to locate changes in radiation on the surface without necessarily correlating them with metal temperature. As stated above, there is a large difference in emissivity between liquid and solid metal which is manifested in the image by strong intensity discontinuities. The main advantage of using IR detectors instead of CCD arrays is to eliminate arc interference. This is shown in Figure 7. It can be seen that at wavelengths greater than 2 μm the energy radiated by the arc light decreases below the radiation emitted by the molten metal. Thus the effect of arc radiation is reduced in comparison to radiation from the metal. However, the use of filters is still necessary in GMA to eliminate the radiation emitted by the metal drops transferred across the arc [Ramsey et al. 63]. The hot tungsten electrode in the GTA process may also cause interference.

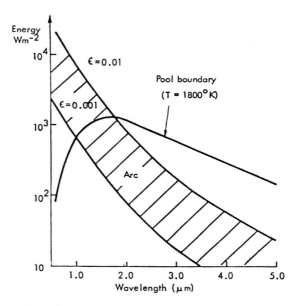

Figure 7 Radiated energy of arc and molten metal (from [Rider 83]).

An array of optical fibers, individually coupled to PbS detectors, was used by [Boillot et al. 85, Bégin and Boillot 83]. The IR array was aimed across the joint in front of the torch. In this case, the relevant geometrical features of the workpiece, as well as the weld pool boundary, are extracted from the variations in the recorded emission profile. This system has the advantage of being compact and light, and is thus well suited for robotic applications.

Analysis of temperatures as opposed to emission in the weld pool area has been extensively studied by [Chin et al. 83]. Using a thermographic image of the weld pool, they were able to detect impurities and variations in joint geometry and penetration. In this approach, the thermal profile is obtained directly. Thus, it is possible to detect the weld pool edges, given the melting temperature of the metal. An asymmetry in the recorded thermal distribution may indicate that the weld pool is not properly centered. Using this information, corrective action can be taken to control the torch lateral position.

A complete knowledge of the thermal distribution, either one or two-dimensional, is not always necessary to control the welding process. For example, temperature sensing at only two points, at an equal distance from the torch on both sides, may be sufficient. The purpose is then to maintain these two points at a constant temperature to ensure a symmetrical fusion with respect to the joint [McCampbell et al. 66, Chin et al. 83]. Moreover, the information recorded can be used as a reference temperature in the application of a mathematical model for welding process control.

2.4 Implementation

The design of a fully automated expert welding system requires much more than the adaptive control of the process. In fact the system must be able to identify the start and stop positions of the weld as well as to detect abnormalities that could occur during the process.

The starting location of the weld can be determined with the help of a vision system [Edling and Porsander 84]. Here a search is made around the starting position of the preprogrammed path. One system uses a tactile sensor to locate this start position [Kotykhin and Antonenko 80]. The stop position could also be recognized in the same fashion.

In order to maintain a good quality weld, the start and stop positions involve the application of special procedures [Kerth 85] [3]. The arc energy must be carefully increased or decreased, as the case may be, while the torch spirals over the seam.

An automatic welding system must also be able to detect any error conditions during the process. Such defects include shortage of electrode wire, shortage of shielding gas and extinguishing of the arc during welding. [Gellie 83].

In many industrial applications, welding requires that the torch follow a weave pattern across the joint. Thus, the torch describes a periodic movement in the lateral direction while traveling along the weld. This feature is particularly useful for the filling passes of wide joints. Most existing welding robots offer this weaving capability, but no commercial system permits the modification of the weaving parameters during the process. It is possible to select weave patterns [Kerth 85] according to their shape (square, sinusoidal), frequency and amplitude. The selection of the weaving pattern is done before the welding begins, based on the geometry of the joint. For instance, the weaving parameters could be selected from the information acquired during a teaching pass. It has also been suggested [Nachev et al. 83] that these parameters be varied on-line based on a 3-D analysis of the joint.

Not all robot manipulators can execute weaving movements in all positions or all planes. The manipulator must be flexible enough to allow precise torch positioning and adequate velocity control. Many studies have dealt with the control of torch position and velocity for welding applications [Alekseev et al. 79, Alekseev et al. 80, Tomizuka et al. 80, Khosla et al. 85].

Because of the great complexity of the welding process, it is important to have control over as many independent input variables as possible. In the GMAW process, wire feed rate and welding current are dependent variables. If the weld geometry is such that at one point

---

[3] A good quality weld requires that the ends of the weld seam be free of craters. These craters are experienced if the torch does not remain over the seam for a sufficient time to allow the metal buildup. Craters are points of high stress concentration that may lead to joint failure.

more metal needs to be deposited to fill the joint, one must increase the wire feed rate or reduce the torch speed. An increase of the wire feed rate causes the current to rise, resulting in a higher amount of heat brought to the process. This energy increase has a significant effect on the penetration and heat distribution in the workpiece, which may not be desirable. Thus it is very difficult to dissociate completely the heat control from the mass control. One way to overcome this problem is to use pulsed GMAW, which offers supplementary control variables such as pulse and background duration of the current [Linden et al. 80]. The amount of deposited metal can be controlled via the pulsing rate while reducing the heat input [Ogilvie and Ogilvy 83].

Among all of the welding robots which were surveyed in 1982 by [Timchenko and Nachev 82], few have adaptive welding process capabilities. Many of them can perform seam tracking as reviewed by [Gellie 83]. In most cases, those that implement adaptive functions for process control use one or two control variables coupled with one type of sensor. The approach presented here suggests the ability to simultaneously control the mass balance and the heat balance through the manipulation of three independent input variables and the use of two different types of sensors.

3. Hierarchical Control

In this section we discuss the physical structure of the expert welding system. A key consideration is modularity and device independence. Since it must be expected that the computer, robot, and sensor technologies will alter with time, the proposed design must be extremely flexible. This has generally not been the case with existing robots; usually, the controller is specifically tailored for a particular hardware configuration. To achieve this objective, MARS has been designed as a hierarchical control system [Albus et al. 81].

The MARS welding station controller is shown in Figure 8, and it consists of four main components. First the sensory processors are dedicated to preprocessing and analyzing the sensor data. Similarly, the robot trajectory and process controllers are responsible for real time control. Finally the fourth component is the heart of the expert system, playing the role of a planner and consultant. Its objective is to define, in conjunction with the welder, the optimal welding parameters for a specific welding task. To achieve this aim, it is necessary to have an appropriate world model, the topic to be discussed in the next section. The welder interaction environment, which is part of the Weld Task Planner, will allow the user of this system to interact with MARS by means of a user-friendly, menu-driven icon based visual programming language [Dorf and Kirschbrown 85, Computer 85]. At this level, the user will define the projected welding task employing offline programming. Ultimately, a CAD/CAM system could be used at this point in conjunction with a global vision system. For testing the trajectory aspects of the task, it is also convenient to include a robot simulator.

A simplified version of the control hierarchy is illustrated in Figure 9, where the communications paths are also shown. At the strategy level, the user (welder) interacts with the system in terms of an application language, in this case, an arc welding language for the specific task at hand. Here the user will plan the complete welding task and specify the appropriate constraints to the real time controller at the next level below (see Figure 10).

At the tactics level, a robot workcell language is used to co-ordinate the dynamic behaviour of all of the elements of the workcell under consideration. Such a system should have the following features:

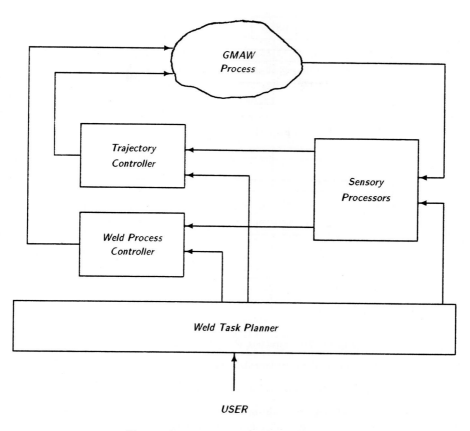

Figure 8 Architecture of the control system

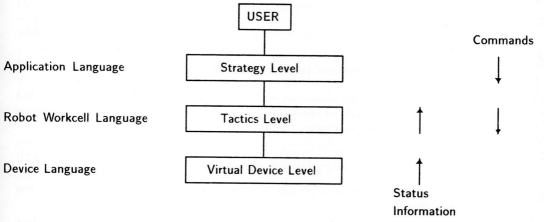

Figure 9 MARS defined as a hierarchical control system. (The terminology is due to Gregory Carayannis at McGill.)

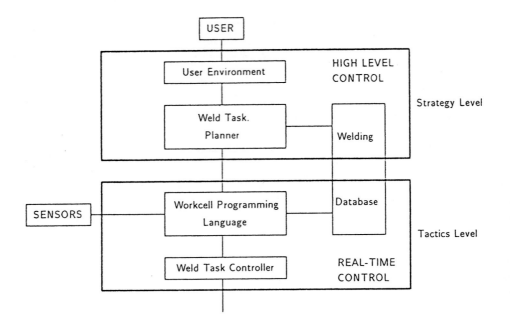

Figure 10 Structure of strategy and tactics levels

(i) It should be a general purpose robot-cell programming language.

(ii) It should be portable, as it will interact with virtual devices (see Figure 9).

(iii) It should interact with a world model (database).

(iv) It must provide for task concurrency. This is necessary because the workcell is a distributed computer system, with perhaps multiple processes running on each computer.

(v) It must provide for dynamic creation and deletion of processes.

(vi) It must have real time decision-making abilities based on the interpretation of sensory data.

(vii) It must handle the collision avoidance problem in real-time.

The basic requirements listed above cover the design of a general purpose controller and do not address the special needs of the welding environment. Design of a third generation welding system involves further additional requirements:

(i) Have control over all weld process parameters (voltage, current, wire-feed, gas, weld pool geometry, weld velocity, torch orientation, weave patterns).

(ii) Provide real time adaptive control over the weld process as a function of process parameters.

(iii) Provide sensory feedback of 3D geometry to permit measurements of those weld parameters not directly accessible (e.g. seam location, weld pool geometry).

(iv) Provide a simple description of welding task to the welding robot system using "smart" teaching. (The set-up time required for current systems makes the application of welding systems unattractive for small production runs. A task description based on weld level descriptions extracted from global and local sensory data would greatly simplify the teaching problem).

Each of these requirements imposes specific constraints on the design of MARS.

The next level down in Figure 9 is the virtual device level. It is desirable is to design MARS so that it is independent of the particular workcell robot being employed. In this way it would be possible to use this welding station controller with different robots from different maufacturers. Figure 11 illustrates the five *types* of hardware devices available. Each element is assumed to be independent and controlled by its own processor. From the point-of-view of the tactics level, each of these elements is considered to be a "black box". This creates a distributed processing environment where the elements must be able to communicate for the purposes of synchronization, control, and monitoring. Figure 12 indicates the logical levels for the virtual devices.

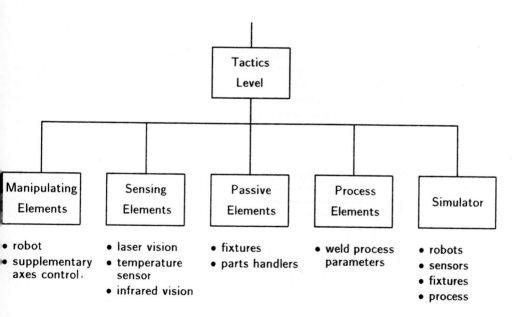

Figure 11 There are five classes of virtual devices and each class may have several elements, each controlled by its own processor.

From the point of view of the *user* in Figure 8, we envisage two nested levels as indicated in Figure 13:

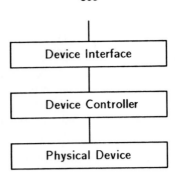

Figure 12 Logical levels of a virtual device.

(i)   Welding Technician Level

End user who tailors existing programs to specific instances. He sets up the production line for a run on a particular work piece. Since approximately 75% of the interaction between the system and user is at this level, particular attention must be paid to the man/machine interface, especially as it impacts the teaching.

(ii)   Engineer Level

End user who designs the software systems used at the technician level. The engineer's programs, written in a high-level robot programming language, will hide many of the system aspects. Nevertheless, they are sufficiently flexible for writing application programs. The robot programming language must allow for complete access to the internal states of the controller as well as full control of the robot functions.

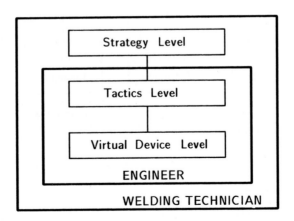

Figure 13 The user point-of-view

4. An Adaptive Welding System

Given the hierarchical control structure discussed in the previous section, in this section, we describe an approach for adaptive process control in welding. The architecture was designed within the scope of a project to develop an expert welding robot. The block diagram of the proposed system architecture is shown in Figure 8. Before explaining how each module works, it is interesting to examine the overall architecture of the system. The information gleaned by the sensors is processed to obtain the desired parameters describing the environment. Subsequently, the results are sent to the *Trajectory Controller* and the *Weld Process Controller* to adjust the manipulator coordinates and the process variables. The *Weld Task Planner* supplies all of the necessary information to the different modules operating in real-time. This information may originate in the welding database or the user, as explained shortly.

The adaptive functions needed to achieve a good quality weld should address the following issues [Boughton et al. 79]:

1. Position of the weld relative to the position of the joint.

2. Size of the weld relative to the size of the joint.

3. Cooling rate of the solid hot metal relative to its composition.

4. Solidification speed of the liquid weld metal relative to its composition.

The first item is taken care of by the *Trajectory Controller* and items 2 and 3 by the *Weld Process Controller*. The last item is not treated here, but an approach to its implementation is briefly discussed.

The variables controlled in real-time are voltage (u), wire feed rate (r) and torch speed (v). For each sample of measurements, the *Weld Process Controller* computes the required values of u, r and v that will give a satisfactory weld for the particular conditions encountered. One must keep in mind that these variables are not monitored. The voltage, wire feed rate and torch speed commands sent to the process are assumed to be maintained at the desired level until the following sample. If this cannot be achieved in practice, a monitoring feature would have to be implemented.

The following sections give detailed explanations of each module in Figure 8.

4.1 Sensory Processing

The measurements dealing with the varying environment include joint geometry (G), cross-sectional area of the joint (S), weld pool width and position (W_w, W_p) and workpiece temperature (T). Two sensors are needed to obtain this information. These environmental parameters are obtained by processing the sensor signals in the *Joint Profile Processor* and the *Weld Pool Processor*. The details of the sensory processing unit are depicted in Figure 14.

4.1.1 The Joint Profile Processor

Joint dimensions are obtained by processing the range data from the laser sensor. The laser sensor, mounted on the robot arm, is a very light and compact device. The light is scanned across the joint in front of the torch and range data of the joint profile are determined using triangulation techniques [Rioux 84]. The laser diodes operate at a wavelength of 830 nm to avoid interference by the arc. In fact, a typical argon-oxygen MIG welding arc produces low illumination in the range of 820 - 835 nm [Bégin et al. 85]. The measured signal is then

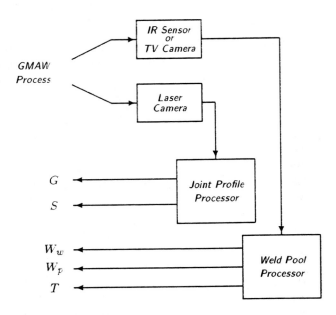

Figure 14 Sensory processing.

sent to the *Joint Profile Processor* which extracts the relevant information from the profile. Figure 15 shows the critical dimensions G for the three types of joint.

These dimensions are susceptible to variation during the process. The joint cross-sectional area S must also be computed in order to determine the amount of filler metal needed. This can be easily done by simple geometrical calculation. Efficient processing of joint information can be achieved by supplying a model of the joint to the *Joint Profile Processor*. Because only small variations are expected, the model is helpful in eliminating the signal noise and rapidly locating variations in the critical dimensions. In the same way, a poor correlation between the signal and the model indicates a significant change in the environment, such as the presence of tack welds or even corners.

The final output of the *Joint Profile Processor* thus comprises a series of parameters defining the joint profile dimensions and area. The number of parameters computed varies from one application to the other, depending on the type of joint.

4.1.2 The Weld Pool Processor

The remaining measurements, namely the width and location of weld pool and the workpiece temperature T, can be determined with the help of an infrared sensor or a television camera. The two approaches are possible but require different processing algorithms. Before discussing each case, some further explanation is needed about the temperature measurement. What is understood by this measurement is in fact the sensing of the plate temperature at a fixed location with respect to the torch. For the same conditions of voltage, wire feed rate and torch speed, the weld pool behavior is substantially different for a hot and a cold plate. Generally the weld pool tends to penetrate deeper in hot plates. No *direct* method of measuring the penetration is provided in this scheme, so that its value must be estimated from

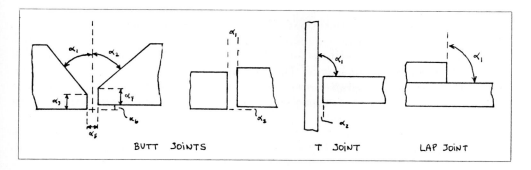

Figure 15 The joint dimensional parameters G.

front-face measurements, as shown shortly. Thus it is necessary to continually determine the temperature to take into account the heating of the workpiece during the process.

The first method uses an infrared sensor to scan the weld pool in front of the torch. It could be a linear array of detectors coupled to optic fibers [Boillot et al. 85] or a scanning device to sweep a point sensor across the weld pool. The former method is more convenient because of its compactness and reliability [4]. In both cases the signal obtained shows the radiation profile across the pool. As shown in Figure 16, the abrupt variations in the radiated energy indicate the weld pool edges. These discontinuities in the radiation profile are caused by the change of emissivity at the boundary between solid and molten metal. The middle point between these edge features represents the weld pool center position. For an adequate detection of these signal variations, the sensor must be located such that the reflection of the arc light from the molten pool is minimized. With regard to the temperature measurement, it can simply be taken at a fixed location in the solid metal region. Using the known value of the emissivity of the appropriate metal, the radiation at that point can be correlated with the metal temperature.

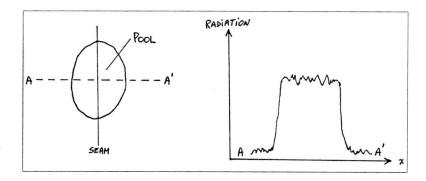

Figure 16 Radiation profile across the weld pool.

A slight variation of the above is infrared thermography which records the radiated energy of the metal and relates it directly to temperature, as shown in Figure 17. The temperature scale must be computed from a knowledge of the exact relationship between the energy

---

[4] To date scanning IR sensors have been too large and heavy to be used for this purpose.

recorded and the temperature. This is possible by knowing the specific values of emissivity in the solid and liquid regions for the type of metal considered. From the temperature distribution obtained, the molten metal-solid metal interface can be identified at the points corresponding to the metal melting point temperature. The distribution presents two inflections at this temperature, as shown in Figure 17. However, these inflections may disappear if the distribution is not symmetrical, in which case the pool edges can only be detected using the known melting temperature of the metal considered.

By exploiting the fact that sudden variations of energy are experienced at pool edges, the analysis of the radiation profile given by an IR sensor, as discussed above, may be sufficient. As only radiation is taken into account, no temperature calibration is needed and less sources of error are introduced.

The use of a television camera provides an alternate method for obtaining the desired information. From an economical point of view, this method is preferable since TV cameras are cheaper than IR detectors. To block the bright arc light, the camera needs a special filter or has to be synchronized with the extinguishing of the arc, as explained in Section 3. The image acquired in this fashion shows a bright region on a dark background representing the weld pool. By processing this image one can extract the weld pool geometry from the contour of the bright region. This two-dimensional analysis of the molten pool offers the advantage of providing extra information such as the weld pool length and area. Acquisition of these features using infrared techniques requires a two-dimensional detector, which is not yet commercially available. The additional parameters of length and area could also be used in a sophisticated model to estimate penetration from the complete pool geometry. Successful processing algorithms of weld pool images have been developed in research laboratories [Richardson and Richardson 83]. However the model that relates penetration to weld pool surface geometry has not been established.

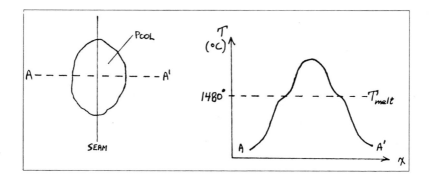

Figure 17 Thermal profile across the weld pool for a symmetric weld.

Since only the weld pool width and position are needed in the model presented here, the processing of a single video line can provide all the information. In fact, the signal on a line across the weld pool shows abrupt changes of intensities at the weld pool edges. By averaging and differentiating the signal [Richardson et al. 82], one can locate the weld pool edges and compute the width from the distance between edge features.

One inconvenience of using a camera in lieu of an infrared sensor is the loss of accuracy in the plate temperature measurement. Because the camera does not supply any information about temperature, the latter must be computed on a theoretical basis. From the locations of the pool edges, which represent the melting point of the workpiece, it is possible to extrapolate temperatures in the solid metal area. The workpiece temperature at a fixed location can thus be estimated, but with the help of a model of the temperature distribution in the weld pool area. The accuracy of the result is then limited by two factors. The first factor is the accuracy of the model itself which must take into account the varying features of the environment, such as joint geometry and arc motion. This model would have to be determined empirically. The other factor is the preciseness in the detection of the actual pool boundaries. In fact it has been known that the temperature gradient in the solid metal around the weld pool is very strong, particularly near the electrode [Fihey et al. 83]. By using an extrapolation method, a small misinterpretation of the weld pool edge location may cause a significant imprecision in the plate temperature estimation.

As in the case of the first sensor, the functioning of the *Weld Pool Processor* needs additional information from the welding database. For instance, physical properties of the metal such as emissivity must be known. These characteristics differ with the type of metal to be welded and must be revised before each welding application.

4.2 The Trajectory Controller

The *Trajectory Controller* takes care of the torch path and its orientation. The latter is defined by the transverse and longitudinal angles, which are set prior to the process. The transverse angle depends on the type of joint. It is normally chosen such that it splits the angle of weld. For example, the fillet welding of two plates forming an angle of 60 degrees requires that the torch be positioned at 30 degrees from each plate. The longitudinal angle also depends on the type of application and can be determined before the process. Generally the torch orientation remains constant during the process, unless significant changes in the conditions occur. In fact, the transverse angle may be adapted to cope with possible distortion in the configuration of the plates. This distortion is normally caused by the intense heat generated by the process. The adaptive control of torch angle could easily be implemented using information about joint geometry supplied by the *Joint Profile Processor*. A torch angle control system is discussed in [Kremers et al. 83].

The torch path is governed by the speed and position commands. For the start and stop procedures, the accelerations and decelerations may also be controlled, but this issue will not be discussed here. The torch speed command is determined by the *Weld Process Controller* to meet the energy and mass transfer requirements imposed by the environment.

The adequate positioning of the torch is a critical issue for successful welding. This position is defined by torch horizontal and vertical locations with respect to the joint center line. The torch height must be such that it provides sufficient space to account for arc length fluctuations. Indeed, voltage variations during the process cause the arc length to vary. The setting of torch height must also take into account the electrode stickout.

The *Trajectory Controller* is depicted in Figure 18. The control of torch lateral position can be achieved from the knowledge of joint geometry and weld pool location. The control law may be expressed as follows, where x_t is the torch lateral position:

$$x_t = f_1(G) + f_2(W_p) \quad cm \tag{1}$$

where G represents the joint profile geometry and W_p the weld pool center position. Functions f_1 and f_2 are supplied by the *Welding Database* in accordance with the type of joint and the nature of the pass performed. For instance, the root pass of a V-groove weld (Figure 19) requires that the torch be located exactly over the center of the gap separating the plates. For successives passes the function f_1 must be such that the torch is guided to the appropriate position.

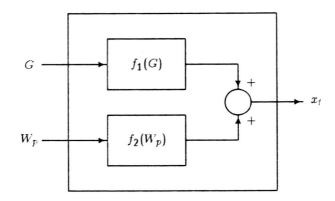

Figure 18   The Trajectory Controller

A simple seam tracking situation is illustrated in Figure 20. In this case, $f_2 = 0$ and torch positioning in the lateral direction is done according to the control law [Khosla et al. 85]:

$$x_t = \frac{u_{xi} + v_{xi}}{2} \qquad (2)$$

where u_{xi} and v_{xi} are the x-coordinates of points u_i and v_i in Figure 20.

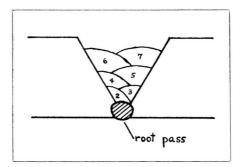

Figure 19   Multipass welding.

The function f_2 accounts for possible deviations of the weld pool center line with respect to the joint. This function must be determined by experimentation. If the pool location is found to be off-track on one side, the torch must then be pulled slightly to the opposite side. The extent to which the torch must be displaced is proportional to the shift of the weld pool position. The accuracy in the determination of f_2 is not that critical, since the sampling period of the control system is much less than the response time of heat propagation in the metal.

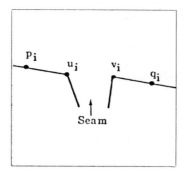

Figure 20 Seam tracking (from [Khosla et al. 85])

A more direct method for centering the molten pool uses temperature measurements at two equally distant locations on both sides of the electrode. If ε_1 is the error signal resulting from the difference between the two measurements, the equation could be stated as:

$$x_t = f_1(G) + k_1\varepsilon_1 \quad cm \qquad (3)$$

where k_1 is a constant. When the pool is symmetrical with respect to the electrode, the measured temperatures are the same and $\varepsilon_1 = 0$. This control law would only apply for symmetrical parts. It could not be used to track a lap joint where the heat propagation is not the same on both sides of the weld [Bégin and Boillot 83].

The operation of the *Trajectory Controller* benefits from the knowledge of a preprogrammed path which can be stored in the *Welding Database*. In fact, the robot initial trajectory could be specified during a teaching phase prior to the process. With the help of a graphics interface and CAD/CAM techniques, the teaching of the robot's trajectory could be done interactively. Once the path is known, adaptive torch guidance could be used to account for small variations in parts positioning and joint preparation. Ultimately, if the torch control system turns out to be very efficient, only the start and stop positions of the robot trajectory would have to be taught.

4.3 The Weld Task Planner

The *Weld Task Planner*, detailed in Figure 21, contains all the necessary knowledge to supervise the welding procedure. It comprises the user interface, a decision module, a set of user and process constraints, and a welding database.

The information acquired through the *User Interface* is twofold. On the one hand, the user defines the fixed parameters such as plate thickness, base material composition and joint type. On the other, he specifies the acceptable quality criteria that must be met for the existing task. These criteria affect the weld bead geometry and the metal cooling rate. As shown in the *User Constraints* block in Figure 21, the penetration P is bounded by ρ_1 and ρ_2 which are usually expressed as percentages of plate thickness. When applicable, the constraint β on bead width B is specified as an absolute value. For some applications, the user may wish to also impose a limit K which then constrains the ratio of penetration to bead width. Finally, the cooling rate C_r is bounded to ensure good metallurgical properties of the resulting weldment.

The *Decision Module* is able to select process-related characteristics that permit the system to meet the quality criteria imposed by the user. This decision is based on the type of application, which is defined by the type of material to be welded, the nature of the pass to perform, the position of parts and the type of joint. Factors such as electrode material and size, shielding gas composition and flow rate, along with the metal transfer type, are selected with the help of charts stored in the *Welding Database*. Once the type of process is entirely defined, one can generate the constraints acting on wire feed rate (R_1, R_2), torch speed (V_1, V_2), voltage (U_1, U_2) and current (I_1, I_2). These *Process Constraints* define a domain of permissible values for the process variables to respect while performing the selected welding procedure.

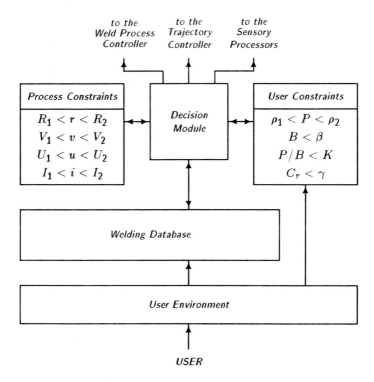

Figure 21 The Weld Task Planner.

The main role of the *Decision Module* is to administer the information in the *Welding Database*. It supplies necessary data to the *Trajectory Controller* and the *Sensory Processors*, as explained previously. Its most important function is to select the set of appropriate functions that define the welding model embedded in the *Weld Process Controller*. The set of equations is extracted among a large number of functional relationships, each one prevailing under particular conditions. This operation is performed before each welding run and is based on information from the user and the sensory processors. Plate and joint characteristics supplied by the user contribute to the determination of the model. Supplementary information, such as initial plate temperature, can be obtained via the sensors.

The *Weld Task Planner* plays a determinant role in this control scheme. It constitutes the link between the user and the process control loop. Further developments in higher-level automation of the welding process should aim at refining this "expert" module.

4.4 The Weld Process Controller

The *Weld Process Controller* is the heart of the real-time feedback loop. It uses the information from the sensory processors to generate the voltage, wire feed rate and torch speed commands that suit the conditions encountered. For each sample of measurements, the computation of the new values of the process variables involves the application of the model. Each result must then be verified by the *Weld Task Planner* to ensure that all user and process constraints are satisfied. If a set of values falls outside acceptable limits, some iteration might be necessary. The functional diagram of the *Weld Process Controller* is shown in Figure 22.

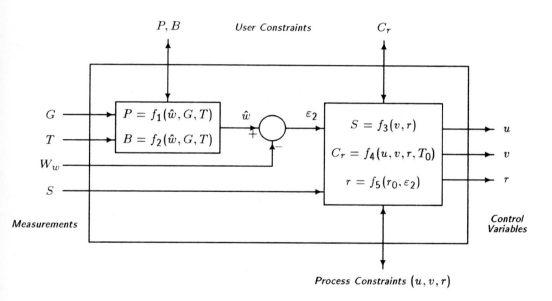

Figure 22 The Weld Process Controller.

The computation is initiated in the left-hand box where the reference weld pool width $\hat{w}$ is evaluated using the following equations:

$$P = f_1(\hat{w}, G, T) \tag{4}$$
$$B = f_2(\hat{w}, G, T) \tag{5}$$

Penetration P and bead width B are expressed as functions of weld pool width $\hat{w}$, joint geometry parameters G, and workpiece temperature T. Using the measurements of joint dimensions and plate temperature obtained by the *Sensory Processors*, we can compute a reference width $\hat{w}$ such that the geometrical constraints imposed on the weld bead will be respected. This is possible because the final configuration of the bead is directly related to the shape taken by the molten pool during the process. The number of equations in this part of the model may vary from one application to the other. In the case of fillet welds for instance, the equations would address the penetration and the two leg lengths. Similarly one may end up with a single constraint on penetration if welding the root pass of a V-groove joint. If more than one equation is present, a common value of $\hat{w}$ must then be found to satisfy all the constraints at the same time.

The actual weld pool width W_w determined by the *Weld Pool Processor* is then compared to the reference width. The difference ε_2 is used to directly control one of the process input variables u, v, or r. As shown in the diagram, the wire feed rate r could be controlled using this principle with a proportional control law for f_5 of the form:

$$r = r_0 + k_2 \varepsilon_2 \quad Kg/sec \tag{6}$$

where r_0 is the original wire feed command and k_2 a gain constant, determined empirically.

Once the wire feed rate has been computed, we are left with two equations and two unknowns, u and v:

$$S = f_3(v, r) \tag{7}$$
$$C_r = f_4(u, v, r, T_0) \tag{8}$$

Equation (7) represents the mass balance of the process. It relates the cross-sectional area of the deposited metal (which must equal the joint area S) to the torch speed and wire feed rate. It can be stated as follows [Lancaster 84]:

$$S = \alpha \frac{r}{\rho v} \quad meter^2 \tag{9}$$

where r is the wire feed rate expressed in Kg/sec, ρ is the mass density of the electrode material in Kg/m^3 and v is the torch speed in m/sec. The factor α represents the metal transfer efficiency for the process. A percentage of filler metal is actually lost due to spatters during the process. This factor could easily be approximated from previous experience. In order to deposit the exact amount of metal into the joint, this equation must be satisfied at all times.

Finally equation (8) deals with the cooling rate C_r of the solidified metal. It can be expressed as a function of voltage u, wire feed rate r, torch speed v and initial workpiece temperature T_0. It also depends on such fixed parameters as plate thickness and joint type, but it is assumed that the proper function f_4 is selected from the database in accordance with the prescribed application. The initial plate temperature can be acquired before the weld run using an infrared sensor or a thermocouple. The cooling rate C_r computed with this equation is subject to the constraint imposed by the user.

As indicated earlier, a complete process control scheme should also address the problem of solidification time of the molten metal. This issue has not been extensively studied for process control purposes. One analytical expression is given by [Fihey et al. 83]:

$$P = \frac{t_s}{k_3} \frac{dT}{dR} \quad cm \tag{10}$$

where P is the penetration, t_s, the solidification time in sec, and dT/dR, the thermal gradient at the liquid-solid interface, expressed in $°C/cm$. Constant k_3 depends on the material composition and thickness. It can be seen that the solidification time could be estimated from the measurement of the thermal gradient and the computed value of penetration.

Except for the mass balance equation (9), all the functional relationships of the *Weld Process Controller* must be determined from empirical data. This can be done efficiently using the "EM2000" system developed at the *Ecole de Technologie Supérieure* in Montréal [Galopin and Boridy 83]. The method was originally developed to determine optimal welding procedures using computer aided experimentation. The apparatus comprises a welding unit mounted on a track to allow displacement of the torch along the weld line. During the test run, a computer varies the process variables and records the measurements. Correlation between measurements and input variations is then performed using statistical analysis. The corresponding curves relating input variables to quality parameters can thus be obtained. The great advantage of this method lies in its capability of reducing the amount of experimentation. This is made possible by a rational plan for each experiment. The analysis performed on the results supplies information about more than one variable. The following experiments are then focused on the variables that have a significant effect on the weld quality.

For the present application, the use of this tool is a valuable asset. Its purpose is to vary the input variables and to measure the resulting weld characteristics. For example, equations (4) and (5) could be determined by varying torch speed, voltage and wire feed rate while measuring the weld pool width and plate temperature. This experiment would be done for different sets of joint dimensions. After the test run, the bead would be cut at the locations where the measurements were taken in order to observe the resulting penetration and bead width. To record the measurements of pool geometry and plate temperature, the welding unit would have to be equipped with an IR sensor or a television camera. The cooling rate could be approximated by considering successive samples of temperature measurements. It is estimated that the determination of the required functions for a specific welding application could be achieved within a few days.

An important characteristic of the set of equations (4-9) forming the model is that they relie only on those parameters that are susceptible to variation during the process. For instance, the penetration is obviously a function of more than the three variables appearing in equation (4). It also depends on such secondary inputs as type of shielding gas, electrode diameter and workpiece composition, just to name a few. Each function is thus directly associated with a known set of secondary inputs and its applicability is limited to these particular conditions. If one of the secondary or unmanipulatable inputs is modified, a new relationship may have to be determined. This way of proceeding may seem awkward but it offers two important advantages. First it assures that the results obtained are accurate within known limits. Second it is well suited for industrial applications where the same welding task is performed many times. The set of equations needs to be determined only once, and this can be done relatively fast using the method presented above.

In this scheme, the role of the *Welding Database* takes primary importance. The memory must be organized such that each function is related to a precise set of process parameters. It will be necessary to rapidly locate and retrieve the appropriate functions that apply to the existing welding procedure.

In summary, the weld process control system can be stated as a set of functions describing the behavior of the existing process. These functions relate the input variables and the measurements to the weld quality criteria. The on-line regulation of input variables requires that all equations forming the model be satisfied, while at the same time respecting the constraints imposed on the weld quality (*User Constraints*) and input variables (*Process Constraints*). This computation must be such that variations undergone by input variables be minimized from one command to the next. Thus a cost function would be useful to optimize the process by according a certain priority to the torch speed, for instance.

5. Conclusions

Features that could enhance the performance of this control system are related to considerations of speed and flexibility. In the first place the sensory processing algorithms should be relatively fast, perhaps 30 Hz. One should aim at a frequency response in the range of 5-10 Hz for the control system. The resulting sampling period would allow sufficient time for sensory processing and be fast enough compared to the process time constants. In fact, expected variations in joint geometry and weld pool behavior are not susceptible to abrupt change. Important environmental changes, such as corners, could be entered by the user before the run and be known to the sensory processors.

System flexibility could be improved by allowing the user to select the input variables of the system. The two sensors proposed for this system should be able to provide all the necessary information about the varying environment. However the set of input variables to be controlled may differ depending on the task to be achieved. For instance, the welder may wish to have control over weaving parameters, such as shape, amplitude and period. One can also adaptively control the electrode stickout or the torch angle. This is possible provided that appropriate functional relationships have been determined beforehand. Finally, in the Gas Tungsten Arc Welding process, we note that the welding current can be used as a fourth input variable. Indeed current and wire feed rate can be controlled independently because the filler metal is provided by a separate wire.

Besides the process model itself, the most important factor to consider in welding automation is the man-machine interface. The main objective when designing this part of the system is to make it easy to learn and use, while providing adequate tools for good communication. It is important to clearly state what the user needs to know for operating the system. In the approach discussed here, the user must provide information regarding the plate and joint geometry, the final quality criteria and the torch trajectory. The latter can be done by interactively entering successive coordinates of the path. The robot controller must then interpolate the complete trajectory. Input of the robot path can also be done by manually guiding the end effector along the weld.

A complete automated welding system should make use of CAD/CAM techniques. This would be useful to facilitate communication between the user and the system and to model the welding environment. A sophisticated graphics interface would permit the user to specify joint dimensions. The system could display a model of the joint to be welded and the user could adjust the dimensions by manipulating a mouse, for example. Using this information and that from the laser sensor, the welding workcell could be modeled in the database with a precise definition of part position and orientation. It would then be possible to determine a complete sequence of operations to achieve complex tasks. Moreover, the initial robot trajectory could even be computed by the system from the knowledge of the part configuration. Modeling of the workcell would also permit detection of the presence of obstacles on the robot path. Using this information, the trajectory could be planned before the run in order to avoid collisions during the process.

The man-machine interface serves as the communication channel for an expert welding system. The role of such a sophisticated system is first to act as a consultant to guide the user in the preparation phase of the welding process. Secondly, the system must be able to plan the welding task in accordance with user specifications. According to the geometry of the parts to be welded, the system can give advice in edge preparation. It can also be useful to position the parts before the run in order to cope with heat distortion during the process. For some applications, it can suggest where to place tack welds to maintain the workpiece in a stable position. Finally, the task planning capability of the system can help to determine the number of passes required to meet restrictions imposed on the final product. It can also give advice for selecting torch orientation and adequate process parameters.

The expert welding system discussed here could enrich its knowledge from the different tasks encountered during operation. Its increasing welding expertise would eventually allow the welder to concentrate on performance optimization by simulating different welding procedures.

The objective of this work was to investigate the nature of the knowledge required to achieve automation of the arc welding process at the level of a human welder. An approach has been proposed for the on-line control of process variables using feedback methods within the scope of an expert welding system. Based on results from previous research in the field, the requirements in terms of sensory equipment, system architecture and software development have been defined.

It can be seen that the mastery of the welding process by machine is feasible, but on a limited scale at present. It is evident that no general mathematical solution for the automatic control of all possible welding tasks can be found. A minimum amount of preliminary experimentation is unavoidable for the determination of adequate control schemes. Although the scope of applicability of these schemes is limited by the conditions of the experiment, they can contribute in the long run to the enlightenment of the complexity of the welding process. A lot of research still has to be done, but the present work should be a useful guideline for further developments towards autonomous robotic arc welding.

6. References

[Agapakis et al. 85] J.E. Agapakis, K. Masubuchi and N. Wittels, "General Visual Sensing Techniques for Automated Welding Fabrication", *Proc. 4th Int. Conf. on Robot Vision and Sensory Controls RoViSeC4*, London, Oct. 1984, pp. 103-114.

[Albus et al. 81] J. Albus, A. Barbera and R. Nagel, "Theory and practice of hierarchical control", *Proc. of the IEEE COMPCON Fall Conf.*, 1981, pp. 231-237.

[Alekseev et al. 79] K.B. Alekseev, V.A. Afonin and M.M. Fishkis, "A two-channel control system for an adaptive welding robot", *Automatic Welding (GB)*, vol. 32, no. 12, pp. 25-28, December 1979.

[Alekseev et al. 80] K.B. Alekseev, M.M. Fishkis and V.V. Fokin, "The adaptive control of a welding robot", *Welding Production (GB)*, vol. 27, no. 9, pp. 4-8, September 1980.

[Althouse et al. 80] A.D. Althouse, C.H. Turnquist and W.A. Bowditch, "Modern Welding", South-Holland, Ill., The Goodheart-Willcox Co., 1980, Chapter 30, pp. 733-742.

[Arata and Inoue 72] Y. Arata and K. Inoue, "Automatic Control of Arc Welding by Monitoring Molten Pool", *Transactions of JWRI*, vol. 1, no. 1, pp. 99-113, 1972.

[Arata and Inoue 73] Y. Arata and K. Inoue, "Automatic Control of Arc Welding (Report II) - Optical Sensing of Joint Configuration", *Transactions of JWRI*, vol. 2, no. 1, pp. 87-101, 1973.

[Arata et al. 76] Y. Arata, K. Inoue, M. Morita and G. Kawasaki, "Automatic Control of Arc Welding (Report V) - Application of Digital Picture Processing Technique to Automatic Control", *Transactions of JWRI*, vol. 5, no. 1, pp. 77-85, 1976.

[Arata et al. 77] Y. Arata, K. Inoue, Y. Shibata, M. Tamaoki and H. Akashi, "Automatic Control of Arc Welding (Report VI) - Recognition for Intersection of Two or Three Planes", *Transactions of JWRI*, vol. 6, no.1, pp. 7-16, 1977.

[Bégin and Boillot 83] G. Bégin and J.-P. Boillot, "Welding adaptive functions performed through infrared (IR) simplified vision schemes", *Proc. Third Int. Conf. on Robot Vision and Sensory Controls RoViSeC3*, Cambridge, Mass., SPIE vol. 449, Nov. 1983, pp. 328-337.

[Bégin et al. 85] G. Bégin, J.-P. Boillot, C. Michel and G. Teubel, "Third Generation Adaptive Robotic Arc Welding Unit", *Proc. Int. Institute of Welding Annual Conf.*, Strasbourg, France, Sept. 1985.

[Boillot et al. 85] J.-P. Boillot, P. Cielo, G. Bégin, C. Michel, M. Lessard, P. Fafard and D. Villemure, "Adaptive Welding by Fiber Optic Thermographic Sensing: An Analysis of Thermal and Instrumental Considerations", *Welding Journal*, vol. 64, no. 7, pp. 209s-217s, July 1985.

[Bonvalet et al. 85] J.C. Bonvalet, Y. Launay and C. Philip, "Application de capteurs de vision pour la commande adaptative du soudage à l'arc", *Soudage et Techniques Connexes*, vol. 39, no. 1-2, pp. 53-58, Janvier-Février 1985.

[Boughton et al. 79] P. Boughton, G. Rider and C.J. Smith, "Towards the automation of arc welding", *CEGB Research*, Central Electricity Generating Board, UK, pp. 33-40, June 1979.

[Chin et al. 83] B.A. Chin, N.H. Madsen and J.S. Goodling, "Infrared Thermography for Sensing the Arc Welding Process", *Welding Journal*, vol. 62, no. 9, pp. 227s-234s, September 1983.

[Clocksin et al. 85] W.F. Clocksin, J.S.E. Bromley, P.G. Davey, A.R. Vidler and C.G. Morgan, "An Implementation of Model-Based Visual Feedback for Robot Arc Welding of Thin Sheet Steel", *The Int. J. of Robotics Research*, vol. 4, no. 1, pp. 13-26, Spring 1985.

[Computer 85] *Computer*, IEEE Computer Society, vol. 18, no. 8, August 1985.

[Cook 80] G.E. Cook, "Feedback control of process variables in arc welding", *Proc. IEEE Joint Automatic Control Conf.*, San Francisco, CA, vol. 2, Aug. 1980, paper FA7-B.

[Cook 81] G.E. Cook, "Feedback and adaptive control in automated arc welding systems", *Metal Construction*, vol. 13, no. 9, pp. 551-556, September 1981.

[Cook 83a] G.E. Cook, "Robotic Arc Welding: Research in Sensory Feedback Control", *IEEE Trans. on Industrial Electronics*, vol. IE-30, no. 3, pp. 252-268, August 1983.

[Corby 85] N.R. Corby Jr., "Machine Vision Algorithms for Vision Guided Robotic Welding", *Proc. 4th Int. Conf. on Robot Vision and Sensory Controls RoViSeC4*, London, Oct. 1984, pp. 137-147.

[Dorf and Kirschbrown 85] R.C. Dorf and R.H. Kirschbrown, "KARMA - A Knowledge-Based Robot Manipulation System", *Robotics*, vol. 1, 1985, pp. 3-12.

[Dornfeld et al. 82] D.A. Dornfeld, M. Tomizuka and G. Langari, "Modeling and Adaptive Control of Arc Welding Processes", *Meas. and Cont. for Batch Manuf.*, Hardt ed., ASME, Nov. 1982, pp. 53-64.

[Dornfeld and Tomizuka 84] D.A. Dornfeld and M. Tomizuka, "Development of a Comprehensive Control Strategy for Gas Metal Arc Welding", *Proc. 11th Conf. on Prod. Research and Technology*, Pittsburgh, PA, Soc. of Manuf. Eng., May 1984, pp. 271-275.

[Dufour and Bégin 83] M. Dufour and G. Bégin, "Adaptive robotic welding using a rapid image pre-processor", *Proc. Third Int. Conf. on Robot Vision and Sensory Controls RoViSeC3*, Cambridge, Mass., SPIE vol. 449, Nov. 1983, pp. 338-345.

[Dufour and Cielo 84] M. Dufour and P. Cielo, "Optical inspection for adaptive welding", *Applied Optics*, vol. 23, no. 4, pp. 1132-1134, April 1984.

[Edling and Porsander 84] G. Edling and T. Porsander, "Adaptive control of torch position and welding parameters in robotic arc welding: Examples and practical use", *Proc. 14th Int. Symp. on Ind. Robots, 7th Int. Conf. on Ind. Robot Tech.*, Gothenburg, Sweden, Oct. 1984, pp. 359-363.

[Fihey et al. 83] J.L. Fihey, P. Cielo and G. Bégin, "On-Line Weld Penetration Measurement Using an Infrared Sensor", *Proc. Int. Conf. Welding in Energy-Related Projects*, Toronto, Canada, Sept. 1983, pp. 177-188.

[Galopin and Boridy 83] M. Galopin and E. Boridy, "Une approche statistique du choix d'un mode opératoire de soudage", *Soudage et Techniques Connexes*, vol. 37, no. 11-12, pp. 403-412, Novembre-Décembre 1983.

[Gellie 83]. R.W. Gellie, "Sensing for Automated Welding", *Proc. 31st Annual Conf. of The Australian Welding Institute*, Sydney, Australia, Oct. 1983, pp. 193-200.

[Inoue 80] K. Inoue, "Image Processing for On-Line Detection of Welding Process (Report II)", *Transactions of JWRI*, vol. 9, no. 1, pp. 27-30, 1980.

[Inoue et al. 80] K. Inoue, H. Akashi, M. Tamaoki, Y. Shibata and Y. Arata, "Automatic Control of Horizontal Narrow Gap Welding (Report II)", *Transactions of JWRI*, vol. 9, no. 1, pp. 31-37, 1980.

[Kawahara 83] M. Kawahara, "Tracking Control System for Complex Shape of Welding Groove Using Image Sensor", *Proc. of the IFAC/IFIP Symposium - Real Time Digital Control Applications*, Guadalajara, Mexico, Jan. 1983, pp. 257-263.

[Kerth 85] W.J. Kerth Jr., "Knowledge-Based Expert Welding", *Proc. Robots9 Conf.*, June 1985, pp. 5-98 - 5-110.

[Khosla et al. 85] P.K. Khosla, C.P. Neuman and F.B. Prinz, "An Algorithm for Seam Tracking Applications", *The International Journal of Robotics Research*, vol. 4, no. 1, pp. 27-41, Spring 1985.

[Kotykhin and Antonenko 80] Yu. P. Kotykhin and V.T. Antonenko. "Finding the Start of a Weld Automatically When Welding with a Robot". *Automatic Welding (GB)*, vol. 33, no. 11, pp. 45-46, November 1980.

[Kremers et al. 83] J.H. Kremers et al., "Development of a Machine-Vision Based Robotic Arc-Welding System". *Proc. of 13th Int. Symp. on Industrial Robots*, Chicago, Ill., April 1983.

[Lancaster 84] "The Physics of Welding". J.F. Lancaster ed., Pergamon Press, Oxford, UK, 1984.

[Lindberg and Braton 76] R.A. Lindberg and N.R. Braton, "Welding and Other Joining Processes", Boston, Allyn and Bacon Inc., 1976.

[Linden et al. 80] G. Linden, G. Lindskog and L. Nilsson, "A control system using optical sensing for metal-inert-gas arc welding", *Developments in Mechanised, Automated and Robotic Welding*, London, The Welding Institute, Nov. 1980, paper 17.

[McCampbell et al. 66] W.M. McCampbell, G.E. Cook, L.E. Nordholt and G.J. Merrick, "The Development of a Weld Intelligence System", *Welding Journal*, vol. 45, no. 3, pp. 139s-144s, March 1966.

[Morgan et al. 83] C.G. Morgan, J.S.E. Bromley, P.G. Davey and A.R. Vidler, "Visual guidance techniques for robot arc welding", *Proc. Third Int. Conf. on Robot Vision and Sensory Controls RoViSeC3*, Cambridge, Mass., SPIE vol. 449, Nov. 1983, pp. 390-399.

[Nachev et al. 83] G. Nachev, B. Petkov and L. Blagoev, "Data processing problems for gas metal arc (GMA) welder", *Proc. Third Int. Conf. on Robot Vision and Sensory Controls RoViSeC3*, Cambridge, Mass., SPIE vol. 449, Nov. 1983, pp. 291-296.

[Nevins et al. 84] J.L. Nevins et al., "Adaptive Control, Learning and Cost Effective Sensor Systems for Robotics or Advanced Automation Systems", Massachusetts Institute of Technology, Aeronautics and Astronautics Departments, First Annual Report, 1984.

[Ogilvie and Ogilvy 83] G.J. Ogilvie and I.M. Ogilvy, "The Pulsed GMA Process in Automatic Welding", *Proc. 31st Annual Conf. of The Australian Welding Institute*, Sydney, Australia, Oct. 1983, pp. 16-19.

[Page et al. 81] C.J. Page et al., "Non-Contact Inspection of Complex Components Using a Range Finder Vision System", *Proc. First Int. Conf. on Robot Vision and Sensory Controls RoViSeC1*, Stratford-Upon-Avon, U.K., Apr. 1981.

[Ramsey et al. 63] P.W. Ramsey, J.J. Chyle, J.N. Kuhr, P.S. Myers, M. Weiss and W. Groth, "Infrared Temperature Sensing Systems for Automatic Fusion Welding", *Welding Journal*, vol. 42, no. 8, pp. 337s-346s, August 1963.

[Richardson et al. 82] R.W. Richardson, D.A. Gutow and S.H. Rao, "A Vision Based System for Arc Weld Pool Size Control", *Measurement and Control for Batch Manufacturing*, Hardt ed., ASME, Nov. 1982, pp. 65-75.

[Richardson and Anderson 83] R.W. Richardson and R.A. Anderson, "Weld Butt Joint Tracking With a Coaxial Viewer Based Weld Vision System", *Control of Manufacturing*

Processes and Robotic Systems, Hardt & Book ed., ASME, Nov. 1983, pp.107-119.

[**Richardson and Richardson 83**] R.D. Richardson and R.W. Richardson, "The Measurement of Two-Dimensional Arc Weld Pool Geometry by Image Analysis", *Control of Manufacturing Processes and Robotic Systems*, Hardt & Book ed., ASME, Nov. 1983, pp.137-148.

[**Rider 83**] G. Rider, "Control of weldpool size and position for automatic and robotic welding", *Proc. Third Int. Conf. on Robot Vision and Sensory Controls RoViSeC3*, Cambridge, Mass., SPIE vol. 449, Nov. 1983, pp. 381-389.

[**Rioux 84**] M. Rioux, "Laser range finder based on synchronized scanners", *Applied Optics*, vol. 23, no. 21, November 1984.

[**Sicard and Levine 86**] P. Sicard and M.D. Levine, "An Approach to an Expert Welding Robot System", Technical Report no. TR-86-1R, Computer Vision and Robotics Laboratory, McGill University, Montréal, March 1986.

[**Smati et al. 85**] Z. Smati, D. Yapp and C.J. Smith, "Laser Guidance System for Robots", *Proc. 4th Int. Conf. on Robot Vision and Sensory Controls RoViSeC4*, London, Oct. 1984, pp. 91-101.

[**Sweet et al. 83**] L.M. Sweet, A.W. Case Jr., N.R. Corby and N.R. Kuchar, "Closed-Loop Joint Tracking, Puddle Centering and Weld Process Control Using an Integrated Weld Torch Vision System", *Control of Manufacturing Processes and Robotic Systems*, Hardt & Book ed., ASME, Nov. 1983, pp. 97-105.

[**Timchenko and Nachev 82**] V.I. Timchenko and G. Nachev, "Current State and Trends in The Development of Welding Robots", *Automatic Welding (GB)*, vol. 35, no. 7, pp. 37-46, July 1982.

[**Tomizuka et al. 80**] M. Tomizuka, D. Dornfeld and M. Purcell, "Application of Microcomputers to Automatic Weld Quality Control", *Transactions of the ASME*, vol. 102, pp. 62-68, June 1980.

The Edinburgh Designer System as a Framework for Robotics

R.J.Popplestone
Computer and Information Sciences,
University of Massachusetts at Amherst.

Abstract

I discuss how the capabilities of the Edinburgh Designer System can be extended and used to support symbolic computation for robotics. I conclude that the Algebra Engine requires to handle temporal constructs, groups and tolerances, that the taxonomy can support *activity modules* and that automatic plan formation would require the creation of a *specialist*.

1 Introduction

The Edinburgh Designer System (EDS) [26] [27] provides a coherent set of *inference engines* which operate upon a *common formalism* appropriate to the representation of engineering designs in general. This formalism stems from the work of Barrow [3], but has been extended in various ways, including to provide for the representation of the evolution of a design. The exploration of the space of possible designs is related to the work of Latombe [21], although conducted in a Prolog framework. The EDS has been implemented at Edinburgh, and has been mounted on a Sun workstation at U.Mass. My present concern at U.Mass is to define extensions of the EDS needed for Robotics.

In the EDS, a design is specified in terms of *modules*, which are engineering functional units (eg a motor or a keyway or a shaft). Thus modules are *not* necessarily rigid bodies, but may be features of bodies (eg an oilway) or assemblies of bodies (eg a gear-box). The extension of the EDS to deal with plan formation requires that modules be regarded as existing in time, for example the activity of drilling of an oilway or the support of a shaft while a gear is fitted to it.

Barrow specified interactions between modules in terms of connections between *ports* of modules. This formalism is appropriate for his domain (the logical analysis of VLSI designs) because interaction between modules always occurs in a standard way through conductors. However in general engineering, interaction between modules can take place in a rich variety of ways, so that knowledge about such interaction cannot be regarded as a static part of the EDS. Considerable conceptual economy can be achieved by treating the interaction between modules as a module in its own right, called an *interface module*.

Such interface modules can be regarded as establishing a relational network or graph upon the *concrete modules*.

In order to describe the formalism and behaviour of the EDS, I will use Prolog terms, typeset in mathematical form, with the Prolog convention that variables begin with capital letters. The EDS represents facts about particular modules which have been postulated to exist in the design with terms which are free of Prolog variables, whereas general facts about classes of modules are represented using Prolog clauses, which will contain Prolog variables.

Modules have *parameters*, which are symbols denoting quantities which are determined at design time, and *variables* which are symbols denoting quantities which may vary while the module is operating. For example a gear has as one of its parameters the pitch radius, and as one of its variables its angular velocity. Interface modules will define constraints upon parameters of the modules connected by the interface. For example the interface module *meshing_gears* equates the pitch line velocities of the two component gears. The parameters and variables of particular modules are *ground level* terms. For example $r\$g1$ may denote the pitch radius of the gear $g1$. In fact module parameters and variables are all *substitution instances* of the Prolog term $V\$M$. If we think of $A\$B$ as being A_B, this notation turns out to be quite in accordance with engineering conventions, for example if $g1$ is a gear and r is its pitch radius, then $r\$g1$ corresponds naturally with r_{g1} which could appear in engineering texts.

Modules are classified in a *taxonomy*, which is part of the formal support of the exploration of possible designs. For example, having decided that $g1$ belongs to the module-class *gear*, a further step of detailing it is to decide that $g1$ belongs to the module-class *spur_gear*. Other important steps in detailing involve deciding on values for parameters, and using one of the engines to determine new relationships.

In the EDS, the representation of shape is supported by a Constructive Solid Geometry (CSG) modeller [28]. The current implementation makes use of the ROBMOD modeller [9], but an interface to the Noname modeller [2] is currently being implemented. It is straightforward to interface the EDS to any CSG modeller which has a well defined textual or procedural interface. Current work at U.Mass includes the development of a ray-tracing modeller which will provide high quality images.

We use the term *inference engine* for any program which applies some body of knowledge relevant to the EDS. Such an engine will typically follow paths in the relational structure placed upon concrete modules by interface modules, and infer new relationships between entities which are at some distance from each other in the original structure, or it may derive new relationships by examining cycles in the structure. The inference engines typically perform forward chaining upon facts about the design: it is necessary to place strict limits on this chaining to avoid an undesirable proliferation of deduced facts.

Thus the RAPT interpreter [13] [24] is used as an inference engine for deducing module locations from defined spatial relations. Spatial relationships hold between geometric features of modules, which are derived from the primitives which make up the shape.

Thus the spatial relationship *fits* may be used to express the relationship between a journal on a shaft and the bush of a plain bearing.

The term "specialist" is used in the EDS context to mean systems which it is intended to provide an integrated treatment of knowledge about some sub-domain of design, for example planning of machining or assembly. specialists will be procedural, but will make use of appropriate declarative knowledge held in the taxonomy.

Any system which supports the activity of design must cope with the fact that designing involves exploring a space of possible designs. The exploratory nature of designing is treated by regarding any statement that the human designer makes to the system as an *assumption*, which may or may not form a part of the final design. For example, a designer may wish to consider the possibility of using an electric motor or a hydraulic motor in a given application, or, at a greater level of detail, the possibility of using a journal bearing versus that of using a rolling element bearing. While some of the design work associated with a given assumption may be lost if an alternative assumption is finally chosen, it is important to avoid the massive loss of effort which would arise if chronological backtracking were employed, an consequently we employ an Assumption Based Truth Maintenance System (ATMS) [14] to manage these assumptions. As well as assumptions about the existence of particular modules, the design process will also involve making assumptions about the values of parameters of modules and the spatial relations between their features.

Specialists likewise may make use of ATMS to support a more automatic exploration of design space in their area of specialism.

The EDS has been implemented in Poplog [16] [30] running under Unix on a Sun workstation. While the basic formalism is expressed in Prolog terms, most of the engines described have been written in Pop-11, or a mixture of Pop-11 and Prolog.

2 The Algebra Engine

Of all the inference engines, the Algebra Engine is the most basic. It is capable of simplifying expressions denoting a number of entities important in engineering and of interest to robotics, including locations, vectors and Constructive Solid Geometry shape descriptions. It also embodies Minipress, a system with many of the capabilities of the Press [7] program for solving equations involving trancendental functions symbolically. Press was developed at Edinburgh by Bundy and his co-workers.

The Algebra Engine was originally conceived of as being an extension of Press. The choice of Press as the basis for the Algebra Engine was determined by the need to solve equations in transcendental functions, since these commonly occur in engineering, and the need for an engine which could be readily interfaced to form a sub-system of the EDS. There are a number of algebraic manipulation packages available on the market, for instance REDUCE and MACSYMA, both of which are LISP based. Both of these packages have a front end which means that users do not have to couch their needs in rebarbative LISP syntax. However they are not readily accessible in a form which makes them ideally suited to be sub-systems of the EDS.

2.1 Simplification

We shall use $T1 \longrightarrow T2$ to mean that the Algebra Engine unconditionally rewrites an instance of term T1 as term T2 during simplification. These rewrite rules do not constitute a complete definition of simplification. The statement $T1 = T2$ means that the Algebra Engine may replace an instance of $T1$ with $T2$, or conversely, depending upon circumstances.

The Algebra Engine can be called upon to simplify a term by executing the Prolog goal $X := Y$ where Y is bound to the term being simplified. For example

$$X := a + b + 2 * a + c - b$$

will bind X to $a * 3 + c + 2$.

Algebraic simplification is not in general a straightforward, or even well defined problem. There is not, in most algebras, a standard "canonical form" into which all expressions can be transformed and which will serve as a basis for doing any operation subsequently required. Thus expression simplifiers for the ordinary algebra over the reals or complex numbers will normally do a number of standard useful operations, leaving others to be performed by specialist code for specialist purposes, or to be specified by options in the simplifier.

The EDS Algebra Engine does not make use of a formal theory of type, but relies upon a use of symbols which is consistent with laws built into the simplifier. Thus, for example it is not possible to use + to mean boolean *or* (or *union*) operation, since the Algebra Engine assumes that + obeys the laws of an abelian group and a boolean algebra does not obey the same laws (eg. $X + Y = X + Z$ does not imply $Y = Z$ if '+' is interpreted as union). In terms of Universal Algebra [12], we restrict any set of entities upon which function symbols operate to be a *variety*. It should be noted that 0 and 1 are to be regarded as nullary operators.

2.2 Datatypes and basic notational conventions

It is necessary to represent integers, real numbers, 3-vectors, locations (ie Euclidean transformations of 3-space), matrices, and shapes defined as subsets of 3-space. In the present implementation of the EDS integers and reals are represented as fixed and floating point quantities respectively. However rational arithmetic is available within Poplog and has definite advantages for the more theoretical aspects of design.

We use $X-> FX$, where FX is a term in X to mean the function which maps X to FX. (cf. $\lambda x.F(x)$). Thus $I-> I \uparrow 2$ is the squaring function.

Other datatypes are represented as terms. Thus vectors are terms with the functor vec. In particular, $vec(X,Y,Z)$ is a 3-vector and $vec(0,0,0) \longrightarrow 0$, since zero is known to the simplifier as the identify of the abelian group operator +.

The unit vectors are denoted by the atoms ii, jj and kk, although these are all rewritten as vec(1,0,0) etc., since these symbols are not regarded as basic operators of the Algebra Engine.

We use the term *location* to mean a rigid transformation of 3 space which preserves handedness of axes, ie a member of the *Euclidean Group*. The best canonical representation of constant locations is a matter of some debate: a presentation of the possiblities is to be found in [5]. At present the EDS uses a 4×3 array representation, derived from RAPT, although a 4×4 matrix representation would have the advantage that locations were no longer special entities.

$X + Y$ means the addition of entities usually added.

$-$ is used for Binary subtraction and unary negation.

Thus
$$X + Y = Y + X, \quad X + 0 \longrightarrow 0,$$
$$X + (-X) \longrightarrow 0, \quad X + (Y + Z) \longrightarrow (X + Y) + Z.$$

This latter is in accordance with Prolog conventions. Binary minus is rewritten by $X - Y \longrightarrow X + Y * (-1)$. $X + Y$ may denote the sum of reals, complex numbers, matrices, etc.

$X*Y$ means commutative multiplication, eg. of numbers, scalar multiplication of vector by number. * has zero 0 and identity 1, so

$$X * Y = Y * X, \ X * (Y + Z) = X * Y + X * Z, \ X * (Y * Z) \longrightarrow (X * Y) * Z$$
$$X * 0 \longrightarrow 0, \ 0 * X \longrightarrow 0, \ X * 1 \longrightarrow X, \ 1 * X \longrightarrow X$$

The division operator is eliminated (except in the special representation of rational functions discussed below).
$$X/Y \longrightarrow X * Y \uparrow (-1).$$

$X \uparrow Y$ means X to the power Y. $X \uparrow 0 \longrightarrow 1$.

$X@Y$ means the associative, non-commutative multiplication of x by y, for example of locations, permutations, matrices, and the concatenation of sequences. Thus @ plays the same role as the "." operator in Macsyma. @ is also used for locating any entity eg. $block(1,2,3)@rot(ii,0.1)$ means "a block of dimensions 1 2 3 rotated by 0.1 radian about the x-axis".

inv is the inverse of @. $X@inv(X) \longrightarrow X$, $1@X \longrightarrow X$. $X@1 \longrightarrow X$.

2.3 Polynomials and rationals

Terms of the form $poly(X, A_0, A_2......A_n)$ (ie. terms of $n+2$ arguments whose functor is $poly$ and whose first argument is X) represent the polynomial $A_0 + A_1X + A_2X^2 + \cdots + A_nX^n$). A rational function is represented as the formal quotient of two polynomials. Note that, except for the case of a polynomial of the above form where X is a constant, conversion to and from these forms is not performed by the := predicate, but is invoked by predicates $poly$ and $rational$.

2.4 Sequences and iterations.

$M..N$ denotes the finite sequence $[M, M+1...N]$, ie. the Prolog list.

$sigma(S, F)$ sums the function F over the sequence S, eg.

$$sigma(1..3, i- > i \uparrow 2) \longrightarrow 1 \uparrow 2 + 2 \uparrow 2 + 3 \uparrow 2 \longrightarrow 14.$$

$pi(S, F)$ similarly takes the product of the function F over the sequence S.

2.5 General operations.

$mod(X)$ is a numeric measure of the size of any entity. Eg. $mod(V)$ is the modulus of the vector V. mod obeys over an abelian group the laws $mod(0) \longrightarrow 0$, $mod(X+Y) =< mod(X) + mod(Y)$.

$mod(M, N)$ is the remainder when the integer M is divided by the integer N.

$unit(X)$ is a version of the entity X which is of unit modulus. Eg. $unit(vec(3,4,0)) \longrightarrow vec(0.6, 0.8, 0)$

2.6 Locations.

$trans(X, Y, Z)$ denotes a translation by the vector $vec(X, Y, Z)$. $trans(V)$ denotes a translation by the vector V.

$rot(V, T)$ denotes a rotation by an amount T about a vector V.

2.7 Equations

Terms involving equality are simplified depending upon the entities being equated. In particular, any equation over the reals which involves only on e symbolic quantity is given to Minipress to solve, and simplifies to the form $Variable = Constant$. Equalities on terms which have a functor known to be *free* may give rise to equations of the arguments of the terms.

2.8 Ordering relations.

The symbols $<$, $=<$ etc. are used to mean a total ordering over the reals, following Prolog conventions. Terms involving them are simplified by collecting constants to the right, non-constants to the left. Any inequality involving just constants is evaluated to the Pop-11 true or false.

The following conventions are used for *partial orderings*:- $<:=$ Less than or equal to in a partial ordering, so that it is reflexive, symmetric and transitive.

$X <:= X$. $X <:= Y$ and $Y <=: X$ implies $X = Y$. $X <:= Y$ and $Y <:= Z$ implies $X <:= Z$.

2.9 Lattices and boolean algebras.

A lattice is an algebra with operators \/ and /\, which are associative an commutative and obey the axioms:–

$A/\backslash(A\backslash/B) = A$. and $A\backslash/(A/\backslash B) = A$.

Boolean algebras are special cases of lattices, and in particular the point sets defined by Constructive Solid Geometry form a Boolean Algebra. Boolean algebras are distributive over both \/ and /\, and admit a subtraction operation \, for which

$(A\backslash B)\backslash/B = A$.

It can be shown that lattices are partially ordered unter the relation

$X <:= Y$ iff $A/\backslash Y = X$

It should be noted that lattices do not in general obey the distributive laws of boolean algebra, so that these cannot be built into the Algebra Engine.

2.10 Shapes

Constructive Solid Geometry is used to represent shapes using the following primitives:–

$block(X, Y, Z)$ denotes a cuboid, centroid at the origin, with the stated dimensions along the coordinate axes.

$cyl(L, R)$ denotes a finite cylinder of length L and radius R, centroid at the origin, axis along the Z-axis, and

$cone(H, R)$ denotes a cone of height H and radius R, base centre at the origin, axis along the Z-axis.

$sph(R)$ denotes a sphere of radius R, centre at the origin.

$tor(R1, R2)$ denotes a torus of minor radius $R1$ and major radius $R2$, centroid at the origin, axis of radial symmetry along the z-axis.

These primitives are combined with the boolean operations \/ and /\ for union and intersection, and with \ denoting set subtraction. The term *Shape@Loc* denotes a *Shape* relocated by the location *Loc*.

It is a convenience to allow *Shape@Vec* to denote the *Shape* translated by the vector *Vec*.

A block has faces called *top, bottom, left, right, front and back*, named under the convention that the X axis is forward, the Y axis is left and the Z axis is upward. These names are preserved under rotation. The lower, upper and curved faces of a cylinder are called the *proximal, distal and curved* faces.

It should be noted that there are some difficulties with the treatment of shape outlined above. A desirable property of any formalism is *referential transparency*, which means that we can always substitute one term for another if the two are equal without changing the sense of what is said. If we regard the CSG primitives as *just* denoting subsets of 3-space, then the face naming conventions, which are needed for specifying spatial relationships, imply a lack of referential transparency. For example if $b1 = block(1, 2, 3)$ and $b2 = block(1, 2, 3)@rot(ii, pi)$ then, as sets, $b1 = b2$, but the *bottom* of $b1$ is the *top* of $b2$. A potential resolution of this difficulty would be to treat shapes not as subsets of space, but as functions from 3-space to a set of *labels*. It is possible to ensure that such *labelled sets* form a boolean algebra by requiring the labels themselves to form a boolean algebra.

3 Bodies

An important property of a module is that of being a *rigid body*, which implies that the nominal relative positions of any sub-modules are determined at design time. The shape of a body module m is determined from the shape of the modules listed in the parameter *parts\$M*, but the way that this shape is computed may depend upon the module-class to which M belongs. In what follows, I shall use "body" to mean a module which happens to be a rigid body, and use "features" loosely to mean a sub-module, or geometric features of the shape of the body.

It should be noted that, during the course of design, the existence of many modules may be established long before their membership of bodies is established, and indeed body membership may change during the course of design. For example, it may be seen to be necessary to make a module such as a valve seat, which requires fine grinding, as an insert rather than be machined as an interior feature of a larger body.

3.1 Group Theory

Many modules will have symmetry, either actual or functional. It is important to extend the EDS to have an explicit representation of these symmetries. In [25] it is shown that the capabilities of RAPT inference engine could be generalised if a group theoretic representation were used, and in particular an exploitation of finite symmetries would be possible.

The basic idea is that any spatial relationship between features determines the relative positions of bodies possessing these features *modulo the symmetry group(s) of these features*. Suppose body $B1$ has feature $F1$, and body $B2$ has feature $F2$, then one possible implication of the statement that $F1$ fits $F2$ (or $F1$ fits $F2$ exactly) is that they have the same symmetry group (up to automorphism), and that the relative location of $B2$ with respect to $B1$ is a coset of that symmetry group. For example, if the socket of a socket wrench has a six fold symmetry (some do, others have twelve-fold symmetry), then its symmetry group is the cyclic group we shall denote by $cy(6)$. To be more precise, in our context $cy(6)$ denotes a particular cyclic subgroup of the euclidean group. If the socket fits a bolt head with six fold symmetry, then the position of the wrench with respect to the bolt is determined to be a member of a coset of the common symmetry group, that is $T1@cy(6)@T2$ where $T1$ and $T2$ are constant transformations. (eg. if the axes are centrally embedded in the bodies, then perhaps $T = trans(0,0,6)$) and $T2 = 1$).

From [25] we see that a basic operation which needs to be performed on groups is the simplification of expressions involving intersections of cosets of the Euclidean group. It is also possible that a representation of groups of permutations could be desirable in treating sets of identical entities.

A suitable notation for these groups needs to be chosen, and a suitable representation for performing the necessary computations. A classification of the infinite groups can be found in [17]. Many of the finite or finitely generated sub-groups of interest are crystallographic groups [19]. A uniform representation of the finitely generated sub-groups is possible simply by listing their generators, preferably in a canonical order. The infinitely generated groups can be specified by defining their generators parametrically. A suitable convention would be to denote groups by gp(List), where List is a list of functions or constants. eg $gp([t-> rot(kk,t)])$ would denote the group of rotations about a single axis.

Calculating the intersections of cosets of such sub-groups from first principles involves the solution of equations, which will either be non-linear equations in real variables, or diophantine equations, or both. The implementation of RAPT described in [13] does in fact deal with the former type of equations, in the form of *location equations*. However, solving such problems from first principles is not efficient — it is better to have a catalog of the solutions to commonly occurring ones eg.

$$cy(N)/\backslash cy(M) \longrightarrow cy(hcf(N,M)).$$

ie. the intersection of two cyclic sub-groups is the cyclic sub-group whose order is the hcf of the orders of the two sub-groups. Following Knuth, Bendix, Huet, [18] it is possible to check such rewrite rules for *confluence*, and add new rules, depending upon the existence of "critical pairs". Thus if we say:

$$gp([rot(kk, pi/N)])/\backslash gp([t->rot(kk,t]) \longrightarrow gp([rot(kk, pi/N)])$$

ie. the intersection of a cyclic subgroup with the sub-group of rotations about the z-axis is the cyclic subgroup, we are creating a need for rewrite rules to convert from the form $cy(N)$ to $gp([rot(kk, pi/N)])$. It is important when producing a final form of a coset expression to choose a standard member of the automorphism class of any sub-group occurring — for example, where a single axis of rotation is involved, that should be about the z-axis.

4 Exploring Design Space

The EDS is not intended to perform fully automatic design: it is a *Designer's Apprentice*, which is capable of performing some Design functions automatically, but is dependent upon human guidance for the more strategic decisions. The EDS is also seen as a vehicle upon which more fully automatic design *specialists* can be built. It will play this role also in Plan Formation.

The process of design is the construction of a design description document (DDD), which is held as consequences within the ATMS. The DDD contains a set of assumptions, and a set of consequences (called *values* by de Kleer). Each assumption corresponds to a design decision. For example, the assumption $g1$: *gear* says that $g1$ is a gear, and the assumption $r\$g1 = 10$ says that the parameter $r\$g1$ has the value 10. There is no requirement for assumptions to be consistent. For example, the assumption $r\$g1 = 20$ might coexist with the assumption that $r\$g1 = 10$. However, a final design is characterised by a set of assumptions which must not be known by the system to be inconsistent. So the two values for $r\$g1$ quoted above can only form part of alternative possible designs.

Consequences are held as Pop-11 records of the form $conseq(N, B, As, J, P)$ where B is a term with a truth functional value (eg. $r\$g1 = 10$) and As is the set of assumptions that the consequence depends on. P, the parents, is the set of consequences that the consequence was derived from. J is the justification — the name of the rule used to derive the consequence.

While a small design exercise could be essayed with no additional structure on the ATMS, serious design requires us to provide the concept of *focus*, whereby a limited set of consequences is in use for forward chaining, corresponding to a particular part of the design that is being considered. For example, in the case of a gearbox design, the initial focus will be on the gear teeth, followed by the location of the bearings on the shafts, the loading of these bearings, the analysis of the shafts for stiffness and strength, the detailing of bearings, housings and gear wheels, etc. An experimental focussing capability is available in the EDS, which allows the user to select particular consequences, or consequences containing particular variables, or which depend on particular assumptions.

The Prolog goal $assume(Term)$ is used to enter assumptions into the ATMS. For example, $:-assume(g1:spur\_gear)$ states that the module $g1$ is a spur gear.

Let us now discuss the interaction between the ATMS and Prolog. When a new fact is entered in the ATMS it can give rise to forward chaining. Possible forward chainings are defined by the prolog predicate *implies*, and by separately encoded equality propagation. The result of a forward chaining is always given to the Algebra Engine to simplify before it is entered in the DDD.

4.1 Equality propagation.

Only consequences of the form $Variable = Constant$ give rise automatically to forward chaining. Any consequence of which contains $Variable$ spawns another consequence with $Constant$ substituted for it. If the new consequence has the same assumption base as the old one, the old consequence is said to be subsumed by the new one, and is deleted. New consequences are simplified by the Algebra Engine, and may give rise to yet more equality substitutions. Recall that the Algebra Engine may invoke Minipress to solve equations during this simplification process, which will thus treat systems of linear equations if their matrix is triangular. A separate engine is being provided to solve non-triangular systems of linear equations.

4.2 Propagation of other consequences

Propagation of other consequences is determined by the predicate *implies*. Implies is defined both for single consequences and for pairs of consequences. Every new consequence is given to *implies*, and is paired up with old consequences and the pair is given to *implies*. For example consequences of the form $M : Mc$ state that M is a member of the module class Mc. Thus $g1 : gear$ states that $g1$ is a *gear*. *implies* deals with this by *consulting* a file containing the definition of the module class (if necessary) and instantiating all of the constraint clauses. This forward chaining can proceed quite far, since a module may have *parts* each of which is stated to be a member of some other module class. However the chaining is normally terminated before the whole design is automatically expanded into the DDD by the fact that some modules will be at too abstract a level in the taxonomy to have parts specified.

5 The Taxonomy

Module class definitions are held in an external form in the taxonomy, and translated by a program called Dracula into prolog clauses which are held textually in files. The external form contains parameter and variable declarations, tagged pieces of English intended for appropriate regurgitation to the user, constraint and table definitions. One view of the taxonomy is that it constitutes something like a generative grammar for designs.

All properties of modules (eg. names and types of variables and parameters) are strictly inherited down the taxonomy, which puts significant consistency constraints upon that structure. Thus we should be in no danger of finding off-white elephants in our trees [4].

Currently the structure of the taxonomy is implemented by textual references in *header files* associated with each module class. This structure is very cumbersome to modify interactively, making it difficult to install new modules, and it is intended to change it.

5.1 Constraints and Tables.

Constraints are relationships between the mathematical entities which characterise a module or modules. The general form of a constraint as a Prolog clause is:–

$$constraint(Mc, M, B, Id).$$

where Mc is the module class to which the constraint applies, B is a boolean term expressing the relationship, and Id is an identifier used to refer to the constraint, so that it can be identified and used by the appropriate inference engines. (Note the EDS has at present an extra place in the *constraint* predicate associated with the implementation of ports, which was at first essayed). M is a prolog variable which occurs in B and which is bound to give an instance of the constraint which will be entered into the DDD.

For example
$$constraint(spur\_gear, G, ft\$G = tor\$G/r\$G, 1)$$

says that the tangential force on a spur gear is equal to the torque divided by the pitch radius. If we say, for example, $g1 : spur\_gear$, then the equation $ft\$g1 = tor\$g1/r\$G1$ is loaded into the DDD as a consequence.

Shapes are defined by means of shape constraints, which have the same form as any other equations. The shapes associated with modules may be positive, that is indicating the presence of material, or negative, that is indicating its absence. The parameter $shape\$M$ is used to denote the nominal positive shape of any module. A module such as an oil-way will have a negative shape, indicating material that must not be present, and a positive shape, indicating material that must be present to provide adequate walls.

In engineering, as well as relationships between parameters an variables being expressed as equations, a tabular or graphical form is often used, either because the mathematical form is felt to be too complex for hand computation, or because the relationship has been established empirically, or to express some arbitary convention. For example, the factor by which the power capacity of a pair of gears is reduced with increasing speed is expressed in the form of a graph, and is input to the EDS as a table.

Thus the EDS is capable of handling constraints in tabular form, which requires a descriptive apparatus for defining what the entries in a table mean. We have used the work of Codd [11] (see also [23]) as a guide to this, although there is a need to extend the concepts of relational databases because they are dependent on the idea of an exact match between components of tuples, whereas some engineering tables require an interpolation to be done. While a detailed account of the treatment of tables is outwith the scope of this paper, suffice it to say that tables are similar to Codd relations with module parameters or variables as column headers, and that they undergo manipulations corresponding to those undergone by constraints, with the relational join operation playing a similar role to the elimination of a variable or parameter between two constraints.

6 Invoking the Inference Engines

Many inference engines will have been written with representation conventions other than those used in the EDS, or indeed may be completely separate programs. Thus their use will in general involve a measure of data conversion.

6.1 Invoking Press

Press [7] is a system which is able to solve symbolic equations, and is used to allow the equations arising in design to be solved for an appropriate variable or parameter, for example to allow it to be eliminated between two equations. Press is written in Prolog, with a compatible representation of terms to that used in the EDS, so that the interfacing task is straightforward. Press itself is however a very large and cumbersome program. The purposes of the EDS are adequately met by Minipress, which makes use of Bundy's methods, and some of the axioms, but is a new program written by the author.

The simplifier automatically makes use of Minipress if it discovers an equation with only one symbolic quantity — the equation is solved and the result substituted for.

Press will be available to the user of the EDS to allow him to solve a named equation (occurring as a consequence) for a named variable or parameter. More complex use is envisaged to allow other inference engines, or specialists, which will trace through relational paths, to eliminate variables of a certain class from equations of a certain class at each step. For example, the input torque to a drive train can be deduced from the output torque by equating the input and output torques of successive stages and applying the torque conversion laws appropriate to each stage.

6.2 Invoking the RAPT inference engine

In the EDS, rigid modules are the conterpart to bodies in RAPT. In any focus of the ATMS, certain spatial relations will be known to hold between features of modules. The RAPT inference engine can be used to derive explicitly more constraining relations from these, and in some cases to infer fixed relative locations for the modules. Features are named by reference to the conventional names of features of the primitives of module shapes.

RAPT is written in POP-2 [8], and so is readily available within the EDS, its invocation requires the building of data structures encoding the relationships known to the EDS, and expressing the locations of module features. These latter are inferred by examining the CSG definition of the module shapes.

6.3 Invoking the Modeller

The modeller exists as a separate process, and communication with it is via some form of Inter Process Communication. The elementary interface which has at present been implemented simply makes use of a printing routine which prints CSG data structures in the syntax required for input to the modeller, and sends the resulting character stream to the modeller through a Unix pipe, and listens through another pipe for the modellers's response. This interface is used both to request the modeller to draw shapes and to query the modeller about matters such as the volume of shapes and whether they intersect. The CSG trees thus transmitted are variable free, so than no attempt is made to use the modeller's capability to handle symbolic descriptions.

More efficient use of the modeller would allow the EDS to to feed variable-containing shape expressions to it and to keep a tally of the values the variables are bound to. Also a procedural interface rather than a textual one may be used.

7 Plan Formation

The approach to plan-formation which I am advocating in this paper is to construct plans as *behaviour modules* in the EDS. The kernel of any element of behaviour arises from the existence of modules in the design itself. If a shaft module $sh1$ exists in the design, then the activity $acquire\$sh1$ can be generated. Whether the acquisition involves buying or manufacturing is not immediately manifest: choosing the acquisition method is a matter of descending the taxonomy. If the "buy" choice is made, the acquisition module will have parameters like $supplier\$sh1$ and $cost\$sh1$, whereas if the "make" choice is made, the acquisition module will become a complex structure where decisions about the formation of various sub-modules are made, for example bearing seatings and gear hub seatings. Potential strategies for forming these will themselves be found in the taxonomy, and will be taxonomically structured as in [31].

Likewise the existence of an interface module in which a gear drives a shaft implies (if they are separate bodies) the existence of an activity of assembling the gear to the shaft. However, the representation of activities does require an extension of the symbolism of the present EDS. For example, we need to be able to say that one activity occurs before another, or during another [1].

The temporal ordering of activities in assembly is strongly dependent upon geometry. It is, of course, important to avoid considering in detail all of the possible assembly sequences, and it is certainly worthwhile analysing any activity module both for its feasibility in isolation, and to examine the temporal dependencies of activities taken pairwise. Koutsou [20] has shown how it is possible to plan trajectories for isolated bodies using two key components of the EDS, the Spatial Relation Engine, RAPT, and the Modeller. This approach can be used to answer questions such as "Suppose I insert shaft $sh2$ in bearing $b1$ before I insert bearing $b2$ in bearing housing $bh2$, can I plan a suitable trajectory for this latter insertion?", thereby obtaining facts about the possible ordering of assembly operations, even if these are not necessarily completely adequate for determining the order. Developing an ordering in this way offers the possibility of treating what the planning literature refers to as "goal interaction", which is particularly difficult in robotics if we are dealing with bodies having a realistic range of shapes because it is not possible to codify the interactions as can be done in the "blocks world".

It is also necessary to treat what are referred to in the planning literature as "preconditions" and "postconditions" (or "effects"). The most important post-condition of an activity module $acquire\$M$, where M is a module which is present in the final design, is simply that M exists, and normally this will be, from the point of view of the EDS, a permanent state of the world, although of course trial assemblies are not unknown. However it will also be necessary to design jigs and fixtures for manufacture, and it is possible that some of these will have to be assembled and disassembled during the course of manufacture. It will be natural to treat the design of a jig or fixture with the same EDS mechanisms as any other design: hence it will be necessary to admit that the main effect of an activity may have a finite future.

The preconditions of an activity can be treated as a requirement for the existence of certain interface modules. For example, suppose bearing $bg1$ is to be inserted in a bearing-housing $bh1$ belonging to an end-plate $frame1$ by an activity $acquire\$[bg1, bh1]$. Let $frame1\_jigged$ be an interface module in which $frame1$ is mounted in a jig. Let $bg1\_gripped$ be an interface module in which $bg1$ is gripped by a robot gripper. Then

$$acquire\$[bg1, bh1]\ during\ frame1\_jigged$$

holds, as does
$$bg\_gripped\ overlaps\ acquire\$[bg1, bh1]$$

It should be noted that in the realm of assembly planning and machining planning the concept of *module identification* (qv.) plays an important role. For example if we are going to perform another insertion into a feature of $frame1$ the interface module $frame1\_jigged\_1$ to exist, then it is probably more efficient to make the identification $frame1\_jigged = frame1\_jigged\_1$ if possible, ie. to use just the one jigging operation. Of course, the result of this identification is that conditions upon the two jigged states are conjoined, and may become unsatisfiable. The automatic generation of such identifications will be the task of *assembly specialists* which will cluster operations for similarity in such properties as direction of approach, so that there will be a strong expectation of satisfiablity of conjoined preconditions over these clusters.

8 User Interface

The present implementation of the EDS has a minimal user interface, requiring the user to input assumptions into the ATMS with the *assume* predicate. Some work has been done in displaying the taxonomy as a visible tree. It is anticipated that a more user friendly system will make use of modern interactive graphical capabilities, where for example an assumption about the existence of a module will be generated by pointing to a token for the module class presented on the display. Alphanumeric input will still be required for other purposes, for example to override the default name assigned by the system to a new module, or to assume a value for a parameter, but we would use the technique familiar in the Mackintosh, where the significance of such input is determined by its location in a box on the screen.

Considerable thought will have to be given to the presentation to the user of the current design state, since an undifferentiated mess of equations and drawings will be of little value to him. This will almost certainly have to make use of a windowing structure in the work station, with data that is in some sense related grouped together in the same window. Some preliminary work to these ends has already been done.

While the Prolog form of input and output of mathematical expressions is considerably more user friendly that that of Lisp, we expect to implement further improvements in typography by providint for the display of greed letters and subscripts and superscripts.

9 Implementation

The EDS has been implemented in a demonstrable form, and has been mounted on Sun workstations at various industrial and academic sites of the Design to Product Demonstrator, and at UMass. Throughout this paper I have distinguished between what has been done and what might be done or is planned to be done by using suitable modal verbs for the latter.

10 Discussion

It should not be assumed that the definition of a design in terms of modules will result in a modular design, since the modules referred to in this paper are basic engineering entities, which can be freely combined. In order to permit the system to contain definitions of fairly complex high level modules, while not coercing a modular form of design, a specific capability, that of *modular identification* needs to be introduced. This allows the designer to decide that two modules which were originally considered to be distinct are in fact identical. For example, an electric motor always has a frame, which supports the working parts, and a transmission likewise always has a frame. However in an optimised design the frame of the motor and of the transmission may be identified.

A number of important issues have not been discussed in this paper. All of the locations and shapes of bodies have been treated as though knowledge of the *nominal location* and *nominal shape* were adequate. Some excellent work has been done on *reasoning with uncertainty*, for example Brooks [6] has adopted an algebraic approach to treating the problem of errors in robot planning. Requicha [29] provides an interpretation of engineers' tolerancing of shape in a CSG context. Fleming [15] has treated the accumulation of location tolerance arising from loose fits, and is studying the accumulation arising from toleranced bodies. Lozano-Perez, Mason and Taylor [22] have studied how a goal can be achieved despite position uncertainty provided the uncertainty envelope lies in a *pre-image* of the goal.

Providing a treatment of tolerance in the EDS promises to be a formidable activity. However I would suggest the following guide-lines.

1. The development of a formalism for specifying tolerance on scalar parameters, and the provision of a tolerance inference engine to propagate tolerances through constraints.

2. The examination of Fleming's techniques for representing location tolerances, with a view to their possible adoption. Fleming uses location tolerance zones which are intervals of the angles, but have a linear relationship between angular tolerance and linear dimensions. These structures, while they may not give as tight a hold on tolerance as is obtainable in theory by Brooks' approach, can be manipulated readily without heavily taxing the symbolic computational capabilities of a system.

3. The adoption of Requicha's scheme for tolerancing shapes. This is likely to mean that the *block* primitive falls out of use, since its sides will need to be independently toleranced. It will be replaced by half spaces.

4. Much knowledge about tolerances will be encoded in modules, rather than derived from first principles, in line with the philosophy of the EDS.

Within the framework of the EDS, simple sensors can be treated as modules which bind the values of certain variables. Dealing with complex sensors, such as an image acquisition system with its associated image segmenting software, could be done by a spacialist, since the EDS does contain constraints upon the scene, and in the ATMS a system for supporting alternative interpretations.

Acknowledgements

The implementation of the EDS forms part of the Alvey Demonstrator Project "Design to Product", funded in part by H.M. Government through her Department of Trade and Industry.

The EDS is an organic part of the Demonstrator, and owes much to the other collaborators GEC, Lucas, Leeds University, Loughborough University and the National Engineering Laboratory. Individuals who have made important contributions to the evolution of the EDS include A.P.Ambler, T.Arai, P.Davey, I.Powell, M.Sabin and J.Streeter. The EDS was realised by the Edinburgh Team — J.Corney, O.Franks, R.Gray, G.Sahar, S.Renton, T.Smithers and S.Todd.

References

[1] Allen, J.F., [1983], Maintaining Knowledge about Temporal Intervals, CACM, Vol 26 No 11.

[2] Armstrong et al [1982], Noname Description and Users Manual, Department of Mechanical Engineering, Leeds University, England.

[3] Barrow H.G. [1983], Proving the Correctness of Digital Hardware Designs, Proceedings of the National Conference on Artificial Intelligence, Washington,DC.

[4] Brachman R.J. [1985], "I lied about the Trees", or Defaults and Definitions in Knowledge Representation. A.I.Magazine, Vol 6 No 3.

[5] Brady M.,Hollerbach J., Johnson T., Lozano-Perez T., and Mason M. [1982], Robot Motion: Planning and Control, MIT Press, Cambridge MA, and London.

[6] Brooks, R.A.,[1982], Symbolic Error Analysis and Robot Planning, IJRR Vol. 1 No. 3, pp 29-68.

[7] Bundy A., Byrd L., Luger G., Mellish C. and Palmer M. [1979], Solving Mechanics Problems using Meta-level Inference, in specialist Systems in the Micro Electronics Age, ed. D.Michie, Edinburgh University Press, Edinburgh.

[8] Burstall, R.M., Collins, J.S. and Popplestone, R.J., [1971], Programming in POP-2, Edinburgh University Press, Edinburgh.

[9] Cameron, S.A. [1984] Modelling Solids in Motion, Ph.D. Thesis, Department of Artificial Intelligence, Edinburgh University.

[10] Clocksin W. and Mellish C. [1981], Programming in Prolog, Springer Verlag N.Y..

[11] Codd E.F. [1970], A Relational Model of Data for Large Shared Data Banks, CACM 13, 377-387.

[12] Cohn, P.M. [1965], Universal Algebra, jointly published by Harper and Row, N.Y., Evanston and London, and by John Weatherhill, Tokyo.

[13] Corner D.F., Ambler A.P., and Popplestone R.J., [1983], Reasoning about the Spatial Relationships Derived from a RAPT Program for Describing Assembly by Robot, Proc 8th. IJCAI, Karlsruhe,DBR.

[14] de Kleer,J. [1984] Choices without Backtracking, Proc Conf AAAI.

[15] Fleming A, [1985], Analysis of Uncertainties in a Structure of Parts, IJCAI 85, pp 1113-1115.

[16] Hardy S., [1984], A New Software Environment for List-Processing and Logic Programming, in Artificial Intelligence, Tools, Techniques and Applications, eds. O'Shea and Eisenstadt, Harper and Row, N.Y..

[17] J.M.Hervé [1978] Analyse Structurelle des Mécanismes par Groupe des Déplacements. Mechanism and Machine Theory, Vol 14 No 4.

[18] Huet G. [1978], Confluent Reductions: Abstract Properties and Applications to Term Rewriting Systems. IRIA Roquenfort France.

[19] Shokichi Iyanaga and Yukiyosi Kawada [1968], The Encyclopaedic Dictionary of Mathematics, English edition M.I.T. Press, Cambridge. MA. and London, Japanese original published by Iwanami Shoten, Tokyo.

[20] Koutsou, A., [1986], Phd Thesis, Department of Artificial Intelligence, Edinburgh University.

[21] Latombe J.C., [1977], Une Application de L'Intelligence Artificielle à la Conception Assisstée par Ordinateur (Tropic), Thèse d'état, L'Université Scientifique et Médicale de Grenoble.

[22] Lozano-Perez, T. Mason, M.T. and Taylor, R.H., [1984], Automatic Synthesis of Fine Motion Strategies for Robots, IJRR Vol 3 No 1, pp 3-24.

[23] Popplestone,R.J., [1979], Relational Programming, in Machine Intelligence 9 (ed D.Michie). Ellis Horwood, Chichester, Sussex.

[24] Popplestone,R.J., Ambler A.P., and Bellos I. [1980], An Interpreter for a Language for Describing Assemblies, Artificial Intelligence Vol 14 No 1, pp 79-107

[25] Popplestone,R.J, [1984], Group Theory and Robotics, Robotics Research: The First International Symposium, (eds Brady and Paul), MIT Press, Cambridge MA and London.

[26] Popplestone R.J. [1984], The Application of Artificial Intelligence Techniques to Design Systems, International Symposium on Design and Synthesis, Japan Socy. Precision Engineering, Tokyo.

[27] Popplestone R.J. [1985], An Integrated Design System for Engineering, Preprints of the 3rd ISRR, Gouvieux, France.

[28] Requicha A.A.G. and Tilove R.B., [1978] Mathematical Foundations of Constructuve Solid Geometry: General Tobology of Closed Regular Sets, TM27A, Production Automation Project, University of Rochester, Rochester N.Y..

[29] Requicha A.A.G. and Tilove, [1983,] Toward a Theory of Geometric Tolerancing, IJRR Vol.2, No 4.

[30] Sloman A. and Hardy S. [1983], POPLOG: A Multi-purpose Multi Language Program Development Environment. AISB Quarterly, No 47, 1983.

[31] Tenenberg J,[1986], Planning with Abstraction, Proc. AAAI-86, pp 76-80.

IMPLEMENTATION OF COMPLEX ROBOT SUBSYSTEMS ON DISTRIBUTED COMPUTING RESOURCES

S.Y. Harmon
Robot Intelligence International
4660 Long Branch Avenue
San Diego, CA 92107

ABSTRACT

An approach to the coordination of a complex robot's subsystems has been developed and implemented. In this approach, the computational load is distributed functionally over several microprocessor systems in both tightly and loosely coupled configurations. Tightly coupled functions communicate through shared memory on the same high speed parallel bus. Loosely coupled functions communicate through a local area network. However, whether tightly or loosely coupled, communications between functional modules appear as if a single blackboard memory is shared. This approach to robot integration has been used to explore various concepts for sensor data fusion. An autonomous mobile robot has provided the experimental environment in which experience with this approach has been gained. The concepts fundamental to this approach have also been extended to coordinate multiple interacting robots.

INTRODUCTION

The present state of computing technology requires that complex robots will be implemented as distributed systems. Relatively few approaches have been explored for distributing functionality among the computational subsystems of a complex robot and for coordinating the interaction between these subsystems. The JPL Rover (1) and HILARE (2), two of the most recent efforts, distribute functionality in a star computing architecture using specialized communications protocols to solve the particular problems imposed by the computing environment.

Distributed problem solving researchers have explored some applicable concepts. The cooperating experts of the HEARSAY II speech understanding system communicate and are controlled through a blackboard (3). The blackboard formalism is useful in reducing the effects of uncertainty and error in a system with distributed knowledge sources (4). Unfortunately, the blackboard communications mechanism ignores most of the limitations imposed by actual implementations of distributed computing systems. A number of message passing implementations of distributed problem solving have been proposed (5,6). Hewitt and Baker have proposed the message passing actor formalism to describe and control the assignment of workload between the processors of a general distributed computing system (7). Unfortunately, the actor formalism as presently developed neglects the issues of real time events, message content and world model representation. Smith and Davis have examined these issues in their exploration of the contract net formalism (6). However, the contract net formalism concentrates on distributed system resource allocation as opposed to efficient and realistic handling and representation of real time problems. Existing research experience has not yet produced generalized concepts for the distribution and coordination of function in complex robots.

This paper describes an architecture to support the distribution and coordination of function in a complex robot system. This architecture for robot design, based upon developments in production systems and distributed problem solving, consists of a concept for function distribution, a high level communications protocol, an interface mechanism and a blackboard mechanism. This approach supports various types of sensor data fusion and some of these have been explored experimentally on an autonomous mobile robot called the Ground Surveillance Robot (GSR). This approach has also been extended to address the problem of coordinating multiple interacting robots.

ARCHITECTURAL DESIGN

In the proposed architecture of a complex robot, the functions of the robot are distributed among processing subsystems of three types: sensor, control and knowledge based subsystems. These subsystems are interconnected through a local area network in a bus communications configuration. Each subsystem has equal control of the bus. Two types of messages are exchanged, plans and reports.

In this architecture, each subsystem is as capable and independent as possible and communication between subsystems occurs at as high a level as possible. For example, sensor subsystems perform all the processing on the sensor data to make symbolic level information available to the other subsystems. Likewise, control subsystems provide all the processing necessary to translate symbolic plans into coordinated effector actions. Knowledge based subsystems including planning and knowledge intensive automated reasoning functions. In this design, knowledge based functions could be divided into several separate coordinated subsystems.

The model of subsystem interaction is shown in Figure 1. This is a blackboard model where each of the subsystems exchange information through blackboard memory. This model provides a very powerful and flexible paradigm for coordinating the interactions between capable subsystems. It also provides a rational means for coordinating the programming activity associated with system development since each subsystem developer has a well defined mechanism through which to exchange information with the other subsystems. Unfortunately, the implementation of the blackboard interaction model on the distributed computing resources required in complex robots is not straightforward.

Figure 2 illustrates the actual computing hardware architecture developed for an autonomous mobile robot, the GSR. In this architecture, the major loosely coupled computing systems interact through a local area network (LAN) (i.e., IEEE 802.3) and the more tightly coupled systems interact through shared memory. Each of the loosely coupled modules are interfaced to the LAN through an Intelligent Communications Interface (ICI), the components of which are illustrated in Figure 3. The ICIs support the blackboard interaction model, illustrated in Figure 1, between a robot's sensor, control and planning subsystems which are implemented in the distributed computing environment illustrated in Figure 2.

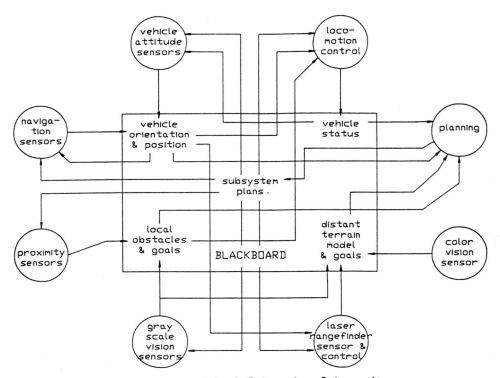

Blackboard Model of Subsystem Interaction

Figure 1

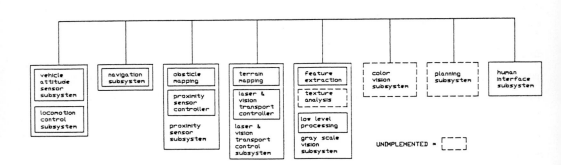

FIGURE 2 - GSR SYSTEM ARCHITECTURE

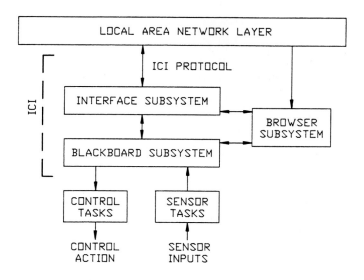

Figure 3
INTELLIGENT COMMUNICATIONS INTERFACE COMPONENTS

The distribution of robot function into sensor, control and planning subsystems has been employed in the JPL Rover (8). Furthermore, simulations of distributed problem solving networks have confirmed the effectiveness of employing very flexible protocols which permit free communication between subsystems but which provide mechanisms for controlling deadlock and race conditions (9). Just such protocols have been developed in the proposed architecture.

Protocols

The communications protocols define the framework for a high level language which governs all communications between distributed intelligent robot subsystems. These protocols enable manipulation of all sensor and actuator states and maintenance of a consistent world model throughout the robot subsystems.

The robot's world model is represented as a structured collection of classes and instances of those classes. Each subsystem represents only that portion of the world model which is relevant to its operation. When the function of one subsystem depends upon the function of another, then their world models must overlap.

All intermodal communication occurs as messages. The structure of the messages exchanged between ICIs is illustrated in Figure 4. Message transport between subsystems is supported by the transport layer of the communications system. A message consists of a source address, a priority and a body. The body can be one of two message types: a plan or a report.

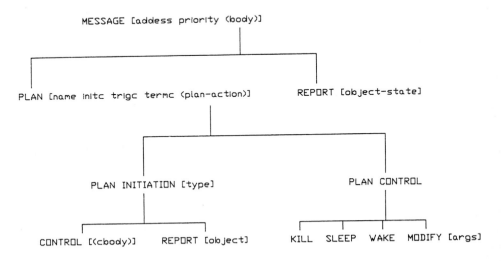

FIGURE 4 - ICI PROTOCOL TAXONOMY

A plan consists of a plan name, an initiation condition, a trigger condition, a termination condition and a plan action. Of this information, only the plan name is necessary. The remaining information can be substituted by the subsystems as default values thus making communication of commonly used plans very efficient. A plan is a form of production rule which has been extended to manage real world continuous processes. The initiation condition specifies when the plan can be first invoked and the termination condition specifies the situation after which the plan will never be invoked again. The trigger condition describes when if the initiation condition is true the plan action should be taken.

All conditions can be represented in a manner similar to the production conditions in the language OPS5 [10]. The trigger condition can also be modified by predicates which permit control of a plan's invocation between the initiation and termination conditions. They are particularly useful for controlling such actions as report generation.

A plan action either initiates a new plan or controls an existing plan. Two types of actions can be initiated: report actions and control actions. A report action results in the generation of one or more report messages describing some part of the local world model. A report action is specified by describing the object attributes of the world model which are desired in the report.

A control action results in the change of some part of the local world model. Control actions are divided into open loop control and closed loop control actions. An open loop control action is described by an optional goal state list and a control type. The two types of open loop control actions are discrete (switches and discrete state actuators) and continuous (actuators which have a continuous range of possible states). Continuous actions have an optional gain state description which specifies the rate at which the actuator state is changed toward the goal. A closed loop control action is described by a goal (specifying not only the closed loop control), a gain description (specifying the rate at which control actions are taken) and a control type (approach or avoid). An approach action exerts the control to reach the specified goal while an avoid action exerts the control to prevent reaching the specified goal. Control actions can be modified by a metaoperator which permits the existence of multiple competing goals. All but one of any competing control actions without the metaoperator are terminated to prevent control action conflicts.

A plan control action can kill, wake, put to sleep or reprogram an existing plan. Reprogramming a plan changes some of the existing plan parameters. These plan control primitives permit external control of plans once they have been instantiated.

Report messages communicate either plan status information or information about some part of the world model. Plan status reports convey information about a plan's execution state to the plan initiator (when the plan has been parsed, initiated, triggered or killed). Any plan will cause one or more plan status reports to be generated depending upon the success or failure of the plan processing. Furthermore, plans are represented in each subsystem as objects in the world model and, therefore, report plans can be created which report plan status information.

Reports containing world model information are responses to report plans. The content of these reports is specified by the plan. World model information is represented in an object/attribute/value form everywhere in the robot.

Interface Mechanism

The ICI interprets the protocols and translates message content to subsystem function (Figure 5). This interface mechanism parses arriving messages, adds report information to the local world model, adds new control actions, modifies existing control actions, interprets local sensor data, controls action processes and searches for true conditions. The local view of the world model is maintained in an active blackboard which enables rapid information sharing between local processes.

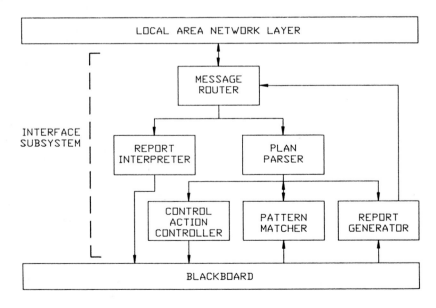

FIGURE 5 - ICI INTERFACE MECHANISM

The report interpreter and report generator translate world model data between report representation and blackboard representation. The report interpreter translates external reports from symbolic form to the blackboard's internal representation. When any report plan's initiation and trigger conditions are true and its termination condition is false, the report generator translates blackboard representation into a symbolic report.

Sensor interpretors, specific to the sensor source, put sensor input into the blackboard. Consistency processes associated with particular object classes can be invoked when making an entry into the blackboard. Simple sensor interpreters can be very fast and make data from sensors which need little processing very quick to update.

The plan parser parses received plans and makes entries into tables of conditions, report formats and distribution information. It passes control action information to the control action controller to make entries into the blackboard and, later, to schedule and manipulate control processes. These control processes access blackboard information directly which reduces all unnecessary processing delays between feedback sensor and control process and makes real time closed loop control realistic.

The pattern matcher controls the initiation, triggering and termination of plans for the entire subsystem. Each of the three conditions associated with a plan (initiation, trigger and termination) is essentially an expression written in terms of the attributes of the objects contained in the blackboard. The pattern matcher evaluates the conditions associated with each plan and initiates the appropriate action when a condition evaluates true.

Conceptually, as with any production system, the pattern matcher runs continuously and instantaneously with different rules firing as the blackboard is continually updated by local sensor information and external reports. Typical production systems fire a rule only once when its condition becomes true. This is the case for plan initiation and termination conditions. However, trigger conditions are continuously evaluated until their plans' termination conditions are true. Since some processes are discrete (most notably, report generation), constraints must be placed upon trigger conditions to generate a determinate number of trigger messages.

These constraints can be represented as a series of special predicate operators. Three predicates have been identified thus far: PERIOD, TOGGLE and DELTA. PERIOD enables simultaneous control of a trigger with both time and a condition. TOGGLE triggers whenever a prescribed condition or predicate changes truth value. DELTA triggers whenever a situation makes a specified transition from one condition to another. These trigger predicates can be nested, thus, providing a considerable degree of flexibility.

Blackboard Mechanism

A blackboard represents the world model for each subsystem of the robot. All sensor and actuator states and the quantities derived from these states (known as composite objects) for which the subsystem has use are represented in the blackboard. The blackboard concept employed here is not the passive blackboard typically employed in other production systems. This blackboard concept contains active processes which maintain data consistency.

The blackboard incorporates a dynamic set of objects and a set of procedures that manipulate those objects in a real time multi-tasking environment. Its structure was designed with both object oriented and data oriented programming paradigms in mind. A set of small tightly coded blackboard interface procedures provides great flexibility of operation at a minimal cost in efficiency to the real time system.

Data contained in the blackboard is grouped into object entities and is structured hierarchically into a class tree. The class tree relates blackboard objects to one another in time and data space. There are two types of blackboard objects; class objects and terminal objects. Class objects have other classes or terminal objects as children and terminal objects have no children at all. Class objects are created only at system initialization to maximize a running system's efficiency and, thus, cannot be created or deleted dynamically. On the other hand, terminal objects can be created either at system initialization or during runtime from class objects through the inheritance mechanism which completely specifies memory allocation and organization, the attribute field formats and their initial values. Therefore, to create a new instance of a class a task writer need only know the instance's parent class and the attribute values which make the instance unique in the class. This keeps the interface narrow and minimizes runtime overhead. The class tree with inheritance properties is a very efficient data representation.

A data oriented programming paradigm synchronizes nearly independent processes and maintains blackboard consistency. Consistency maintenance processing may be triggered when an instance's attributes are accessed. The blackboard interface procedures permit separate processes to be invoked both before and after reading and writing. This design feature increases the blackboard mechanism's flexibility. Two methods for time stamping attribute values are provided by the blackboard. The default method associates a time stamp generated by the host operating system with each write access to an instance attribute. The second method allows the task writer to specify the time stamp value. This feature makes possible arbitrarily complex data dependent temporal sequencing. Time stamping is crucial for sensor data fusion.

This approach to robot system integration has been applied to three very different systems, an autonomous mobile robot (i.e. the GSR), an intelligent welding work cell and coordinated teleoperated vehicles. The interface processing is equally general and flexible. The blackboard mechanism internal to each subsystem provides a good interface between different designers working on various components of a single subsystem. The use of a multitasking operating system has permitted both the interface and blackboard processing to be implemented as separate intercommunicating tasks. Most of the communications between these tasks goes through the blackboard. These design features will ease the future transition of the architecture to highly parallel (i.e. non von Neumann) computers.

SENSOR DATA FUSION

Today, multiple, redundant and disparate sensors are abundant. Many existing complex robots employ multiple sensors (e.g. TV cameras, acoustic, infrared and laser rangefinders, absolute and relative navigation sensors, force and tactile sensors, etc.). As the capabilities of complex robots increase, they can undertake tasks of greater complexity and perform more robustly in uncertain and unknown task environments. Sensor data fusion is a logical step towards increasing robot capability. It lets a robot make best use of all its sensor resources to solve complex tasks. Sensor data fusion can improve a complex robot's performance by providing hybrid information, by reducing sensor errors and by maximizing sensor use.

Hybrid information is information derived from the combination of sensor data which would not be available from a single sensor source. It can represent new physical properties about known objects in the task environment [13,14] or identification of objects previously unknown to the robot [15,16]. Allen fuses stereo vision and active tactile sensing data for object recognition [15].

Data fusion can reduce sensor errors generated by inaccurate sensor data interpretation models, by sensors driven beyond their bounds of known accuracy and by sensor failures. Flynn investigated the use of multiple sensors with complementary characteristics to reduce the error inherent in sensing mechanisms [17]. Her system also exploited sensor redundancy to recover from sensor failure. The multiplicity of data available from redundant sensors is a useful feature which can go under utilized if a coherent methodology is not developed to analyze sensor performance. This analysis will almost always involve the fusion of data from many sensors. Sensor use can be maximized by using data fusion to optimally control sensor application, to aid in sensor calibration and to filter noisy sensor data. Kent uses fast infrared proximity detectors and a video camera to grasp dynamic objects [16]. In his system the vision system generates realtime calibration parameters for the proximity system. Brooks [18] and Iberall [19] have developed methods for hiding sensory data by representing it in a filtered form which depends upon the perceptive processing needs. Perceptive processing needs can also guide the control of sensors in data gathering activities 19 . In another example, the GSR uses steered acoustic rangefinders and a vision system mounted on a three degree of freedom platform for target tracking 13 . Each sensor system relays target bearing information to the other to help servo its field of view in the proper direction.

Techniques

In the simplest situation, a robot with multiple sensors uses them independently. As the data from multiple sensors is merged, the coupling complexity increases the design complexity significantly. The first step toward dealing with this complexity is to use a uniform sensor data representation. This agreement in representation is necessary to permit economical communication between distributed sensor subsystems. Data from like sensors are the simplest to fuse since they have the same dynamic responses, update rates, calibration and error conditions. Often like sensor data can be most effectively fused at the lowest levels of the computing hierarchy. Data fusion problems increase further when the information from disparate sensors is merged. If the physical domains of disparate sensors intersect then data fusion requires knowledge of the observation times, the sensor responses (both static calibration and error) and the fusion method (i.e. what model is used to equate two different measurements for the same phenomenon). Several methods can be used to fuse disparate sensor data. Of course, a representation and a collection of methods are not sufficient for practical sensor data fusion. Timely effective data fusion requires a flexible computing architecture such as the one described in this paper. Several examples of sensor data fusion onboard the GSR are discussed to illustrate the power of the distributed blackboard technique.

Representation

A robot task environment can be modelled as a finite unordered set of objects, each with an associated set of single valued properties. Unfortunately, knowledge of the environment cannot be obtained directly but must be measured through sensors plagued by the inaccuracies of the real world. Sensors introduce errors for a variety of reasons including inherent uncertainties, discrete sampling and quantification errors, inaccurate calibration and device failures. If no device failures have occurred, a sensor's accuracy can usually be calculated from empirical and analytic sources. If a sensor's accuracy is dependent upon information from other sensors, its computation can involve data fusion. Often good estimates of a sensor's accuracy made in a controlled environment can aid sensor performance analysis in unknown environments. Most error recovery systems proposed for robots depend at least in part on prior knowledge of the task environment [20]. However, many mobile robot applications make this assumption impractical [13,21]. Dynamic redundant sensors may permit identification of sensor failure in complex unknown environments. For these reasons and others, estimates of property value accuracies are necessary for fusion of real world estimates of object property values. Thus, each sensor obtained value for which an accuracy estimate is available adds an additional value for estimated error to the object-property tuple.

Another notion useful to this data representation is the confidence in the interpretation of a property's value. Confidence measures have been proposed as a primary means for fusing data about the same object properties from different sensors [14,15]. Usually measures of confidence represent the sensor source's faith in the model it uses to interpret property values and error estimates. Most simply, a confidence measure can be used as the criterion to choose between conflicting sensor estimates of the same property value. It can also be used for setting thresholds and weights in various averaging schemes. In this scheme, confidence measures are assigned at the moment the information is made available to other consumers. As a result, these measures can become inaccurate in dynamic situations. Consistent confidence measure scaling is critical if confidence is used in sensor data fusion. Nevertheless, confidence is a key aspect of sensor data representation.

Unfortunately, this still remains an inadequate representation in most task domains because it takes no account of the changing property values of a dynamic world and fails to provide any basis for temporal reasoning. A static representation of unchanging or slowly changing (relative to the length of the task) properties is sufficient. However, many properties change too rapidly to ignore temporal considerations. Often a significant amount of time passes between the time a measurement was made by a sensor and the time it is used. Knowledge of this time interval is necessary for data fusion in dynamic situations so decisions can be made as to how best to use the information and so estimates of current values can be predicted (provided the nature of the dynamics are known). In a distributed computing environment, allowing access to different sensor data on multiple occasions by different consumers (e.g. other sensor modules, controllers and planners) decouples complex processing and lowers communication needs. The knowledge of the time elapsed since property measurement can be recorded as the observation timestamp made on some absolute time scale (usually defined by a system clock although this function can also be distributed between several computing modules if some synchronization scheme is used). Therefore, complete task environment representation must include the observation timestamp with each object-property tuple.

Some error is to be expected in the timestamp itself due to inaccurate synchronization, delays introduced by the sensor and any processing overhead in acquiring the data and associating the timestamp. Timestamping sensor data should take place as near to actual data acquisition as possible and known delays should be taken into account in order to minimize these errors. Occasionally, calculation of the sensing delay is dependent upon the present state of the task environment and, thus, requires knowledge of other sensed values. These cases represent other examples when sensor data fusion can increase the precision of property value estimates.

Methods

Several methods exist for fusing sensor data depending upon the situation. The simplest situations represent complementary sensor measurements. If two sensor observations, A and B, are complementary and describe similar (i.e. same measurement units) but independent situations (i.e. no intersection between measurements) then they can be logically added to the total environment description without concern for conflict. Each measurement also contributes its own associated values for accuracy and confidence. If these observations are complementary but coupled (i.e. $C = f(A,B)$) then the dependent quantity can be computed from the prevailing model (i.e. $f()$) and input measurements (i.e. A and B). The dependent quantity's accuracy and confidence can also be computed from this model.

More complex situations arise when sensor observations intersect (i.e. represent information about the same property value of the same object). In these situations, the information must be merged. If intersecting measurements represent observations at different times or at different but continuous spatial positions, then simple interpolation is the most straightforward merging method. Accuracies and confidences associated with interpolated values can also be interpolated although more sophisticated models. for computing these parameters can be used if necessary. Several merging methods are available if independent sensor observations are completely intersecting (i.e. same object, same property, same spatial location and approximately the same time) including some form of averaging, a decision between two intersecting sensor observations and the act of guiding the processing of one observation with another.

Filtering. This method, along with guiding method mentioned below, uses confidence measures in the spirit that they are usually intended [14,15]. This is a simple percentage calculation based on confidences which are then used as weights in an averaging of sensor data values. Other more complex methods for computing confidence have been proposed including using error manifolds [22], heuristics [23] and probability theory [24]. The fused timestamp can be computed in many different ways. The latest or earliest timestamps could be chosen, a cluster of timestamps could be interpolated or the mean could be computed.

Deciding. This fusion method simply makes a discrete choise between several different measurements of the same property. This choice must be based upon some method to priority order the different measurements. The confidence associated with each value in the representation discussed above provides one means to make this ordering. In the simplest situation, the value with the highest confidence is chosen. However, more sophisticated heuristics can be used which may depend upon knowledge of the state of other parts of the task environment. In general, the confidence and timestamps associated with the fused values are those originally associated with those values. However, in many situations, Bayes rule can be used to compute the resultant confidences of fused values.

Guiding. In these fusion methods, the values from one sensor are used to guide the control or processing of other more accurate estimates from another sensor [16]. As implied, coarser estimates guide the definition of more accurate estimates. Guiding the thresholding of sensor input is one example of guiding the processing of sensor data. Thresholding is usually applied in combination with another method such as averaging to filter either the inputs or the output of the fusion process. A threshold value can be obtained either from an a priori constant, a previous sensor value or confidence or the resultant value or confidence of a fusion process. Thresholding the fusion output effectively filters multiple sensor readings and may be used used as a measure of sensor performance [14,15]. Guiding can also apply to such processes as windowing to decrease search and steering controllable directional sensors (such as range-finders and vision sensors with restricted fields of view).

Implementation

To this point, we have discussed a data representation, and a collection of data fusion methods. These are just a collection of tools by which to accomplish sensor data fusion. In the following section, we discuss the application of those tools to the problem to sensor data fusion for an autonomous mobile robot, the GSR. The computing resources architecture on the GSR is configured to support the data fusion at several different levels while minimizing consumption of costly internal communications bandwidth.

Each subsystem's world model is defined by and designed into its own blackboard which resides in part of the subsystem's shared memory. In this architecture, sensor data fusion can be done within a single processor (usually representing a single sensor group) between processors within a subsystem (usually closely interacting sensor and control groups) through the subsystem's blackboard memory and between subsystems (usually loosely coupled subsystems) through ICIs.

These concepts are currently being implemented and tested in the GSR, an experimental autonomous robot vehicle 13 . The GSR is a M114 armored personnel carrier which has been modified for computer control. It has sensors and processing for vehicle attitude, vision, navigation and proximity as well as computer control of all aspects of vehicle locomotion. An onboard planning subsystem is currently under development and a long range laser rangefinder is being added. With this sensor complement, the GSR provides a fertile testbed for sensor data fusion research. The underlying architecture permits data fusion to occur at a number of levels. While the blackboard is tree structured, the GSR's functional structure is quite monolithic (see Figure 2) so any subsystem can interact with any other subsystem without regard for architecturally imposed limitations. This implementation philosophy has provided a flexible framework within which to experiment with a variety of data fusion methods. Several examples of these are discussed below.

The vision subsystem is mounted on a three degree of freedom transport platform and will eventually incorporate both a solid state TV camera and the laser range finder for obstacle location, moving target tracking, terrain modelling, surface trafficability analysis, and landmark navigation on the moving vehicle. Obviously, sensor data fusion will be necessary for these functions. Figure 6 represents the interactions between various sensor subsystems for terrain modelling. The camera and laser must interact to build the terrain map by using the contrast map obtained from the camera output to guide the discrete sampling of

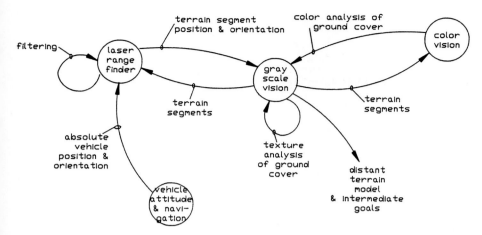

FIGURE 6: SENSOR COUPLING FOR TERRAIN MODELLING

the computer steered laser rangefinder. Region range properties can be merged with relative depth cues from camera imagery to produce the terrain map. Trafficability estimates for different regions derived scene texture and laser range variability analysis can then be combined with this map for route planning purposes. Camera, laser and transport platform position data must be combined and merged to obtain landmark position to enhance vehicle absolute position estimates.

The proximity sensor subsystem uses acoustic ranging sensors to provide short range obstacle position and target tracking information. Some of these sensors are steered azimuthally to within one degree of accuracy while others are fixed to the robot. Figure 7 illustrates the interactions between the proximity, gray scale vision, vehicle attitude, navigation and locomotion subsystems required to accomplish robust obstacle avoidance. The proximity sensor subsystem must fuse the data from this array of nearly identical sensors into consistent target and obstacle position and velocity vectors. This is done largely using different averaging techniques. Unfortunately, the acoustic ranging sensors have very poor angular resolution. One technique to overcome the problem inherent to these sensors is to use multiple sensors to trangulate upon the position of the

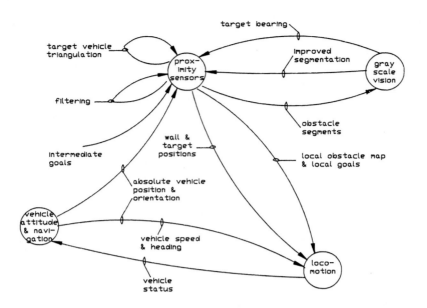

FIGURE 7: SENSOR AND CONTROL COUPLING FOR OBSTACLE AVOIDANCE

target. This is another example of filtering methods for data fusion of information from like sensors. Further, during target tracking, vision estimates of target bearing can be fused with proximity estimates to improve the knowledge of target angular position and motion critical to accurate vehicle response. Likewise, coarse proximity information can be used to guide the camera steering and processing of the camera's complex visual information rapidly enough to permit timely vehicle control.

The vehicle attitude sensor subsystem monitors absolute vehicle speed, track speed, heading, pitch and roll angles. Figure 8 illustrates the sensor coupling used to develop and maintain accurate and continuous estimates of vehicle orientation and position. A fused value of vehicle speed is derived from a true ground speed sensor using doppler radar and a track speedometer using a confidence based decision method to take advantage of the greater low speed accuracy of the track speedometer yet use doppler speed for high speed or other times when track slippage conditions may prevail. Measurements of vehicle speed

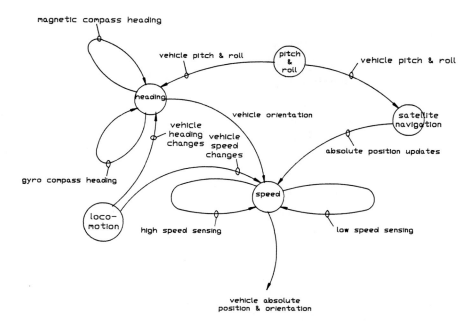

FIGURE 8: SENSOR COUPLING FOR VEHICLE ORIENTATION AND POSITION

are combined with those of vehicle angular motion to estimate relative vehicle position. Then decision fusion methods are used to combine instantaneous vehicle position estimates from dead reckoning sensors with the intermittent absolute estimates from satellite and landmark navigation processes.

These are just a few examples of the sensor data fusion possibilities being explored on the GSR using the technique proposed in this paper. Further details on this sensor data fusion technique can be found in Ref. [27].

MULTIPLE ROBOT COORDINATION

Multiple robots can be coordinated to achieve greater diversity and to obtain greater parallelism. Diversity provides multiple robots with the abilities for spatial extension and resource tailoring. Spatial extension permits multiple robots to address tasks which require simultaneous coordinated actions at two or more separate locations. Resource tailoring lets a mix of tasks with differing requirements to be most economically addressed by assigning to each task only the resources required. This capability could lead to the implementation of robot families with various specialized skills and the assignment of the appropriate robot team to perform each task. Cooperative behavior between redundant robots can be used to detect and correct subsystem failures and processing errors. This ability enables multiple robots to ensure the success of critical tasks where maintenance and repair services are not available despite the occurrence repeated equipment failures (e.g. extended space missions). Parallelism also provides the opportunity for multiple robots to increase overall system throughput. Even though a single robot may be entirely capable of performing a large task, the application of multiple robots to the same task can speed its completion. In this situation, coordination is necessary to ensure that effects are not duplicative or counterproductive. Eventually, multiple robots will be employed for any and all of the above purposes.

Unlike the problem of coordinating a single robot where some guidelines for system structuring could be suggested, coordination of multiple robots is by its very nature an expansive problem and very little can be said about the structure of a multiple robot system other than it must be flexible. Even though the organization of cooperating robots depends primarily upon the nature of the application some general observations can be made. Multiple robots should be designed to benefit from both diversity and parallelism. This implies that each robot's sensor and effector domains should overlap another domain somewhat to accommodate inevitable device failures and to be able to exploit task parallelism when it is available. Further, complex tasks usually demand more capability than available from a single robot so complete functional overlap while desireable is often impossible. Coordination of multiple robots for practical applications requires a very flexible system structure. This requirement places challenging demands upon the world model representation.

World Modelling

In a collection of multiple robots, each with its own resident knowledge, each robot must model only that portion of the task environment which is relevant to its functions. Robot world models must logically overlap only when the function of one robot depends upon the function of another. For cooperating robots, the additional awareness of the system context must be added to the world model [26]. Now information represented by a single robot's world model may be either local (i.e. sensed directly by self) or remote (i.e. sensed by some other robot in the system). The system context is much like that of self but it represents the perceptions and intentions of the other robots in the collection. Obviously, the system context includes only those robots with which there are communications. No cooperation is expected from the objects from the surroundings thus if noncommunicating robots are present in the task environment then they are simply represented as components of the surroundings. Cooperating robots can be viewed as set of capable knowledge based entities acting on the entire self, system and surroundings knowledge base. Knowledge in all of these contexts can be represented as collections of blackboard objects implying that no substantial changes are needed in the single robot representation described above to cope with the demands of multiple coordinated robots.

As discussed earlier, a robot can be modelled as a collection of sensor, effector and reasoning capabilities. Sensor knowledge includes all the physical and logical state information available to the robot about the self, system and surroundings contexts. Effector knowledge includes information about all the physical and logical actions available to the robot in the self and system. Reasoning capabilities include the knowledge based expertise and planning available to the robot from both the self and the system contexts. Active functions and plans associate blackboard object accesses with actions of sensors and effectors. Local sensor and effector knowledge is provided by drivers writing to the local blackboard through the standard blackboard interface procedures.

Nonlocal sensor and effector objects, while treated no differently than local objects, require communication with the sources and destinations of their information. This communication can be transparent to the individual robot but the world model of each cooperating entity should have the same structure and must have enough overlap in content to permit communications (i.e. cooperating robots must communicate in common languages). The amountation through which the communication is conducted, by the difference in levels between the communicating robots, by relationship between the communicating entities and by the structural differences in the world models. If necessary, effector objects can represent plans just as well as past actions and the history of results. No longer must these spaces represent only the possible actions which might be appropriate for the prevailing sensed situation. This modification permits multiple robot cooperation as well as enabling learning behavior in single robots.

Communications

The message passing paradigm employed to coordinate the subsystems of a single autonomous robot can be applied almost directly to the coordination of multiple robots. One small difference is that messages must now be sent through a distributed message routing system. For coordinated robots, a message consists of source and destination addresses (either local procedures, local modules or separate robots) and an informational body. A distributed routing mechanism designed into the ICI mechanism provides the necessary routing functions transparent to the individual robots. As described above, the body can be either a report or a plan and it can be limited to very terse symbolic representations. Structuring the world model as a class tree permits very sparse communications since the greatest level of generalization can be used by taking advantage of inheritance properties and active functions. Plan actions now represent alterations to world models as well as control actions.

Implementation Issues

Programming. This technique provides a high level set of tools and, more importantly, a set of standard interface specifications which allows the system designer to concentrate on WHAT information needs to be represented by the robots' world model and WHAT information is needed for communication not HOW to store the information or HOW to communicate that information. The designer may validate his system first using only the robot's local blackboard and later integrate the robot into the system context. This technique gives the designer freedom from exhaustively programming the robot for every possible situation. This is a key advantage in programming distributed robots since their inherent flexibility makes deterministic programming a very difficult job. This technique also permits much of the system behavior to be changed very easily by changing the plans which are communicated between robots. Software must be designed as modules which accomplish a certain task usually in parallel with other activities and whenever the situation demands. In non-robot applications, this kind of software is usually written as systems support software. The difference is that now software user (the robot themselves) will have no opportunity to visit a systems guru if the module generates difficulty.

Examples. This technique for implementing distributed robot systems has been applied to the design of two different multiple robot situations: a concept for integrating multiple robots and other intelligent components of an automated factory and a concept for integrating multiple teleoperated vehicles driven by multiple interacting operators. The integrated flexible welding system consists of an automated welding workcell with several complex sensors and several associated planning components. The ICIs and the distributed blackboard are used to coordinate the activity of the planners toward generating a final welding plan. This plan is then communicated to an execution controller which decomposes the welding plan into subplans for each of the sensor and control components in the workcell. The teleoperated vehicle system consists of three remotely controlled land vehicles operated from a control vehicle. In this system, the actions of the operators and information from the vehicles are coordinated through distributed blackboards by ICIs. This arrangement permits multiple operators to coordinate combined function on a single vehicle as well as facilitating multiple vehicle coordination by a single operator. The concepts introduced in this paper have proven invaluable for making design and implementation of these complex multiple robot systems possible. Further details on this technique for integrating distributed robot systems can be found in Ref. [28].

CONCLUSIONS

While this technique has been implemented on an autonomous mobile robot, it is general enough to be applied to any complex robot system or system of robots which must be implemented using distributed computing. The representation discussed above is flexible, is able to cope with a dynamic environment and supports the use of a number of different methods of data fusion. The computing architecture discussed above facilitates data fusion at a number of different levels by providing consistent access to system information.

While many different forms of sensor data fusion are currently implemented onboard the GSR, many avenues of data fusion have not yet been explored. In the future, data fusion mechanisms will be further automated by employing the data fusion driven programming available through the active functions supported by the ICI. Through this capability, appropriate data fusion methods can be switched dynamically. This hides higher level fusion from disinterested sensor subsystems thus further reducing complex robot design complexity.

The architectural concepts described here have several useful attributes. First, their generality has been demonstrated by application to three very different problem domains, an autonomous ground vehicle, an autonomous welding workcell, and multiple teleoperated vehicles. Second, the processing mechanisms of this architecture should be easily transportable to a highly parallel computing environment. Third, the inherent modularity of the scheme permits individual components of complex robot systems to be upgraded as new technology allows. And finally, high level symbolic communications simplifies interfacing to complex knowledge based processes such as planners and expert systems. These attributes make the implementation of the next generation of complex robots feasible and practical.

ACKNOWLEDGEMENTS

This work was performed while the author was employed by the Naval Ocean Systems Center, Code S442, San Diego, CA 92152. The author is extremely grateful to the U.S. Marine Corps, the Naval Sea Systems Command, and the Naval Ocean Systems Center, the sponsors of this work, and to the many people at the Naval Ocean Systems Center who have contributed technically to this work.

REFERENCES

[1] A. M. Thompson, "The Navigation System of the JPL Robot", Proc. of the 5th International Joint Conf. on Artificial Intelligence, Cambridge, MA, 22-25 August 1977, pp749-757.

[2] G. Giralt, R. Sobek & R. Chatila, "A Multilevel Planning and Navigation System for a Mobile Robot: A First Approach to HILARE", Proc. of the 6th International Joint Conf. on Artificial Intelligence, Tokyo, Japan, 20-23 August 1979, pp335-37.

[3] L. D. Erman & V. R. Lesser, "A Multi-Level Organization for Problem Solving Using Many, Diverse Cooperating Sources of Knowledge", Proc. of the 4th International Joint Conf. on Artificial Intelligence, Tbilisi, Georgia, USSR, 3-8 Sept 1975, p483.

[4] V. R. Lesser & D. D. Corkhill, "Functionally Accurate, Cooperative Distributed Systems", IEEE Trans. on Systems, Man and Cybernetics, 11(1) 1981, pp81-96.

[5] C. Hewitt, "Viewing Control Structures as Patterns of Passing Messages", Artificial Intelligence, 8, 1977, pp323-364.

[6] R. G. Smith & R. Davis, "Frameworks for Cooperation in Distributed Problem-Solving", IEEE Trans. on Systems, Man and Cybernetics, 11(1) 1981, pp61-70.

[7] C. Hewitt & H. Baker, "Laws for Communicating Parallel Processes", Information Processing 77, B. Gilchrist, ed., North-Holland Publishing Co., Amsterdam, The Netherlands, 1977, p987.

[8] M. Weinstein, "Structured Robotics", Proc. of the 4th International Joint Conf. on Artificial Intelligence, Tbilisi, Georgia, USSR, 3-8 Sept 1975, p609.

[9] R. Wesson, et al., "Network Structures for Distributed Situation Assessment", IEEE Trans. on Systems, Man and Cybernetics, 11(1) 1981, pp5-23.

[10] C. L. Forgy, OPS5 Programmers Reference Manual, Carnegie Mellon University, Pittsburg, PA, 1981.

[11] S. Y. Harmon & M. R. Solorzano, "Information Processing Architecture for an Autonomous Robot System", Proc. of the Oakland Conf. on Artificial Intelligence, Rochester, MI 26-27 April 1983.

[12] S. Y. Harmon, "Coordination between Control and Knowledge Based Systems for Autonomous Vehicle Guidance", Proc. of IEEE Trends & Applications 1983, Gaithersburg, MD, 25-26 May 1983, pp8-11.

[13] S. Harmon, "USMC Ground Surveillance Robot: A Testbed for Autonomous Vehicle Research", Proc. of the 4th UAH/UAB Robotics Conf., Huntsville, AL, Apr. 1984.

[14] J. W. Lowrie, M. Thomas, K. Gremban, M. Turk, "The Autonomous Land Vehicle (ALV) Preliminary Road Following Demonstration", Proc. of the SPIE Conf. on Computer Vision and Intelligent Robots, Cambridge, MA, 1985.

[15] P. Allen, R. Bajcsy, "Object Recognition Using Vision and Touch" Proc. of the 9th Int. Joint Conference on Artificial Intelligence, Los Angeles, CA, Aug. 1985, p1131-1135.

[16] E. Kent, T. Wheatley, M. Nashman, "Real-time Cooperative Interaction between Structured-Light and Reflectance Ranging for Robot Guidance", Robotica, 3, 1985, p7-11.

[17] A. Flynn, Redundant Sensors for Mobile Robot Navigation, Dept. of Electrical Engineering and Computer Science, MIT, Cambridge, MA, July 1985.

[18] R. Brooks, "A Robust Layered Control System for a Mobile Robot", A.I. Memo 864, MIT, Cambridge, MA, Sept. 1985.

[19] T. Iberall, D. Lyons, Towards Perceptual Robotics, COINS T.R. 84-17, Laboratory for Perceptual Robotics, Univ. of Mass, Amherst, MA, Aug. 1984.

[20] M. Gini, et al., Symbolic Reasoning as a Basis for Automatic Error Recovery in Robots, Report 85-11, Univ. of Minnesota, Minneapolis, MN, 1985.

[21] S. Harmon, "Autonomous Robot Submersibles: The Future of Unmanned Submersibles", Proc. of the 2nd ASME Computer Engineering Conf., San Diego, CA, Aug. 1982, p33-36.

[22] R. Brooks, "Visual Map Making for a Mobile Robot", Proc. of the IEEE Conf. on Robotics and Automation, St. Louis, MO, Mar. 1985, p824-829.

[23] J. Crowley, "Navigation of an Intelligent Mobile Robot", IEEE J. on Robotics and Automation, RA-1(1), Mar. 1985, p31-41.

[24] R. Chatila, J-P. Laumond, "Position Referencing and Consistent World Modelling for Mobile Robots", Proc. of the IEEE Conf. on Robotics and Automation, St. Louis, MO, Mar. 1985, p138-145.

[25] S. Harmon, D. Gage, W. Aviles, G. Bianchini, "Coordination of Intelligent Subsystems in Complex Robots", Proc. of the 1st IEEE Conf. on Artificial Intelligence Applications, Denver, CO, Dec. 1984, p64-69.

[26] S. Harmon & D. Gage, "Protocols for Robot Communications", Proc. of the IEEE Conf. on Cybernetics and Society, October 1981.

[27] S. Harmon, G. Bianchini & B. Pinz, "Sensor Data Fusion on an Autonomous Mobile Robot through a Distributed Blackboard" Proc. of the IEEE Conf. on Robotics and Automation, San Francisco, April 1986, p1449-1454.

[28] S. Harmon, W. Aviles & D. Gage, "A Technique for Coordinating Autonomous Robots", Proc. of the IEEE Conf. on Robotics and Automation, San Francisco, April 1986, p2029-2034.

AUTONOMOUS ROBOT OF THE UNIVERSITY OF KARLSRUHE

Ulrich Rembold and Rüdiger Dillmann

Institut für Informatik III
Robotics Research Group
Project Director: Prof. Dr.-Ing. U. Rembold
Universität Karlsruhe
7500 Karlsruhe 1
Federal Republic of Germany

This paper discusses the autonomous research robot which is being developed at the University of Karlsruhe. The device will perform simple assembly operations in the laboratory. The purpose of the project is to develop new technologies for advanced robotic machines for industrial application. The robot contains a mobile platform, a complex sensor system, two manipulators, hierarchical controls and an expert system. Programming will be done by task-oriented instructions. A considerable amount of the fundamental technology which will be integrated in the robot has already been developed at the institute, including a vision system, a robot hand containing 5 different sensors, a programming system and a hierarchical computer architecture.

1. Introduction

The application of mobile robots becomes increasingly of importance in areas where work is hazardous and dangerous for humans. Experience with teleoperators in nuclear power plants, space applications and under water operations has been gathered since the early sixties. For this purpose human controlled master slave systems with several degrees of freedom were developed. Terrain exploration and the use of mobile robots in agriculture are current research areas. In the field of manufacturing mobile robots are used to transport workpieces and tools. In some cases such robots load machine tools and perform assembly tasks. There are examples of travelling assembly robots [1] which are guided along inductive or marked paths [2,3]. Most of todays mobile robots are mounted on a platform and have three or four wheels. Hovering and carterpillar equipped platforms are also under development.

A special class whithin the family of mobile machines are the walking machines. They can operate in difficult terrains. Systems with one, two, four and even six legs are being studied and realized in research laboratories [4,5,6]. For the control of mobile robots master slave techniques with a human operator as master have been developed [7,8]. Guidance controls using inductive loops, orientation marks or world model oriented maps [9,10] are being conceived and developed. The world models for the navigation use a priori knowledge in combination with sensor information and learning algorithms. Thus the approximation and the refinement of the map is possible. Mobile robots can be subdivided into-semi autonomous and autonomous systems. Some examples of autonomous robots are the Stanford Cart [11], the CMU Rover [10] and the System HILARE (Heuristiques Integriciel et aux Automatismes dans un Robot Evolutif) [12].

At the University of Karlsruhe, the Robotics Research Group is developing an autonomous mobile 2-arm robot. This AI project is based on a dynamic multi-sensor world model. The control architecture of the robot is hierarchically organized. It operates on an internal dynamic world model. Sensor- and control operators are applied to the world model to perform the navigation and/or manipulation in an optimal way. A network of system modules performs operations like interpretation of situation and logic inference for task decomposition. Other expert functions coordinate the two arms or support the pilot to perform the navigation. The expert modules exchange data among each other. They cooperate to execute a sequence of subtasks, considering the actual situation of the robot world and the goal to be reached. Each module performs a dedicated task, like sensor processing, navigation, axis control, decision making or execution of operations. The hierarchical control principle is based on a method which subdivides complex problems in an ordered number of solvable subproblems [13-16]. The output of the hierarchical control is an executable action sequence which operates the robot in real time. In the following the structure of the autonomous mobile robot is presented.

2. Structure of the Autonomous Mobile Robot System

The mobile robot performs two basic operations:
- The navigation of the platform between different working places, positioning of the manipulators in the work areas (dynamic docking)
- Manipulation of workpieces (assembly), transport of tools and workpieces

The task of navigation of the platform includes route planning, it's execution and it's supervision. It is subdivided into a hierarchical structure of three tiers. The levels are as follows, (Fig. 1):

1. The path planning level contains of a global planning module operating on a 3-dimensional map of the laboratory [19].
   The output is a collision free macro path

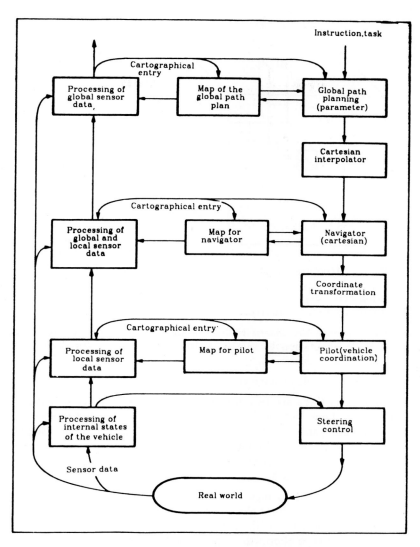

Fig. 1: Control of the mobile platform

A cube based presentation of the free space is proposed to solve the problem. This method considers three types of obstacles/objects: They can be classified as follows:

Stationary, but static obstacles (e.g: wall, building)
Stationary, but dynamic obstacles (e.g: door)
Moving obstacles/objects (e.g: mobile robot, human)

The presentation of free space is started with the presentation of geometric data of the stationary obstacles. Such geometric data could be accessed from a CAD system. The data structure is mapped by a suitable algorithm into a cube structure. Thereby a stationary object is approximated by cubes (Fig. 2). A global cube represents the universe of the robot environment. The intersection between the global cube and the

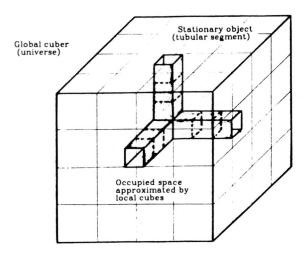

Fig. 2: Structure of cube based presentation of free space

approximation cubes depicts the free space of the robot environment. From this cube list the actually route list is derived.

2. The navigator refines the macro path with the aid of locally aquired sensor data. The result is an adaptation to the local situations and thus the docking process is supported

3. The pilot executes the elementary control operations.

For simple navigation tasks there are two dimensional maps [20, 21] which contain information about obstacle edges, lines, path width as well as path branches and intersections. They aid the generation of linear and circular path and curves of higher order. Because of the dynamic docking maneuver at the worktable a 3-dimensional map consisting of cubic volume elements is used. The path planner checks the volumes which are occupied by obstacles and generates the collision free macro path.

The navigator has the macro path as input. Recognized obstacles cause a modification of the path to avoid collision or a request for a change of the global path. Occupied volumes initiate a path correction to adapt the path to the actual situation. The refined and corrected path which is based on circular interpolation will be executed by the pilot.

Problems may arise in the case of dynamic docking maneuvers, where no fixturing devices like positioning dowels (cones) are provided to hold the vehicle in a known position. In this case the exact relative position between the platform frame and the worktable frame must be known. This can be done by guiding the gripper with a vision system to a set of fixed known points. The robot coordinates are then used to calculate the frame relationships.

The manipulator control system is also hierarchically organized, (Fig. 3). Each arm has a local controller with local trajectory generation capabilities and local sensor processing functions. The two arms are supervised by a higher level two arm coordinator which plans the parallel arm movement [22, 23]. A collision free work space for the parallel motion of both arm is calculated on this coordination level. The input of this level is a command sequence which defines and parameterizes the manipulator trajectories. Vehicle, arm and sensor control is of hierarchical architecture and is integrated into a corresponding control architecture.

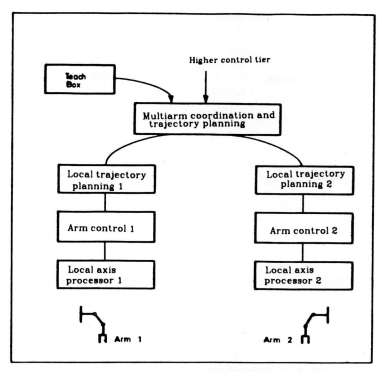

Fig. 3: Schema of a two arm robot control system

3. Hierarchical Control Architecture

The robot architecture is defined by the operation principle of the control modules. The operation principle describes the functional behaviour of the architecture, defining an information structure and a control structure. The information structure of the robot architecture is determined by the information processing components whithin the robot. Whereas the control structure of robot architecture is configured according to the structure of the algorithms for the interpretation and transformation of the information components of the robot.

The mobile robot control architecture is composed of three basic types of modules:

- The dynamic world processor which operates on the real world model containing the actual states of the robot environment (map, frame oriented object description)
- The planning processor which decomposes the task sequences and supervises execution
- The sensor processor which aquires and processes the real world information.

These basic types of modules are horizontally and vertically interconnected. A horizontal layer represents a control tier, whereas a vertical layer presents the interrelation between different control tiers.

The real world model has information about the environment of the mobile robot as reference data. Further it contains the internal states of the robot and reference data about the operating goals to be reached. Operators are applied to the world model which interprete sensor and reference data and which analyze possible conflicts. Thereby expected situations are compared with actual situations and divergencies are determined. The planner has to generate an action sequence and to supervise its execution according the given input tasks or commands. The task sequences (task decomposition) are determined with the help of decision algorithms and tables or problem solving techniques, using the analysed world state data as reference input. The execution of the resulting task sequence is done either on the local level or on the next lower control tier. According to the task sequence a sensor plan is generated as input to the sensor processor to control the task execution (sensor hypothesis). The control plan is hierarchically generated. Similar to processing a decision tree the highest level is considered first then the intermediate level and then the lower one. Thereby the degree of abstraction decreases with each lower level until only atomic functions are executed. The sensor hierarchy has an increasing degree of abstraction at the higher levels for both the pattern recognition operations and classification tasks. The sensor processors cooperate with other types of processor of the hierarchy to perform syntactic and semantic analyses.

4. Autonomous Decision Making, Planning and Execution of Tasks

Classical control systems for industrial robots are designed for repetitive operations, assuming only small changes in the robot environment. Open and closed loop control methods as well as adaptive sensor guided control strategies are used to optimize the robot motions according to a quality index or an other defined goal. The operation plan is usually defined by a linear program with a low number of branches and is executed cyclically. In the case of a mobile autonomous robot the world is only partly known a priori and characterized by time variant change. The planning and execution of the navigation and manipulation makes the interpretation of the actual situation and the analysis of errors and emergency situations necessary. With this knowledge the operation plan can be adapted and optimized in regard to the actual situation to reach the given goal. Let us assume that two robot arms have the task to connect two workpieces with each other. The right arm tries to orient part one at a fixed location and the left arm tries to connect part two according given rules. Now if one of the workpieces is wrongly oriented it has to be reoriented before assembly. Another situation is given when the workpiece is slipping between the fingers or falling out of the gripper. Then the reaction to this situation is quite different to the one mentioned before. It makes the generation of a new action sequence necessary. In general the problem to be solved can be decomposed into smaller ones which can be processed on a lower planner level. In order to the situation the distance between the actual state and the goal or final state has to be evaluated. The processor which tries to solve this problem makes use of the actual world model and its situation interpreter. Knowing the scenario the planner can generate an operation plan consisting of a chain of transition operators (T-operators). The T-operator maps the actual states via intermediate states (expected states) into the final states. With the information of the expected intermediate and final states a sensor plan is generated which allows to predict that the states will be reached by the robot. After the generation of the sensor plan, the operation sequence can be executed and supervised. The evaluation of the performance of the intermediate and final states and the detection and analysis of errors allows a change or an adaptation of the operation plan. Thus substituting the old T-operators for new modified chain of T-operations. In order to generate the T-operator chain there must be knowledge available about a set of rules for goal oriented behavior, alternative strategies and ways of handling missing a priori knowledge. The overall T-operators form a so called T-space which represents the problem and search space of the planner. Other expert functions perform situation interpretation and conflictanalysis, referencing the dynamic world model, (Fig. 4). The dynamic world model can be implemented as a blackboard. The sensor processor, the situation interpreter and the planner have access to the blackboard via read and/or write functions. Each level of the control hierarchy has its own blackboard. Pairs of consumer and producer, such as the axis control and joint sensors, navigator and approach sensors, gripper control and tactile sensors share dedicated areas on the blackboard. The blackboard is managed by its own processor (monitor). The interpretation of the actual situation is done by the access of the blackboard by the analysis and interpretation experts. Expected results calculated by the T-operators are also written on the blackboard. The blackboard represents in the overall system an own autonomous process.

5. The Karlsruhe Mobile Robot System

The mobile robot system being development at the University of Karlsruhe consists of an electrically driven platform, equipped with two PUMA 260 and a sensor platform with two degrees of freedom. The drive system to navigate the platform has three degrees of freedom. The robot will be used for assembly operations in the robot laboratory and will work at locally distributed worktables. A benchmark from the Cranfield Institute of Technology [24] is used for this purpose. The purpose of building the system and performing the experiment is to study and develop AI tools for autonomous robots. The individual subjects under investigation are:

- Real world modelling
- Conception of decision strategies and action planning
- Sensor processing
- Collision free navigation
- Assembly
- Processing of knowledge

The design of the hardware is done under the following aspects:

- Conception of a flexible and modular structure for the mechanical, sensor and control subsystems as an online part for the autonomous operation, located on the platform
- An off-line control part for the strategical computational operations and planning tasks (host computer with telemetric interface)
- Providence of the system with multisensor capabilities (supplementing or redundant)

Fig. 5 shows a schema of the system architecture of the mobile robot system. The platform contains the controls for the wheel drives, the steering unit and the axis controls for the two PUMA 260. The operators in the blocks of this schema are named according Albus's notation [25]. They are implemented on several 68000 microprocessors, which are interconnected via a VMS bus. The use of the IEEE 896 backplane bus is under discussion. The coupling of the

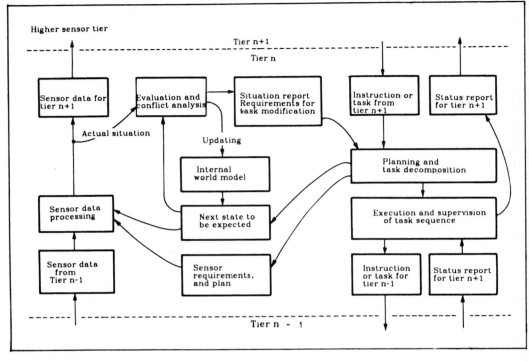

Fig. 4: Elements of model-based control tier

host computer (a MICROVAX II) is performed via a telemetry channel. The vision system on the vehicle will be using an own video channel to interface with the image processing system of the host computer. Processing of other sensor data from ultrasonic sensors, proximity sensors, tactile sensors, and gripper sensors is done on board. Three control tiers can be distinguished on the vehicle.

Level 1, primitive control level for the drives of the platform and the arms

Level 2, navigation and manipulator control in joint specific coordinates and supervision of gripper and sensor

Level 3, navigation and manipulator in local cartesion coordinates and local collision avoidance

The additional levels 4-6 have supervisory function they are located in the host computer. A communication processor performs the communication via telemetric channels. The communication between the processors is based on a decentral asynchronous mechanisms for inter-taskcommunication [26].

6. Expert Functions of the Autonomous Mobile Robot

To make autonomous decisions and to obtain a goal oriented behavior of the mobile robot makes necessary the use of expert functions.

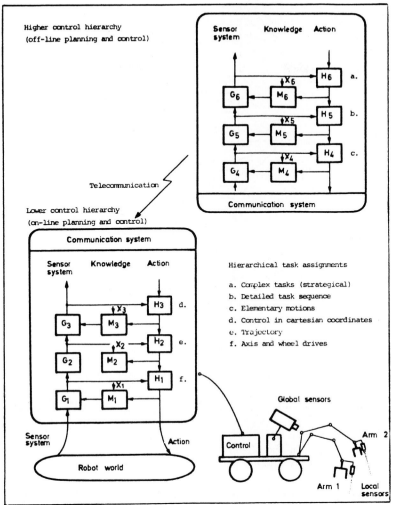

Fig. 5: Hierarchical control concept of the Karlsruhe mobile robot

The expert functions support planning, decision making and analysis tasks by using rules and knowledge about the process. Typical expert functions for autonomous robots are the interpretation of the situation and conflict analysis to support planning of the navigation and manipulation. Knowledge about geometry, technology, context, expected states, and goal states (Fig. 6) has to be made available for the expert functions. Decision rules, goal oriented rules for the system behavior and rules for the task decomposition (T- operators) will support the problem solving. The actual information from the robot world is aquired by the sensor expert functions. They support feature extraction and classification tasks with the aid of sensor data. The context oriented analysis of the actual situation is a basic expert task. In addition the decomposition of sensors and the proper use of pattern matching algorithms has to be performed. Problem solving strategies can be formalized using search trees [27]. "Breadth-first search" and alternatively "depth-first search" are the strategies for finding a

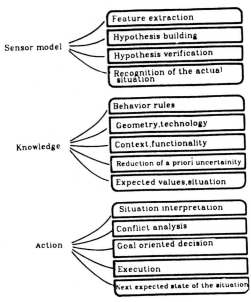

Fig. 6: Conception of the sensor model, Knowledge presentation and Action planning to control the autonomous robot

solution. CMAC state space and decision tables [25] will be considered for realizing the problem and solution space. The development of the expert functions and their connection with the control and sensor processors and the internal world model is supported by an emulation and simulation system. The overall emulated robot control functions form a virtual robot which is an ideal robot model. Single system functions and complex networks of the functions can be tested using the simulator, before testing them with the real experiment, (Fig. 7). For this purpose a simulation system (ROSI 2 [28]) is being developed at the Robotics Institute of the University of Karlsruhe. In the off-line programming phase (implicit programming) the user defines first the problem. A planner generates then the operation sequence using application and system specific knowledge and rules. An essential aspect of this project is to build a shell, to study knowledge aquisition and to produce rules for the autonomous robot. The on-line expert functions for the robot control are developed under real time aspects. Fig. 8 shows model versions of the mobile robot.

7. Implicit or Task Oriented Programming of the Robot Arms

At the University of Karlsruhe a task oriented programming system is being developed for the autonomous robot. The user communicates with the robot by speech or formalized input in a task oriented mode. The comprehensive expert system will contain the following knowledge.

1. How to interprete the task oriented instructions and how to convert these into an explicit program
2. How to describe to the robot, its own physical capabilities and limitations
3. How to describe to the robot, the workpiece, its workplace and tools

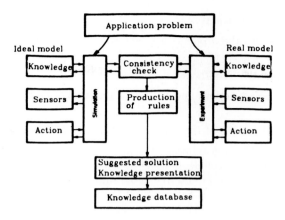

Fig. 7: Schema for the Knowledge acquisition and generation of production rules for the autonomous robot

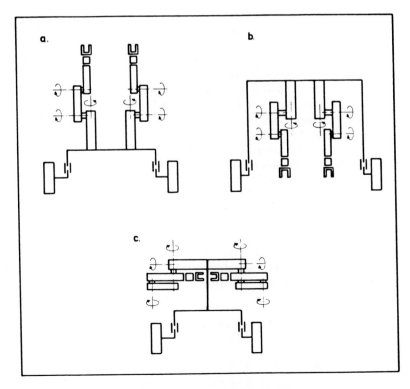

Fig. 8: Different model variants of the mobile robot

4. How to plan assembly operations
5. How to cooperate with other robots, machine tools and material movement equipment
6. How to recognize and solve conflicts
7. How to observe the robot actions and change of the workpiece by using different sensors and vision
8. How to coordinate all actions of the system and the use of the knowledge bases

The concept of the task oriented programming system is shown in Fig. 9 [29]. It consists of eight basic parts, viz., the planning module, a CAD data base, a world model, the SRL compiler and test aids, the simulator, the interactive programming and teach-in module, the interpreter, and the monitor. The system is activated by the task oriented instructions and generates interpretative code for the robot. Knowledge about the workpiece is obtained from a CAD database and knowledge about the robot and its workplace from the world model. Interpretation of the motion instructions is supervised by the monitor with the help of online sensor information. This monitor also performs corrective actions in case of conflicts. The different modules outlined above are discussed in more detail in the following paragraphs.

7.1 The CAD Module

The graphical information about the workpiece, the workstation and the tools and fixtures is entered into the CAD module. This information is accessed by the planning module to identify the workpiece, to locate a suitable grip position, and to plan the assembly. The workpiece must be stored in the database by a solid model presentation which can be easily interpreted by the planner. It is necessary to store only that information which adequately represents all objects of the workplace and which is needed to plan the robot motion and the assembly operations. Furthermore the workpiece representation should be such that it can be accessed by the vision system of the robot in order to identify objects in the viewed scene. It is the task of the designer to render all information needed of the workpiece to the CAD module. For this reason he must be familiar with the assembly operations, the capabilities of the robot and effector and the fixturing requirements. The designer has to implement an object library for all standard parts, robots, work stations, auxiliary components and fixtures. Standard information can be retrieved from this object library.

7.2 The World Model

The structural world model contains the extensions of the CAD data bank and its contents come from two sources. First, the user describes to the world model textually or interactively the robot world. The following information is required:

- The objects of the work cell and their geometric relations
- The geometric relations of the gripper with regard to the workpiece, including stable positions of the workpiece and suitable grip positions
- Technical information about the workpiece, such as weight, mass and tolerances
- The interaction of tolerances of workpiece components during mating operations

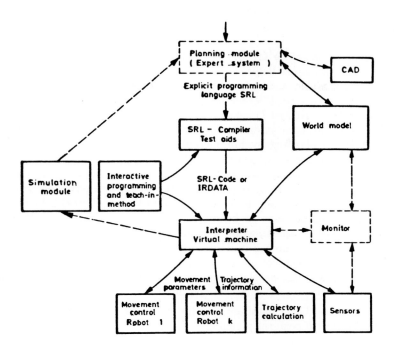

Fig. 9: Hierarchical task oriented programming structure for industrial robots

- Assembly parameters, such as preferred approach and deproach directions, compliance routines, pressfits, torques and forces
- Relations between the manipulator, gripper, auxiliary components, feeders, conveyor belts and fixtures
- The sensor hypotheses for the different assembly operations
- Assembly primitives and standard assembly routines

The objects and their relations are described in the database by a tree structure, whereby the objects are the nodes and the relations are the edges.

The second source of information is entered online by the sensor system of the robot. The sensors constantly observe the robot actions and report the prevailing scenario to the dynamic world model which is part of the world model. For example, new objects may have entered the workplace or a workpiece may have changed its location and orientation. The dynamic world model processes the following information.

- Evaluation of the online sensor information
- Detection of changes of the real world
- Correction of the world model

7.3 The Planning Module

This module is the planning interface for the programmer with the implicit programming system. Two types of information are entered via the interactive programming module. First the description of the robot world and second the description of the robot task. These two different programming tasks require different programming modules. The instructions may be entered by natural or task oriented language. The input instructions are analyzed by an expert system which extracts all basic information needed for describing the robot assignment and the work place. A global planning module consults an expert system and under the assumption of a virtual assembly system selects the robot configuration, the robot arrangement, the assembly line, special work stations, the gripper, tools, fixtures and auxiliary equipment. The other task is planning of the assembly sequence and the coarse motions of the robot. This requires a collision check for all arm and part movements in the robot world. Possible collision between cooperating arms has also to be investigated. The next stage is fine motion planning by an expert module for the assembly operations. In this phase, bolting, insertion, bonding and mating operation are planned in detail. Since the assembly is done under sensor supervision, the different sensor configurations are determined by an expert module which selects the sensor strategies. During the planning phase, the error recovery routines have to be chosen which help the system to solve conflicts and unforeseen assembly problems.

This module has access to the CAD database and the world model. It extracts from these sources all information needed to perform planning.

The reconfiguration of the assembly system or entry of new parameters is done via the interactive programming module.

The output of the planning module forms the code of the structured explicit robot programming language SRL. This code is entered into the SRL compiler. Fig. 10 shows the software levels and interfaces for a robot programming system [29].

7.4 SRL Language Module

The explicit programming language SRL is a powerful tool with general and flexible facilities. Its self-explanatory features, readability, and documentation capabilities are very important and help the user to understand the generated programs and to modify the program if necessary.

The SRL system has been developed at the University of Karlsruhe based on experience with the AL language from the Stanford University and PASCAL. The main features of SRL are:

1. The conception of structured data types was taken from PASCAL. This allows very flexible adoption of the user defined objects.

2. Integration of the predefined geometric data types, e.g., vector, rotation, and frame into structured data types. Positions and orientations of move points or objects can be described off-line.

3. With the help of arithmetic and geometric operations the user can formulate his geometric calculations, e.g. an orientation of a gripper perpendicular to the curved surface of an object.

4. Block structure for a limited scope of variables. This is helpful to reduce memory allocation and to produce different program parts by different programmers.

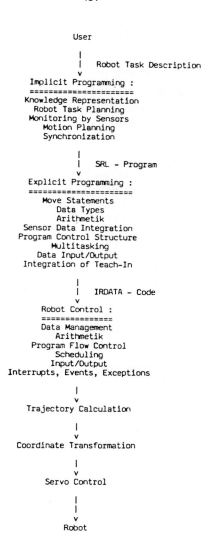

Fig. 10: Software-levels and interfaces for robot programming

5. Syntactical compound statement to handle a statement sequence as one syntactical unit.

6. Multitasking for parallel, cyclic, or time delayed execution of program parts. Tasks (called 'sections') can have parameters with a well defined and limited scope, like variables; this supports a clear program structure. Parallel task execution is possible on one processor or a microcomputer for the robot independent part of the IRDATA interpreter. The task will be executed in a pseudo - parallel (e.g. calculations) and an actual - parallel mode (e.g. move control of two robots or robot and sensor control). This is useful for sensor integration, background calculation or input/output, multirobot control and others.

7. Features for interrupt and condition monitoring allow to stop a robot move if a sensor value limit is exceeded.
8. Input/output to digital or analog ports and sensors, terminals, printers, and other peripherals.
9. For adapting SRL programs to different hardware facilities and for self-documentation a system specification part can precede a SRL program. System components like robots, sensors, effectors, ports, interrupts, or files can be described.
10. Several move statements can describe different kinds of interpolation of robot moves. The user can specify move parameters like velocity, duration, acceleration, precision, intermediate position of a trajectory, forces, approach/deproach frames and others. Several single move statements can be combined to a complicated trajectory by a move specification statement. The robot control must include the needed interpolation modules for calculating the robot angles. A special instruction starts the move execution.
11. The effector statements allow the control of electric or pneumatic grippers and of different tools. In addition the programmer can specify parameters like velocity or grip-force.
12. All features of modern program flow control are included, e.g. conditional statement, loops, CASE-distributor, and procedures. To stop the execution of a task, a procedure or a loop etc. the EXIT statements is defined.

There are other features like synchronization or move stop, which are not listed above. The SRL-language is defined orthogonally; a language construct can be placed into any other construct if the result is of the expected type. The definition of the SRL language is done by extended BNF for the syntax and by the ALADIN definition language for semantics. The formal SRL definition is processed by the GAG - system (Compiler Generator Based on Attributed Grammars). Therefore the syntax and semantic definitions are complete and consistent.

The SRL language has a very advanced concept and should be considered as a standard for a general higher programming language for robots. A sample program is shown in Fig. 11 [29].

The output of the SRL compiler is the standardized IRDATA code. It can be used on any robot which has a IRDATA compatible interface, Fig. 12 [30].

7.5 The Software Interface IRDATA

There are many different robot manufacturers, each of them using his own programming system with a special robot programming language. These languages differ in various aspects concerning syntax, program structure and features. This fact results in the following disadvantages to the user of different robots:

1. Every robot system used in a facility needs an especially trained programmer.
2. The process of transferring an existing robot program to another system is equivalent to implementing the robot task a second time, even if the target robot has similar kinematic features.
3. There is no possibility to transfer taught locations from one robot control to another one.

```
            U N I V E R S I T Y    O F    K A R L S R U H E
**********************************************************************
*                                                                    *
*                              S R L                                 *
*                                                                    *
* PROTOCOL of the SRL-COMPILER                                       *
*                                                                    *
*                                                                    *
* DATE :    4 Apr 85                                                 *
*                                                                    *
* NUMBER of MESSAGES:                                                *
*                                                                    *
*      INFORMATIONS   :   0 (I)                                      *
*      ERRORS         :   0 (E)                                      *
*      SYSTEM ERRORS  :   0 (S)                                      *
*      SYSTEM LIMITS  :   0 (L)                                      *
*                                                                    *
*                                                                    *
**********************************************************************

PROGRAM sample(INFILE,OUTFILE);
VAR magazin,waitposition :    FRAME;
     part_angle:              ROTATION;
     cyclenumber,i:           INTEGER;
     YAXIS,obstacle,save_pos: VECTOR;

BEGIN_PROGRAM
  YAXIS:= VECTORC(0,1,0);
  save_pos:= VECTORC(10,0,75);
  part_angle:= ROTC(YAXIS,130);
  waitposition:= FRAMEC(VECTORC(30,-20,50),part_angle);
  WRITELN(' Give cycle number :');
  READLN(cyclenumber);
  IF ((cyclenumber >= 1) AND ( cyclenumber <= 10))
   THEN
     FOR i:=1 TO cyclenumber DO
       PTPMOVE ARM(1) TO waitposition;
       CLOSE GRIPPER(1);
       SMOVE ARM(1) TO magazin
         WITH VIAFRAMES (obstacle,save_pos)
         WITH VELOCITY=0.50
         WITH ACCELERATION=0.33;
       OPEN GRIPPER(1);
       WRITELN('Part ',i:2);
     END_FOR
   ELSE WRITELN('Wrong part number !') END_IF
END_PROGRAM.
```

```
    U N I V E R S I T Y    O F    K A R L S R U H E

SRL - Codegenerator for IRDATA :

1,PBEG,2,0,36,sample,2,INFILE,OUTFILE;
2,BLBEG,0,1,1,1,FRA,1,VEC,1,VEC,1,VEC,1,1,1,1,1,ROT,1,FRA,1,FRA;
3,GEN,VEC,%ST,#,D,0,#,D,1,#,D,0;
4,MOVDAT,%BR,0,2,VEC,%ST;
5,GEN,VEC,%ST,#,D,10,#,D,0,#,D,75;
6,MOVDAT,%BR,0,4,VEC,%ST;
7,GEN,ROT,%ST,%BR,0,2,#,D,130;
8,MOVDAT,%BR,0,7,ROT,%ST;
9,MUL,I,%ST,I,#,D,-1,I,#,D,20;
10,TYPCON,R,%ST,I,%ST;
11,GEN,VEC,%ST,#,D,30,%ST,#,D,50;
12,GEN,FRA,%ST,%ST,%BR,0,7;
13,MOVDAT,%BR,0,9,ROT,%ST;
14,DATOUT,0,STR," Give cycle number :";
15,DATIN,0,1,%BR,0,5;
16,GE,B,%ST,I,#,D,1,I,%BR,0,5;
17,LE,B,%ST,I,#,D,10,I,%BR,0,5;
18,AND,B,%ST,B,%ST,B,%ST;
19,IF,%ST,34,0;
20,MOVDAT,%BR,0,6,I,#,D,1;
21,FOR,0,6,#,D,1,I,%BR,0,5,33,0;
22,FEDRAT,1,R,#,D,1;
23,ACCEL,1,R,#,D,1;
24,MOVE,V,A,1,%BR,0,9;
25,CLOSE,#,D,1;
26,FEDRAT,1,R,#,D,5.00000000000000e-01;
27,ACCEL,1,R,#,D,3.30000000000000e-01;
28,GEN,FRA,%ST,%BR,0,1,FRAORI,%BR,0,3;
29,GEN,FRA,%ST,%BR,0,1,FRAORI,%BR,0,4;
30,MOVE,L,V,A,3,%ST,%ST,%BR,0,8;
31,OPEN,#,D,1;
32,DATOUT,0,STR,"Part ";
33,DATOUT,0,I,%BR,0,6;
34,FOREND,21;
35,GOTO,35,0;
36,DATOUT,0,STR,"Wrong part number !";
37,BLEND;
38,PEND;
```

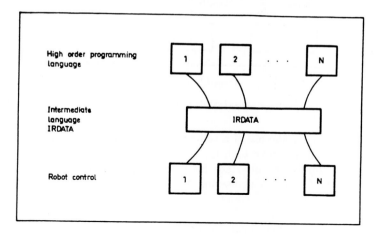

Fig. 11: IRDATA generated by the SRL-Compiler listing of a sample SRL-program

Fig. 12: The IRDATA code

A solution to this problem can be obtained by standardizing the interface between the robot programming system and the robot control system as shown in Fig. 13. This would provide the following advantages:

1. Programs written on a special programming system can be used to control robots of different types.
2. A robot might receive programs or data from any programming system.
3. An expert for programming of a special robot is also a general expert for robot programming.

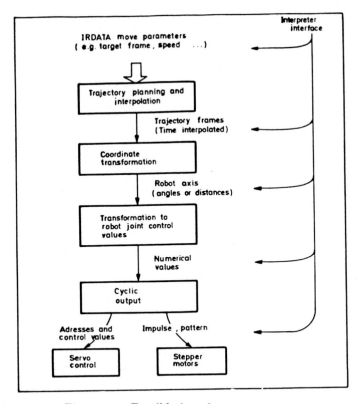

Fig. 13: Possible interfaces to underlying robot control

The resulting flexibility in the use of a robot system and the possibility to reconfigure a robot cell in an easy way will improve both the efficiency and the performance of a robot installation. In order to simplify the interface between the programming system and the robot, a working group of the German Engineering Assocoation VDI, has been formed with an object of specifying a common national standard. It is called IRDATA and is based on the CLDATA standard used for controlling NC-machines. Since the design of IRDATA has not yet been concluded, only the salient characteristics of the standard are discussed in the following paragraphs:

An IRDATA program is a sequence of records which contain a certain number of arguments. The records are classified according to the special requirement of robot programming. Some typical classifications are as follows:

1. The description of
    a) Robots
    b) Tools
    c) Sensors
    d) Working space
    e) Frame lists

2. Motion specification and execution
3. Arithmetic operations
4. Program flow control
5. Input/output operations

To achieve flexibility, the arguments may either be constants or variables. The following data types are provided:

1. Boolean
2. Integer
3. Real
4. Vector (three real numbers for x, y and z-coordinates)
5. Orientation (three real numbers representing angles)
6. World (a vector and an orientation)
7. Joint (a set of real numbers representing robot angles)
8. Character
9. String (a set of character)
10. Pointer

A type conversion between several different types is provided and a large set of standard operations has been defined to handle boolean, arithmetic and geometric calculations. This allows an universal description of the robot position and movement for different types of robots.

The output of the IRDATA interface is a robot dependent code, and is interpreted by the interpreter located in the robot controller.

7.6 The Monitor

The monitor supervises the operation of the robot. Each line of IRDATA code is scanned by the monitor to observe inconsistencies with the actual world. The sensor system furnishes an image of the actual world. This image is compared with the expected scenario described by the program. If a deviation is detected the monitor attempts to plan a corrective action with the aid of a local expert system. This expert system has access to the local parameterized world model which was previously down loaded by the planning system. This system also provides limited error recognition and recovery capabilities. Only common errors are corrected. If an abnormal situation appears, it recognizes and reports as a problem and the assembly terminates. In case the robot has an umbilical cord with the master computer, the planning system tries to correct the problem at a higher level. A more detailed description of the monitor will be given later.

7.7 The Interactive Programming and Teach-in Module.

This module is an interface with the programmer to communicate with the programming system. In principle it has three functions.

1. The system modules can be activated or deactivated and the operating parameters can be entered and changed.

2. It allows the programmer to interactively program the robot motions. Operator guidance is provided by a question and answer dialog. This greatly facilitate programming. It is also possible to invoke the simulation system to check the performance of the entire assembly system or of subsystems.
3. It serves as an interface to the teach-in module. The latter is needed to teach-in the important operating parameters for the assembly, e.g., the exact location of a bolt hole in the three-dimensional assembly space and the position and orientation of the gripper to insert the bolt. For these locations, the SRL program provides non-parameterized variables, which are parameterized during teach-in.

7.8 The Interpreter

The interpreter receives the IRDATA code, interprets it line by line and converts it to robot control commands. An additional function of the interpreter is the trajectory calculation and the coordinate transformation. It also has to schedule the different tasks, process interrupts and control the input and output of performance data.

7.9 The Simulation System

To facilitate the development of programs for the operation of an assembly call, a comfortable programming system is provided. It can simulate the operation of the entire assembly cell or a small component of it, such as the work of a gripper. The simulation program can be invoked either through code generated by the SRL compiler or the interpreter. The principle of the simulation system is shown schematically in Fig. 14.

1. The modeler is used to describe the manufacturing cell and the piecepart including the geometric, physical and functional properties. The CAD-system ROMULUS is used as a tool for geometric modeling of the robots, tools, workpieces and sensors. During future development stages, the generated CAD-data structures may be extended with additional non-geometric and dynamic attributes of the robots, conveyors and fixtures. Functional properties of sensors, axis control loops and relations between objects within the cell can also be defined.
2. The emulator contains the methods and functions which are necessary to plan and emulate the behavior of the robot within a cell and to define its operation. It is also used to build a virtual cell controller to be used for the simulation. Emulated functions are trajectory planning, coordinate transformation, sensor operation as well as control- and decision functions. The virtual cell controller executes the programs defined by the programming module.
3. The programming module permits the programmer to develop programs by different modes. The dialog allows interactive textual and/or graphic specification of program statements. An experimental mode is used for the development and testing of individual trajectory sequences and of basic operations. It is also possible to combine these functions in a robot program. The purpose of this mode is to provide a user-friendly problem-oriented program interface.
4. The simulator is used for the validation and verification of the generated program. The execution of the manufacturing program to the virual cell is displayed on the

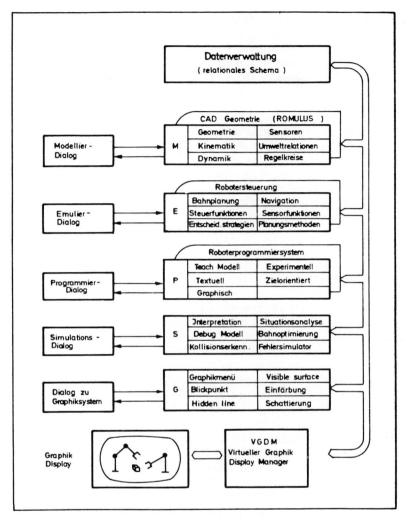

Fig. 14: Structure of an interactive programming system for manufacturing cells

graphic screen. A performance index (criterion) and a number of analysis functions, such as collision check, error recovery procedure, tolerance analysis etc.. allow the programmer to analyze and validate the program. If the simulator detects errors, sources of errors or critical states are displayed as graphic or alphanumeric information on the screen.

The graphic module facilitates the interpretation of the simulated operation. It offers methods, to enhance the information on the screen with the aid of graphic animation subroutines, like hidden lines, visible surfaces, coloring, shading, zooming, view point transformations etc..

6. The interface between the simulator and the graphic system is realised via a VGDM (Virtual Graphic Display Manager). The VGDM makes the system independant of the type of the graphic workstation used. It offers the simulator pseudo-graphic instructions for the construction and manipulation of graphical pictures. With these commands each module can change the structure of the graphic image. Workstation dependant software drivers adapt the VGDM pseudo-display file to a workstation specific control structure.

Each module has access to the central data base. Furthermore, the methods and data are clearly separated to assure consistency of the data to be processed. The data management is based on a relational data model, where objects of the same type are grouped together to object classes. Each object has a structure which is described by attributes. Relations between the objects are indicated by attributes, the value of which represent object names. An example of a virtual robot is shown in Fig. 15.

8. Computer Architecture

A polyprocessor architecture for the control of the intelligent multiarm robot system is being developed, (Fig. 16) [26]. The concept is based on hierarchical control principles. It allows to control different robot types, intelligent sensors and peripherals. The architecture is reconfigurable depending on the robot application. The overall task of the robot will be analyzed by a supervisory computer and decomposed to subtasks. These subtasks will be distributed for execution among the nodes of the polyprocessor. Execution of the individual subtasks is data flow driven. There are dedicated processors for the numerous sensor and control functions and for emergency detection and analysis. Decision making is done with the aid of an intelligent data base system. It helps to perform realtime adaptive control. Several features such as extendability, fault tolerance, autonomous consumer producer relations, distributed operating system functions and function redundancy are being implemented. The interfaces to being IRDATA and SRL allow the use of a high robot programming language. Additional interfaces to a relational robot data base and a graphical simulation system are considered. This will allow to integrate the control system into the robotic software environment. It is intended to extend the polyprocessor system to general purpose robot applications. This makes necessary the definition of general language elements for parallel processing and for constructs to aid programming. Deadlock free processing is assured by the layout of a consumer producer concept during the time of process generation. A prototype of the polyprocessor is being developed with available processors such as the 68000 VME, VAX module and the IEEE P-896 Backplane Bus. The use of UNIX in local nodes is under discussion. To embed PROLOG for programming of the intelligent robot control functions are a second research topic.

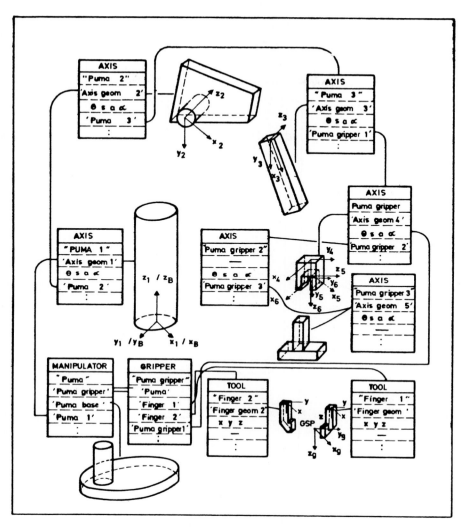

Fig. 15: Modelling and composition of basic arm elements

9. Conclusion

In this paper a project was presented in which an autonomous robot will be built. The robot will be capable to assemble with its two arms under sensor guidance simple workpieces. To make the robot mobile it will be equipped with a selfpropelled platform. Components for this machine have been prepared for several years and they are now being integrated to a selfcontained system. Expert systems are being prepared for navigation, assembly planning and supervision, implicit programming, obstacle avoidance and error recovery. The work is being supported by the Deutsche Forschungsgemeinschaft.

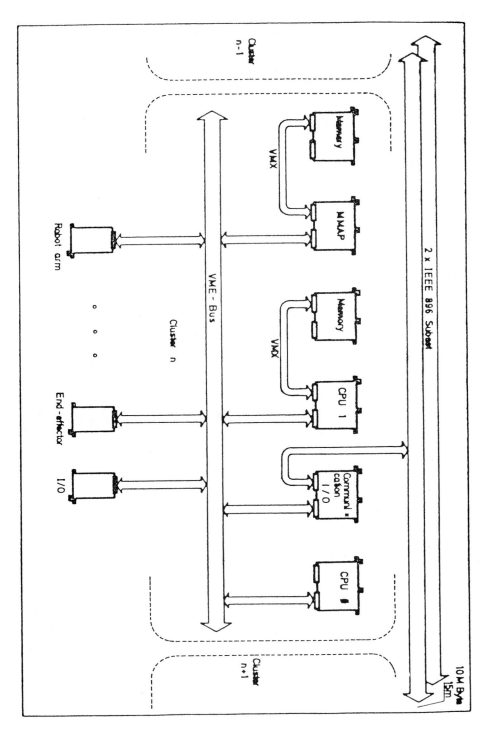

Fig. 16   Hardware architecture of a cluster (node) of the Karlsruhe polyprocessor

10. Bibliography

[1] Hirabayashi, H., Hamada, T., Akaiwa, M., Kikuchi, K., "Travelling Assembly Robot", 13, ISIR-Robot 7, Chicago, April 1983, Proceedings S. 20, 19-20.31.
[2] Kono, H., "Moving Robot Applied Mark Tracing", 9. IFAC World Congress, Budapest, July 1984, Proceedings (Preprints, Band VI, S.33-38).
[3] Warnecke, H.-J., Schuler, J., "Flexible Mehrstellenhandhabung mit mobilem Industrieroboter", Robotersysteme, Nr. 1, April 1985.
[4] Raibert, M.H., Brown, H.B., Chepponis, M., "Experiments in Balance with 3-D One-Legged Hopping Machine", Robotics Research, Vol. 2, N. 2, 1984, S-75-92.
[5] Miura, H., Shimoyama, I., "Dynamik Walk of a Biped", Robotics Research, Vol. 2, N. 2, 1984, S. 60-74.
[6] Raibert, M.H., Sutherland, I.E., "Machines that Walk", Scientific American, N. 1, 1983, S.44-53.
[7] Dobroting, B.,Lewis, R., "A Practical Manipulator System", Proceedings der IJCAI 77, S.723-732.
[8] Bejczy, A., Brooks, T., "Advanced Control Techniques for Teleoperation in Earth Orbit", Proc. der AUVS-80 Conference, Daytona, Ohio, June 1980.
[9] Nilson, N.J., "A Mobile Automation: An Application of Artificial Intelligence Techniques", Proc. der IJCAI 69, S.509-520.
[10] Elfes, A., Talukdar, S.N., "A Distributed Control System for the CMU Rover", Proc. der IJCAI 83, Karlsruhe, Aug. 1983, S. 830-833.
[11] Moravec, H.P., "Obstacle Avoidance and Navigation in the Real World by a Seeing Robot Rover", Diss. an der Stanford Uni., Sept. 1980, veroffentlicht durch UMI Research Press unter "Rover Visual Navigation", Ann Arbor, Michigan, 1981.
[12] Laumond, J.-P., "Model Structuring and Concept Recognition: Two Aspects of Learning for a Mobile Robot", Proc. der IJCAI 83, Karlsruhe, Aug. 1983, S. 839-841.
[13] Nilson, N.J., "A Hierarchical Robot Planning and Execution System", Artificial Intelligence Center, TN-76, SRI-Int., 1973.
[14] Saridis, G.N., Stephanon, H.E., "Hierarchically Intelligent Control of a Bionic Arm", Proc. der Conf. on Decision and Control, Dec. 1975, Houston, Texas.
[15] Albus, J.S., et al, "Theory and Practice of Hierarchical Control", 23 IEEE, Sept. 13-17, 1981, Washington, D.C.
[16] Shin, K.G., Malin, S.B., "A Hierarchical System Structure for Coordinated Control of Industrial Manipulators", Proc. of the Int'l Conf. on Robotics, Atlanta, March 1984, pp. 609-619.
[17] Keirsey, D.M., Koch, E., McKisson, J., Meystel, A.M., Mitchell, J.S.B., "Algorithm of Navigation for a Mobile Robot", Proc. of the Int'l Conf. on Robotics, Atlanta, March 1984, pp.574-583.
[18] Meystel, A., "Intelligent Control of a Multiactuator System", Proc. of 4th IFAC/IFIP Symposium on Information Control Problems in Manufacturing Technology, Washington, D.C., 1982, pp. 126-133.
[19] Soetadji, T., "Cube Based Presentation of Free Space for the Navigation of an Autonomous Mobile Robot", Internal report of the Inst. of Informatik III, Universitat Karlsruhe, 1985, will be published at a later date.

[20] Tsumura, T., et al, "An Experimental System for Automatic Guidance of Roboted Vehicle Following the Route Stored in Memory, Proc. of the 11th Int'l Symposium on Industrial Robots, Tokyo, Oct. 1981, pp. 187-194.
[21] Siy, P., "Road Map Production System for Intelligent Mobile Robot", Proc. of the Int'l Conf. on Robotics, Atlanta, March 13-15, 1984, pp. 562-570.
[22] Park, W.T., "State-Space Representations for Coordination of Multiple Manipulators", Proc. of the 14th Int'l Symposium on Industrial Robots, Gothenburg, Oct. 1984, pp. 397-405.
[23] Alford, C.O., Belyeu, S.M., "Coordinated Control of Two Robot Arms", Proc. of the Int'l Conf. on Robotics, Atlanta, March 1984, pp. 468-473.
[24] Collings, K.K., Palmer, A.Z., Rathmill, K., "The Development of a European Benchmark for the Comparison of Assembly Robot Programming Systems", Proc. of the 1st Robotics Europe Conf., Brussels, June 27-28, 1984, pp. 187-199.
[25] Albus, J.S., "Brains, Behavior and Robotics", Byte Books, McGraw-Hill, Peterborough, U.S.A. 1982.
[26] Kordecki, C., Dillmann, R., "Conceptual Design of Adaptive Miltiarm Control", Proc. der 1984 ASME Int'l Computer in Engineering Conf., Aug. 12-16, 1984, Las Vegas.
[27] Nilson, N.J., "Principles of Artificial Intelligence", Tioga Pub. Co., 1982.
[28] Dillmann, R., Huck, M., "Ein Softwaresystem zur Simulation von robotergestutzten Fertigungsprozessen", Robotersysteme, Nr. 2, 1985.
[29] Blume, C., and Frommherz, B.J., "The Proposed Robot Software Interface SRL and IRDATA", Workshop on Robot Standard, June 6-7, 1985, Detroit, MI. Sponsored by the U.S. Dept. of Commerce, NBS.
[30] Rembold, U., Blume, C., and Dillmann, R., "Computer Integrated Manufacturing Technology and Systems, Marcel Dekker Inc., New York, 1985.

ROBOTICS RESEARCH AT THE LABORATORY FOR INTELLIGENT SYSTEMS NATIONAL RESEARCH COUNCIL OF CANADA

S. Elgazzar
Division of Electrical Engineering
National Research Council of Canada
Ottawa, Ontario, Canada   K1A 0R8

ABSTRACT

This paper summarizes the research and development activities in the field of intelligent robotics at the Laboratory for Intelligent Systems of the National Research Council of Canada. The Council's objective is to use its long term research to provide a base of knowledge from which to attack short term problems in support of the Canadian industry. The Laboratory's work on three-dimensional vision, sensory based control, multiprocessor system architecture and applications of artificial intelligence is presented.

INTRODUCTION

The NRCC is Canada's largest government research and development organization. The overall government objective is to complement industrial research and to perform R & D that would not otherwise be performed. Survey studies of the assembly operations in Canadian industry revealed that a large proportion is batch intensive, hence any automated system must be flexible and adaptable. The approach taken, therefore, has been to investigate advanced robotics systems that can acquire, interpret and respond to changes in the work environment. Thus, our focus in robotics is on sensor based control of complex realtime systems.

The research activities that are pursued at our laboratory are described in the following.

THREE DIMENSIONAL VISION

a) Laser Range Finder Based on Synchronized Scanners:
    A new optical arrangement was developed to improve performances of optical triangulation. Two scanners in

synchronization allow a linear position sensor to be used for surface topography measurement. Besides a speed of measurement in the megahertz range, the new geometry allows considerable reduction of the optical head size compared with the usual geometries, which reduces the shadow effects proportionally. This geometry is insensitive to light interference (multiple reflection, specular reflection, etc) due to a very small instantaneous field of view. It also provides a mean to obtain very large field of view without compromising on resolution.

Geometrical analysis and experimental results can be found in [11,7]. Figure 1 shows the geometry of the camera using a multifacet pyramidal mirror. Outputs of the 3D camera for two different inputs are shown in fig. 2.

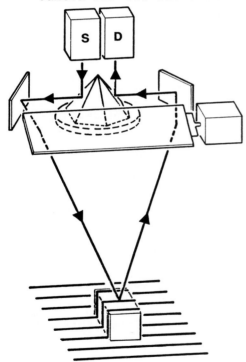

Fig. 1. Autosynchronized geometry using multifacet pyramidal mirror.

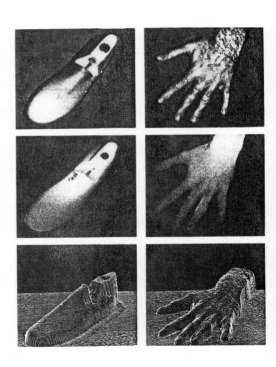

Fig. 2. 3-D camera outputs; top photographs show intensity output, while the four others show the 3-D data.

b) Preprocessing and Recognition of Objects in Range Camera Images:

A scanned 3D scene is represented by an image $Z(X,Y)$ of elevation values Z from the XY-plane as seen from above, fig. 3. The objects in the scene are expected to have smooth surfaces, but otherwise no specific constraints are placed either on the object nor on the contents of the scene. The image is preprocessed for surface normals and curvature at each pixel, the individual surfaces are extracted, and the edges and corners between the surfaces and the borders are detected, fig. 4a and 4b. The output of this stage consists of a stack of registered images, each containing specific information about the scene [6].

These outputs are used in object learning by using the "cross-angle transform" which is invariant to view angle [10]. A description of each 3D object can be formed from multiple views of the same object, such that each surface is seen in the context of its neighbours. Object recognition consists of comparing the preprocessed but unknown scene with descriptions of objects obtained during the learning phase.

Another approach to object recognition uses CAD type models which describe polyhedral objects [1]. The preprocessed results are compared against model descriptions by using Hough transforms on the angle relationships between the surfaces. The best matching CAD models are translated, rotated, scaled and projected onto the image plane for detailed comparisons. A study of object recognition based on edge information and using dynamic programming techniques is underway. Preliminary results can be found in [13].

EXPERIMENTS IN SENSORY BASED ROBOTICS

Experiments with sensory based robotics have been undertaken to better understand their requirements. An important element in these studies has been the investigation of the advantages of parallel processing technology to robotics as provided by the Harmony operating system, which is described in a later section.

Fig. 3. 3-D data to be recognized.

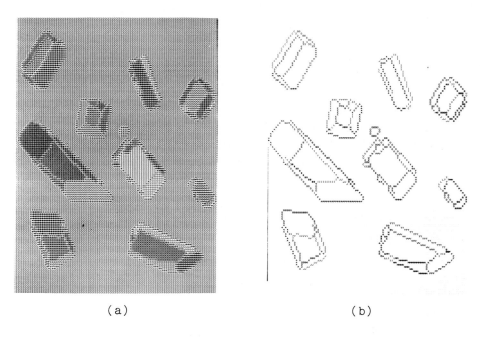

(a)   (b)

Fig. 4 a) Extracted planar surfaces; b) extracted edges and corners

Previous experiments have included a demonstration of a sensor-based robot task using a binary vision system, implemented at our laboratory. The vision task [3] was to locate and recognize regular shaped blocks which are then inserted into matching sockets, fig. 5a. Also experiments with the Instrumented Remote Centre of Compliance (IRCC) of Charles Draper Laboratory were undertaken. The IRCC was used in experiments for edge following and for the insertion of prismatic shaped blocks into their tight fitting sockets, fig. 5b.

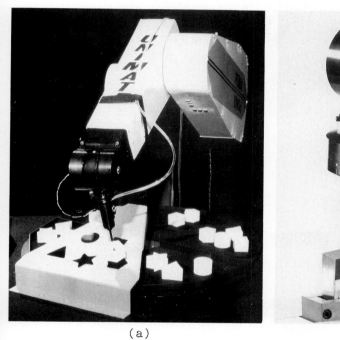

(a) (b)

Fig. 5 a) Experiments with binary vision; b) Experiments with local displacement sensing (IRCC) for the insertion of prismatic shaped block.

The on-going activities include:
a) Design and Implementation of a Profile Scanner with an Integrated Gripper

A compact profile scanner, fig. 6, designed to be mounted on the wrist of a robot, is being built at our laboratory.

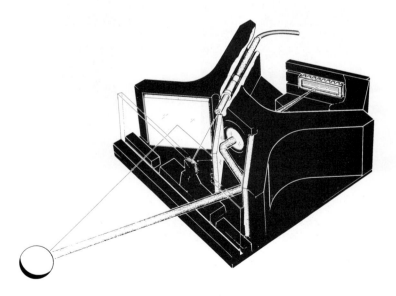

Fig. 6. Profile scanner.

The principle is the same as in the 3D camera. The unit comprises two single profile scanners that generates two parallel beams, one just above and one just below the gripper jaws, fig. 7. Resolution is range dependent. The unit has been designed to provide accurate profile measurements from the grasping range to about 60 cm, but still will provide useful data to locate objects at distances up to one meter. Depth resolution is expected to be better than 0.2% over the design range, and the resolution in the scan angle direction is 256 elements. This unit will deliver range profiles, an example of which is shown in fig.8, at 40 ms intervals. The parallel gripper is driven by a linear actuator and has been designed to minimize the distance between the scanning beams without obstructing the sensor's field of view.

Fig. 7. Two profile scanners with an integrated gripper.

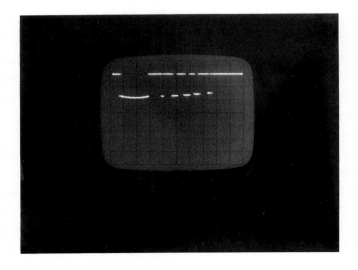

Fig. 6. Range profile for a cup and fingers.

b) Application of the Wrist Mounted Scanner for:
   i) Object recognition and pose determination
      Success will require robust strategies in which a hypothesis is formulated based on a minimum number of profiles. Verification of the hypothesis will commence

from the earliest stages of data acquisition by relating acquired data to models of the objects. This information will be used to direct the scanner to a viewpoint which will contribute most to the validation of the hypothesis. Strategies for this kind of sensor should rely on the surface data more than the edge data since the data is less reliable at the discontinuities.

ii) Bin picking

An approach to the bin picking problem has been tested by simulating the profile scanner using scenes produced by the full range scanner. This approach tries to find good locations, called holdsites, at which to grasp a part by using simple heuristics without actually recognizing individual parts.

All the holdsites that are geometrically acceptable in one plane of the sensor data are calculated using spatial planning methods based on configuration spaces. Both the stability and the safety of a holdsite in the plane are evaluated. The stability is defined as the potential slippage in the holdsite and the safety is the degree to which this holdsite can tolerate translational uncertainty in the part location and still be geometrically acceptable.

A number of parallel planes provided by the scanner is processed similarly and the results combined to select the best holdsite, fig. 9, in three dimensions.

MODEL-BASED OBJECT MANIPULATION SYSTEM FOR SPATIAL AND TEMPORAL REASONING

Autonomous plan generation and execution of assembly tasks requires the implementation of spatial and temporal reasoning concepts. Much of this involves the use of symbolic reasoning processes.

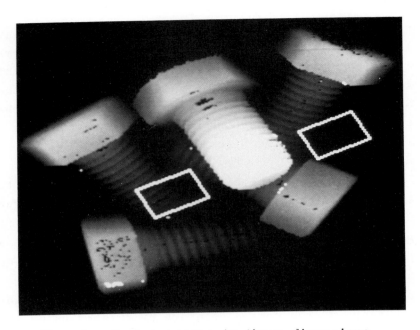

Fig. 9. Best holdsite in three dimensions.

Using a LISP-based object-oriented paradigm (FLAVORS), methods were written for creating, affixing, manipulating objects in 3D space. Objects are described using a volumetric representation based on a subset of generalized cylinders. Each object is characterized by a cross-sectional shape, a spine, and a sweeping rule. Affixment is used to create more complex objects. Upon creation objects are able to compute, and communicate to a MASTER-BUILDER or to other objects, scalar data such as their orientation, volume, surface area, minimum distance to surrounding objects as well as array data such as the cross-section of the object when cut by an arbitrarily located z-plane or any plane perpendicular to its spine. See fig. 10a and 10b for examples. Any object can be selected, rotated, translated, etc. in 3D space. During manipulation the object updates its own attributes.

When complete, this framework will be used to investigate robot path planning and assembly strategies for a task where the scanner-gripper unit will be used to assemble the submodules which comprise it.

Fig. 10a. Examples of invoking procedures for minimum distance to surrounding objects, silhouette extraction, object slices in z-plane, and object slides normal to spine.

Fig. 10b. Successive slices of world in z-space of object at extreme right; used in path planning.

MULTIPROCESSOR SYSTEM ARCHITECTURE: THE HARMONY OPERATING SYSTEM

Harmony is a multitasking, multiprocessing, realtime operating system [4]. In Harmony, resources like memory allocation, task creation and destruction, and connections to servers are dynamic. Message passing is used for intertask communication and for task synchronization. Normally, a task is assigned a particular function and performs that function when required, usually passing its results on to another task on request. Message passing between tasks is the same whether the tasks all reside on one processor or are distributed across a number of processors. This makes system expansion by adding more processors a simple process. Harmony is an open system: to meet the needs of the application, the user can add software and hardware not originally part of the system.

As Harmony is intended strictly for runtime applications, system program development is performed on a separate facility. A core image must be generated for each processor in the system. Harmony is portable across development hosts and target processors.

One of Harmony's advantages is configurability: only those servers used by a program need be included in it. Currently a number of servers have been written for the system including: a terminal server, a file device server, a file system and a clock server. In addition, some debugging tools exist and new tools are being developed.

Originally Harmony has been implemented on a Chorus multi-processor system [5] based on commercially available 68000-based single board microcomputers. The multibus based system could accomodate up to sixteen (16) processors. Current activities are seeing the development of a VME bus implementation using 68010 and 68020-based single board microcomputers.

INTERACTIVE DESIGN ENVIRONMENT FOR ROBOT TASK PLANNING

The focus of this activity is on the development of a multi-tasking, multiprocessor workstation for robotics. Its objective is to provide an appropriate man-machine interface for planning and programming complex robotics tasks for unstructured environment. Its main characteristics are:

i) an object modeller for interactive description of the robot and its environment and the generation of hierarchical data stucture

ii) a near-real-time graphics package processes and displays the data structures on a multi-window display [8]

iii) a user interface supports simultaneously active input devices [9,12]

iv) a research oriented robot controller implements an efficient solution for the kinematics positions, velocities and accelerations for the PUMA 560 robot [2]

v) simple use of simulated "ideal" sensors.

SUMMARY

Intelligent robotics is a large multidisciplinary field. Our goal in robotics technology is towards advanced automated systems. Through the use of sensory perception, these systems will carry out the necessary planning and reasoning processes to interpret their environment and to guide their actions in performing a task.

REFERENCES

[1] Dhome, M., and Kasvand, T., "Hierarchical Approach for Polyhedra Recognition by Hypothesis Accumulation", to appear in the 8th ICPR, Paris, Oct. 1986.

[2] Elgazzar, S., "Efficient Kinematic Transformation for the PUMA 560 Robot", IEEE J. Robotics and Automation, RA-1, (3), pp. 142-151, Sept. 1985.

[3] Elgazzar, S., Green, D., and O'Hara, D., "A Vision-Based Robot System Using a Multiprocessing controller", NRC/ERB-971, June 1984, 97 pages, NRCC No. 23485.

[4] Gentleman, W.M., "Using the Harmony Operating System", NRC/ERB-966, Dec. 1983, revised May 1985, 53 pages, NRCC No. 24685.

[5] Green, D., "Chorus - A Multiprocessor Architecture for Real-Time Control Applications", NRC/ERB-964, Dec. 1983, revised Dec. 1985, 16 pages, NRCC No. 23031.

[6] Kasvand, T., Oka, R., Dhome, M., and Rioux, M., "Segmentation, Learning, and Recognition of Objects in Range Camera Images", Canadian CAD/CAM and Robotics Conference and Exposition, Toronto, June 1985.

[7] Livingstone, F.R., and Rioux, M.,"Development of a Large Field of View 3-D Vision System", SPIE No. 665, June 1986.

[8] Loo, R., "ARIA : A Near-Real-Time Graphics Package", M.Sc. Thesis, University of Waterloo, 1986.

[9] Mackay, S.A., and Tanner, P., "Graphics Tools in Adagio, A Robotics Multitasking Multiprocessor Workstation", Graphics Interface'86, May 1986.

[10] Oka, R., Kasvand, T., and Rioux, M., "Cross-angle Transform for Viewer-Independent Recognition of 3-D Objects", IEEE Proceedings on Computer Vision and Pattern Recognition, San Francisco, pp. 470-475, June 1985.

[11] Rioux, M., "Laser Range Finder Based on Synchronized Scanners", Appl. Opt. 23 (21), pp. 3837-3844, Nov. 1984.

[12] Tanner, P., Mackay, S.A., Stewart, D.A., and Wein, M., "Multi-tasking Switchboard Approach to User Interface Management", to appear in Proc. ACM/SIGGRAPH Conf., Dallas, August 1986.

[13] Yamada, H., Hospital, M., and Kasvand, T., "Rotation-Invariant Contour DP Matching Method for 3-D Object Recognition", to appear in the 1986 IEEE Int. Conf. SMC, Atlanta, Georgia, October 1986.

FUTURE DIRECTIONS IN KNOWLEDGE-BASED ROBOTICS

Deborah A. Stacey
Andrew K.C. Wong
Systems Design Engineering
University of Waterloo
Waterloo, Ontario
N2L 3G1

On the final day, the participants gathered to discuss and summarize the major issues raised and to look at future research directions. Each of the six sections of the panel discussion was aptly led by a researcher who first reviewed the observations made during the workshop.

The first section was on vision and sensing. Nitzan saw the major issues here as being the recovery of two- and three-dimensional information from images as well as the improvement of techniques for range and orientation measurement. Problems include specularity and speed for stereo vision and laser systems. As for monocular vision, it still suffers from the many calculations needed for inference of range and orientation.

All techniques have their niches, but as yet no systematic comparison has been done to assist users in chosing the technique appropriate for their application. At present we can obtain points and lines from captured images, but it is real knowledge and constraints which are needed from a sensing system. A major goal is thus the integration of vision, range information and other sensing data. Progress to date has been slow and we still have a long way to go. Long term prospects will include the introduction of intelligence into sensing modules and the integration of the results into a knowledge base. Vision is an appropriate research area since it demonstrates many general sensor problems. It should be emphasized that more knowledge must be brought into sensory analysis if long term goals are to be achieved.

The next topic was knowledge-based systems. Saridis noted that there is a division between theory and applications. Two main issues are the integration of sensors and knowledge bases and the debate between proponents of real-time planning and organization methods and those of off-line techniques. Theorists tend to be pro real-time. Both approaches are valid and should be explored and evaluated as to cost-effectiveness. Real-time planning is extremely important and we must start now to develop new, effective techniques.

During discussion, Nitzan observed that since knowledged used for real-time planning is usually present from the start, why not incorporate it into the original plan. Lee countered and pointed out that unique and rare events necessitate that planning take place at the time of occurrence. In response, Nitzan suggested the strategy of having generic routines (produced offline) executed in real-time. Sensors would detect unique situations and activate the appropriate routine.

A knowledge representation should be flexible and easy to construct, modify and manipulate. This statement provided the introduction to Wong's summary of knowledge representation. Wong observed that there is a need for autonomous knowledge acquisition and synthesis. Problems have arisen in this field due to the fact that the real world is complex and there is a great variability of knowledge from task to task. The challenge is to retain realism in applications while still achieving generality in the knowledge representation. We need to allow for tolerances and uncertainty and still provide a general, flexible formalism. One goal may be the development of a high-level language for implementation of knowledge representations.

Aggarwal added that there is a need for multiple types of representations. Presently, it is not easy to go from one to another; in fact it is very difficult. But, this approach also raises the question as to whether multiple representations of the same information will be stored or computed when needed. Wong suggested that a general representation may accomplish or provide a medium for the transformation between types. Nitzan brought up the point that two types of knowledge need to be stored and manipulated - bulk, general knowledge and task specific knowledge. He speculated that the former could be explicitly contained in some form of knowledge base of library while the latter, since it is usually less voluminous, could be computed or inferred when needed

Dynamic robot motion control is a major focus for research in vision and motion according to Aggarwal. The goals of this field are recognition of objects, collision avoidance and retrieval of objects. The major problem is that we are still dealing with points, lines and contours instead of high-level object representations. Challenges in this area are multi-sensors and their integration, noise management and parallel processing. Parallel processing is not being dealt with currently but it may provide the answer for the handling of large amounts of data in a reasonable time. Also, search problems may require parallelism. Knowledge representation is very important for integration and conversion from one system to another and is also critical as the field moves into parallel processing. To summarize, short term problems include parallel processing, parallel searches, multiple features and multiple sensors while long term goals are centered around systems to handle both moving objects and moving robots.

Jain observed that most techniques are still in static mode, i.e. they are only working with 2 or 3 frames. Researchers must start to work on continuous data. Saridis offered the opinion that parallelism is important but how is it to be coordinated? Are we to look at hierarchies, networks, master/slave systems, etc. - this is a central issue. Aggarwal commented that the master/slave arrangement is not very good since it limits bandwidth. Harmon added that advantages and disadvantages of tightly-coupled versus loosely-coupled parallel systems are two separate issues to be resolved. Harmon summarizes the problem as being in two stages:

1. no a priori knowledge (cold start)
2. use of successive images to build up knowledge after cold start

For these two stages, parallelism is not the only solution; we must have improved algorithms. This led to discussion of the knowledge based approach which Jain considers worthy of study since general techniques do not seem to be sufficient. But this brings us back to the knowledge representation problem according to Aggarwal. There is no easy way to use present systems to accomplish our goals in the area of vision and motion.

Harmon's autonomous land vehicles talk once again emphasized the point that data fusion (multi-views and multi-sensors) is a vital key to sensory work. The fundamental issues arising from fusion are knowledge representation and the handling of uncertainty. There is also a need for generic planning. At present there are no generic planning engines available. It was suggested these engines may help solve multiple representation problems. The problem with a tool-box approach is in the selection of the appropriate tool - the key is intelligence. This intelligence must be distributed. High-level planning will need intelligent subsystems near the sensors. If intelligence is pushed lower, it may simplify some representation and computation problems. Once again it was stressed that theory is nice but there is a need for more practical experience to expose difficulties in present approaches. In the short term, sensing and interpretation techniques must be improved, and integration and demonstrations are needed. The implementation of intelligent control is urgent and it must not be forgotten that mobility is still an issue. Long term goals emphasize planning (temporal reasoning, uncertainty handling, incremental planning), advanced computer architectures, the need for a formal theory of intelligent machines and the need for learning machines. In the area of architecture, parallelism must be applied appropriately. An architecture should be mapped onto ALV problems. Problem areas include communications between processors, which is always expensive, and the choice of architectural granularity - the massively parallel (the Connection Machine) versus the "smaller-scale" parallel (hypercube, butterfly). Way out future solutions might have to rely on biological computation for the handling of large amounts of sense data.

Parallelism, or the mapping of task to hardware, is also a software concern according to Popplestone. He offered the Imperial College approach which consists of writing in strictly applicable languages which are then interpreted by parallel machines. This is suitable for high-level tasks but not for low-level ones which need much more contact with the hardware and individual CPU's.

A sad note from the software side is that debugging will always be with us.

Harmon contributed some areas for thought and consideration; operating systems for distributed environments, the unresolved issue of dynamic loading and the distribution of processes to processors especially in unknown real environments. In response, Shadia Elgazzar from the National Research Council of Canada reported that NRC's Harmony operating system successfully addresses some of these issues.

The workshop concluded on an optimistic note, with researchers targeting sensor fusion, advanced computer architectures and an increase of machine intelligence in all systems (knowledge bases and sensors) as areas to be actively pursued.

LIST OF SPEAKERS

Dr. Jake Aggarwal
Laboratory for Image & Signal Analysis
Department of Electrical Engineering
University of Texas at Austin
Austin, Texas   78712

Dr. Raja Chatila
Lab d'Automatique et d'Analyse des Systemes
Centre National de la Recherche Scientifique
Toulouse, France

Dr. Nick V. Findler
Computer Science Department
Arizona State University
Tempe, Arizona   05287

Dr. R. Haralick
Machine Vision International
Burlington Center
325 East Eisenhower
Ann Arbor, Michigan   48104

Dr. Scott Harmon
Robot Intelligence International
4660 Long Branch Avenue
San Diego, CA   92107

Dr. Thomas Huang
Coordinated Science Laboratory
University of Illinois at Urbana-Champaign
Urbana, Illinois   61801

Dr. Ramesh Jain
Department of Electrical Engineering and Computer Science
University of Michigan
Ann Arbor, Michigan   48109

Dr. Mark Lee
Department of Computer Science
The University College of Wales
Penglais
Aberystwyth   SY23 3BZ

Dr. Martin Levine
Department of Electrical Engineering
McGill University
Montreal, Quebec
H3A 2T5

Dr. David Nitzan
SRI International
333 Ravenswood
Menlo Park, CA   94025

Dr. L. Pau
Technical University of Denmark
Rormosen 56, Kaarup
DK 4540 Faarevejle
Denmark

Dr. Robin Popplestone
Department of Computer and Information Science
University of Massachusetts at Amherst
Amherst, MA   01003

Mr. Klaus Hoermann
University of Karlsruhe
P.O. Box 6380
D-7500 Karlsruhe 1
West Germany

Dr. A. Rosenfeld
Center for Automation Research
University of Maryland,
College Park, MD   20742

Dr. A. Sanderson
Electrical and Computer Engineering Department
  and The Robotics Institute
Carnegie-Mellon University
Pittsburgh, PA   15213

Dr. G. Saridis
Robotics & Automation Laboratory
Department of Electrical, Computer and Systems Engineering
Rensselaer Polytechnic Institute
Troy, New York   12181

Dr. J. Tou
Centre for Information Research
University of Florida
339 Larsen Hall
Gainesville, Florida   32611

Dr. A. Wong
Systems Design Engineering
University of Waterloo
Waterloo, Ontario
N2L 3G1

Dr. Alan Pugh
Electronic Engineering
University of Hull
Hull   HU6 7RX

LIST OF PARTICIPANTS

Dr. W. Wilson
Electrical Engineering Department
University of Waterloo
Waterloo, Ontario
N2L 3G1

Dr. T. Gajdicar
Computing Devices Company
Development Division
P.O. Box 8508
Ottawa, Ontario
K1G 3M9

Dr. A. Honne
Center for Industrial Research
Box 350
Blindern, N-0314
Oslo 3
Norway

Dr. S. Elgazzar
National Research Council of Canada
Division of Electrical Engineering
Montreal Road
Ottawa, Ontario
K1J 0R8

Mr. T.S. Cinotti, Ing.
Dept. of Electronica
University of Bologna
Bologna, Italy

Mr. L. Froslev-Nielsen
Technical University of Denmark
The Control Engineering Institute
Building 424
2800 LYNGBY
Denmark

Miss D. Stacey
Systems Design Engineering
University of Waterloo
Waterloo, Ontario
N2L 3G1

Mr. A. Sonmez
University Engineering Dept.
Cambridge University
Trumpington Street
Cambridge CB2 1PZ
England

Mr. D. Johnson
Electronic Engineering
University of Hull
Hull   HU6 7RX

Mr. M.D. Wybrow
Department of Production Engineering and Production Management
University of Nottingham
Nottingham, England
N97 2RD

Mr. L. Carrioli
Dept. Informatica vie Abbiategrasso
Pavia University
Pavia, Italy

NATO ASI Series F

Vol. 1: Issues in Acoustic Signal – Image Processing and Recognition. Edited by C. H. Chen. VIII, 333 pages. 1983.

Vol. 2: Image Sequence Processing and Dynamic Scene Analysis. Edited by T. S. Huang. IX, 749 pages. 1983.

Vol. 3: Electronic Systems Effectiveness and Life Cycle Costing. Edited by J. K. Skwirzynski. XVII, 732 pages. 1983.

Vol. 4: Pictorial Data Analysis. Edited by R. M. Haralick. VIII, 468 pages. 1983.

Vol. 5: International Calibration Study of Traffic Conflict Techniques. Edited by E. Asmussen. VII, 229 pages. 1984.

Vol. 6: Information Technology and the Computer Network. Edited by K. G. Beauchamp. VIII, 271 pages. 1984.

Vol. 7: High-Speed Computation. Edited by J. S. Kowalik. IX, 441 pages. 1984.

Vol. 8: Program Transformation and Programming Environments. Report on an Workshop directed by F. L. Bauer and H. Remus. Edited by P. Pepper. XIV, 378 pages. 1984.

Vol. 9: Computer Aided Analysis and Optimization of Mechanical System Dynamics. Edited by E. J. Haug. XXII, 700 pages. 1984.

Vol. 10: Simulation and Model-Based Methodologies: An Integrative View. Edited by T. I. Ören, B. P. Zeigler, M. S. Elzas. XIII, 651 pages. 1984.

Vol. 11: Robotics and Artificial Intelligence. Edited by M. Brady, L. A. Gerhardt, H. F. Davidson. XVII, 693 pages. 1984.

Vol. 12: Combinatorial Algorithms on Words. Edited by A. Apostolico, Z. Galil. VIII, 361 pages. 1985.

Vol. 13: Logics and Models of Concurrent Systems. Edited by K. R. Apt. VIII, 498 pages. 1985.

Vol. 14: Control Flow and Data Flow: Concepts of Distributed Programming. Edited by M. Broy. VIII, 525 pages. 1985.

Vol. 15: Computational Mathematical Programming. Edited by K. Schittkowski. VIII, 451 pages. 1985.

Vol. 16: New Systems and Architectures for Automatic Speech Recognition and Synthesis. Edited by R. De Mori, C.Y. Suen. XIII, 630 pages. 1985.

Vol. 17: Fundamental Algorithms for Computer Graphics. Edited by R.A. Earnshaw. XVI, 1042 pages. 1985.

Vol. 18: Computer Architectures for Spatially Distributed Data. Edited by H. Freeman and G. G. Pieroni. VIII, 391 pages. 1985.

Vol. 19: Pictorial Information Systems in Medicine. Edited by K. H. Höhne. XII, 525 pages. 1986.

Vol. 20: Disordered Systems and Biological Organization. Edited by E. Bienenstock, F. Fogelman Soulié, G. Weisbuch. XXI, 405 pages.1986.

Vol. 21: Intelligent Decision Support in Process Environments. Edited by E. Hollnagel, G. Mancini, D.D. Woods. XV, 524 pages. 1986.

Vol. 22: Software System Design Methods. The Challenge of Advanced Computing Technology. Edited by J.K. Skwirzynski. XIII, 747 pages. 1986.

NATO ASI Series F

Vol. 23: Designing Computer-Based Learning Materials. Edited by H. Weinstock and A. Bork. IX, 285 pages. 1986.

Vol. 24: Database Machines. Modern Trends and Applications. Edited by A. K. Sood and A. H. Qureshi. VIII, 570 pages. 1986.

Vol. 25: Pyramidal Systems for Computer Vision. Edited by V. Cantoni and S. Levialdi. VIII, 392 pages. 1986.

Vol. 26: Modelling and Analysis in Arms Control. Edited by R. Avenhaus, R. K. Huber and J. D. Kettelle. VIII, 488 pages. 1986.

Vol. 27: Computer Aided Optimal Design: Structural and Mechanical Systems. Edited by C. A. Mota Soares. XIII, 1029 pages. 1987.

Vol. 28: Distributed Operating Systems. Theory und Practice. Edited by Y. Paker, J.-P. Banatre and M. Bozyiğit. X, 379 pages. 1987.

Vol. 29: Languages for Sensor-Based Control in Robotics. Edited by U. Rembold and K. Hörmann. IX, 625 pages. 1987.

Vol. 30: Pattern Recognition Theory and Applications. Edited by P. A. Devijver and J. Kittler. XI, 543 pages. 1987.

Vol. 31: Decision Support Systems: Theory and Application. Edited by C. W. Holsapple and A. B. Whinston. X, 500 pages. 1987.

Vol. 32: Information Systems: Failure Analysis. Edited by J. A. Wise and A. Debons. XV, 338 pages. 1987.

Vol. 33: Machine Intelligence and Knowledge Engineering for Robotic Applications. Edited by A. K. C. Wong and A. Pugh. XIV, 486 pages. 1987.